DIAGRAM FOR THE CLASSIFICATION OF WORLD LIFE ZONES OR PLANT FORMATIONS

Altitudinal belts	Mean annual biotemperature, °C*	Mean annual potential evapotranspiration, millimeters	Latitudinal regions
NIVAL	----	— (58.93)	POLAR
	1.5 —		
ALPINE	—	— 118	SUBPOLAR
	3 —	— 177	
SUBALPINE		— 236	BOREAL
		— 294	
	6 —	— 353	
		— 412	
MONTANE		— 471	COOL TEMPERATE
		— 530	
		— 589	
		— 648	
	12 —	— 707	
		— 825	
LOWER MONTANE		-- 943	WARM TEMPERATE
PREMONTANE		— 1060	SUBTROPICAL
		— 1178	
		— 1296	
	24 —	— 1414	
		— 1532	
		— 1650	
		— 1768	
		— 1886	TROPICAL

25   0.0625   0.03125

SEMISATURATED SUBSATURATED   SATURATED

*t^{bio} = Mean of unit-period temperatures with substitution of zero for all temperature values below 0°C and above 30°C, respectively (this formula is tentative pending further investigation)

† = In tropical subalpine only

Redrawn from L. R. Holdridge, Tropical Science Center, Costa Rica, reproduced in "Life Zone Ecology," 1967.

principles of environmental science

principles of environmental science

KENNETH E. F. WATT
Department of Zoology and
Institute of Ecology
University of California, Davis

McGRAW-HILL BOOK COMPANY

New York   St. Louis   San Francisco   Düsseldorf   Johannesburg   Kuala Lumpur   London
Mexico   Montreal   New Delhi   Panama   Rio de Janeiro   Singapore   Sydney   Toronto

PRINCIPLES OF ENVIRONMENTAL SCIENCE

34567890VHVH7987654

Library of Congress Cataloging in Publication Data

Watt, Kenneth E   F   1929–
    Principles of environmental science.

    Includes bibliographies.
    1. Ecology. 2. Human ecology. 3. Environmental
policy. I. Title.
QH541.W28   301.31   72-5344
ISBN 0-07-068575-4

This book was set in Trade Gothic by Progressive Typographers.
The editors were James R. Young, Jr. and Carol First;
the designer was Barbara Ellwood;
and the production supervisor was John A. Sabella.
The drawings were done by Vantage Art, Inc.
The printer and binder was Von Hoffmann Press, Inc.

for Genevieve, Tanis, and Tara

contents

A glance at the table of contents will indicate that this book represents a somewhat unusual grouping of subjects and a view of ecology, or the new, broader field, environmental science, considerably different from that held by many ecologists. Therefore, some words of explanation are in order.

The author feels that traditional plant and animal ecology is really a natural, integral part of a broader subject, environmental science, in which such disciplines as urban and regional planning, economic geography, epidemiology, community medicine, meteorology, and much of agriculture, fisheries, forestry, and range management also belong. This book is intended to demonstrate the existence of a deep underlying core of principles to which all aspects of environmental science can contribute and from which each can draw. Therefore, it is hoped that courses based on this book will provide a foundation of theory for students in many parts of the large university. In short, it is expected that a rigorous scientific foundation can be laid that will serve as one of the main foci for modern scientific effort over a wide range of disciplines, as physics and chemistry do, for example.

Specifically, it appears that fragmentation of the environmental sciences into many disciplines is a result of the way they have been studied. Research interest has focused on numbers of events, description of situations, and specific causes and effects. I propose to shift the emphasis toward an integrated consideration of five fundamental categories of variables: energy, matter, space, time, and diversity. When this is done, it will be apparent that many disciplines which at first inspection appear quite different are in fact studying the same fundamental *processes*. The last word is the key to this book. All the environmental sciences must put principal emphasis on processes in order to attain the most penetrating possible insight into phenomena which they study. Most particularly, this book reflects the recent strong shift of emphasis in ecology to examination of processes in terms of the efficiency of energy transfer in systems. Also, it is now clear that the distribution of entities in space, whether plants, animals, cars, buildings, or cities, has implications for the efficiency of energy flow. Matter is involved, because this is the prerequisite material on which energy operates. Time is important because a process will not reach the threshold level at which it works at all without sufficient time during which conditions have been suitable. Thus, time is a resource. Diversity is important because new research has shown that it is the uninterrupted long-term accumulation of diversity in the form of genetic and species diversity, which determines much of the dynamics of communities of species populations.

This book also differs from existing texts in that many applied subjects are treated at length. There are several reasons for this. First, we wish to point out to all students, pure and applied, that economically important applied problems are typically the best source of data for the theoretician to work with, simply because their economic importance has ensured massive data collection programs. Second, we wish to break down artificial barriers between pure and applied science. These barriers are nowadays purely matters of tradition and prestige because in fact much of the highest-quality basic research is done on practical problems. Third, important applied problems often constitute the best available examples of important theoretical principles. Finally, many practical problems in environmental science are reaching crisis dimensions for all of mankind, and the attention of our most talented youth should be directed to them. As one leading ecologist remarked to me recently, perhaps it is true that pure ecological problems are better in some sense, but if his children and their contemporaries do not do something dramatically effective about air pollution and its consequences, his grandchildren might not be able to study pure ecological problems. Also, if we want to study mature ecosystems, now is the time. We may not be able to study undisturbed coral reefs, rain forests, or African big game in a few years.

We have made a great effort to examine each branch of ecology and environmental science to see what bodies of theory it could contribute to the entire field. Thus, we have made an all-out effort to integrate plant with animal ecology, and forestry, fisheries, agriculture, and range management with both of them.

Chapter 1 introduces essential background material, so that it will not be necessary to break trains of thought in other chapters by defining terms or explaining fundamental concepts.

Chapter 2 is the key to the design of the book. In this chapter fourteen principles are presented as constituting the core theory of environmental science, as the subject is currently understood. Other principles are introduced in subsequent chapters, but those presented in Chap. 2 constitute the foundation for everything that follows.

Chapters 3, 4, and 5 present the core of classical ecological theory, although with certain differences in organization and conceptualization introduced because time and diversity are considered as resources. Chapter 5 puts considerable emphasis on the biological consequences of variations in upper atmosphere turbidity produced by volcanoes, because of the importance of this phenomenon in accounting for certain cycles, and also because it illustrates effects of air pollution.

Chapter 6 stresses the role of man in ecological systems, particularly with respect to the fundamental differences between the effects of man and natural phenomena, in the light of the theory presented in Chap. 2.

Chapters 7 to 11 treat a series of ecosystems, about each of which a great deal is known. Each is selected as a model system peculiarly suitable to illustrate certain types of principles and problems. Other ecosystems could have been chosen, such as the ocean, the coral reef, the high desert, or the rain forest. However, the ecosystems selected have been well studied, are within the range of experience of most students, and are convenient for collection of material (field laboratories or trips), and the five chapters fit together to form a logical sequence. Some of the other types of ecosystems will be dealt with in part, incidentally, in Chaps. 3 to 6 and 14 and 15. The reason for treating five ecosystems, rather than one, is to illustrate a wider variety of principles and also to impress the student with the importance of the comparative method of scientific reasoning.

Biological control is discussed in Chap. 12 to demonstrate that where natural balances are imperfect, or immature, or have been disturbed by man, it may sometimes be possible to restore them through application of ecological theory, by means of biological control.

Disease is treated in some detail in Chap. 13 because the separation of the disease-oriented sciences, particularly epidemiology, from the remainder of ecology has been a damaging loss to both. Particularly with the great new rise of community medicine in the interests of understanding the environmental factors operating on cardiovascular diseases, stronger links should be forged between environmental science and medicine.

Chapters 14 and 15 are intended to show how the basic science of ecology can be applied to a fundamental understanding of man's current position in the world. The startling contrasts with natural systems revealed by considering our situation with respect to the five basic categories of resources reveals the unstable nature of our current situation. The analysis leads to a global strategy for mankind in Chap. 15.

Because of the interdisciplinary character of the problems in this field and the intellectual nature of its demands, we propose that the field has extraordinary cultural value in teaching students techniques of thinking that will be applicable elsewhere. Like physics, mathematics, or genetics, ecology can teach a rigorous, yet imaginative, type of thought that is of value in all walks of life. Particularly, we wish to inculcate in the student an enthusiasm for posing penetrating questions about ordinary phenomena and situations he sees in the world around him. Further, we wish to train the student to think about parallels between different kinds of phenomena by looking at the surrounding environment in terms of processes and asking questions about how a process in one situation relates to corresponding processes in other situations. Does the distribution of ants around an ant nest tell us anything about city planning? What can be learned about the effects of climate on plants by looking at differences in the plant community at different heights up mountainsides, or differences on the lee and windward sides? What is the relation between the composition of plant communities and typical distribution patterns of cloud and fog? What can we tell about the effects of grazing and browsing on plant community succession by looking at the plants on both sides of a fence to keep livestock in or deer out? What can we discover about the effects of pollution on vegetation by looking at the area around a smelting plant when flying over it? Why are there no fish or mammals or mollusks bigger than a sturgeon in large lakes, whereas there are giant squids, whales, sharks, marlins, and other very large organisms in the ocean? Why don't all the frighteningly efficient parasites, predators, and diseases in the world kill off everything else, then die themselves? Why is there so much violence in our big cities? Does this have any counterpart in the fighting that occurs in crowded populations of animals?

This book is intended for courses that can be taken by any university student as soon as he enters the university. It assumes no prerequisites in college biology, social sciences, physical sciences, or mathematics. Although some fairly complex ideas are presented, in all cases an attempt has been made to present them as photographs, graphs, or diagrams.

In a few instances, elementary college-level mathematics has been used to illustrate certain ideas. This has been kept to a minimum, however, and the central flow of the argument will rarely be lost by the student if he does not know this material.

However, the student is advised that for advanced work in this or any other field, failure to obtain a strong mathematical background condemns him to work with phenomena that will be largely incomprehensible to him.

Because of the wide range of topics covered in this book, I have been very dependent on help from friends and colleagues. William J. Hamilton III, a constant source of stimulation for eight years, read an early draft of Chap. 2. Richard H. Wagner, Michael Rosenzweig, and William Murdoch read an earlier draft, and the final version benefited greatly from their labors. After years of trying to get high-quality pictures of tropical fish in coral seas, it turned out that it was virtually impossible to get pictures in the ocean equivalent in quality to those shot in tanks. I am deeply indebted to James O'Neill of the Kona Aquarium, Inc., for permission to take pictures in the aquarium. The U.S. National Science Foundation supported computer rental for the work on insect populations reported in Chap. 5, with grant GB-2889. Richard M. Prentice and the staff of the Canadian Forest Insect Survey supplied me with the data used as input for that analysis. I am extremely indebted to a great friend and colleague, Dr. David Deamer, who first attracted my attention to the role of volcanic eruptions as determinants of geophysical changes. Dr. Leonard Myrup helped me into the literature on the urban heat island. Professors R. J. Smeed of London, Earl Finbar Murphy of Ohio State, and Dr. R. J. Hickey have helped me enormously with their vast knowledge of transportation, natural resources, and the public health consequences of pollution, respectively. Drs. Lloyd C. Hulbert of Kansas State University and William M. Longhurst of California took me on field trips that helped me understand rangeland problems. Robert Boyd made helpful suggestions about the chapter on watersheds. James Young, my editor, has been a great source of encouragement for many years. My wife Genevieve took the pictures of Russia and Switzerland. I am very much indebted to Miss Theresa Batkin and Mrs. Genevieve Tozzi, who helped type the final draft of the manuscript. Finally, I am extremely indebted to Miss Charlene Sonia, who typed much of the final draft and helped in innumerable other ways: correspondence related to rights and permissions, correspondence with the publisher, and with me, and so on. Without her, I might have drowned in the project.

KENNETH E. F. WATT

1 some essential background from biology, mathematics, and scientific methodology

BIOLOGY One of the central concepts in environmental science is the *population*. When this term is used without qualification, it invariably refers to a group of individuals all of whom are of the same species. But how many individuals are in the population, and how much space does it occupy? In some instances, this is an easy question to answer. For example, there are rare species, such as blind cave salamanders, all of which live in one cave. Here all the individuals in the species are clearly part of a single population. In other instances, a species is split up into a number of totally isolated groups of individuals, as in the case of whitefish in northern Canadian lakes which have totally separate drainage basins and between which no migration is possible. Here the population is the group of individuals in one of the lakes.

Defining a population is more difficult when the members of a species are spread over a very large area, with distance being the only effective barrier between different individuals. This is true of the North American white-tailed deer, the Asiatic tiger, the mallard duck, and man. In these cases, the most meaningful way to define population is in terms of the effect that one individual can have on another. Thus, we recognize that a population is a dynamic system of interacting individuals. A population consists of all those individuals within a species that are potentially capable of interbreeding with each other. Consequently, although a white-tailed deer in eastern Oregon is of the same species as a white-tailed deer in New England, they are isolated by distance, cannot interbreed, and are in different populations.

The individuals in a population have various interactions besides those related to mating. If the number of individuals per unit area becomes much larger than normal for some reason, we say there has been an increase in *population density* (the number per unit area). If the increased density creates a demand for food, shelter, and other needs beyond what is available in the environment, then *competition* occurs. Competition takes place among two or more individuals when their combined need for one or more resources exceeds the available supply of resources.

Competition has two types of effects on the competing individuals: short-term, or *ecological*, effects, and long-term, or *evolutionary*, effects. In either case, there is a struggle for existence that is won by some individuals and lost by others.

Four types of short-term effects of competition are important for the population: the birth, survival, or growth rate may be depressed, or the emigration rate may increase.

Competition may involve actual physical interaction between individuals. For example, the reproductive rate will be lowered if the density of males is so great that they spend too much time fighting over females, compared with the time spent mating. Or a female may have difficulty laying eggs if the oviposition site is swarming with females, and there is *interference* among them. However, competition need not involve any physical contact. If one individual starves because other individuals have already eaten all available food, competition has occurred even though the starving individual was never touched or even approached by one of its competitors for food.

Competition has an effect on evolution within a population; for example, the individual that is smaller than others and has reduced longevity, owing to starvation, is less likely to reproduce than individuals which win out in competition. Thus, those individuals which win out in the struggle for existence will leave a disproportionately larger number of progeny to populate the next generation than those individuals which lose. Because of the survival of the fittest, the vigor of the population is maintained.

Two kinds of environmental factors kill individuals, or lower growth and reproductive rates: *density-dependent* and *density-independent*

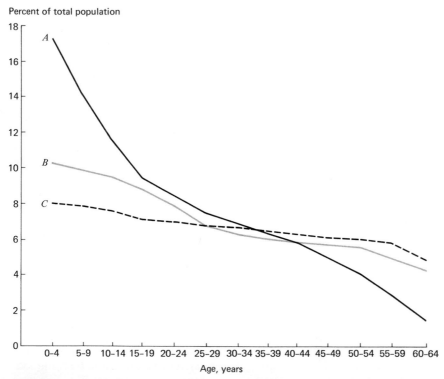

FIGURE 1-1 The effect of population growth rate on population age structure in the human population. Population A has been growing for many years at 2 percent per year; this is typical of many undeveloped countries. Population B has been growing at 1 percent per year; this age structure is what we would now have in the United States but for the Depression, World War II, and their aftereffects on the population age structure. Population C has been growing for many years at 0.6 percent per annum; this is the United Kingdom at present. Note that when the populations at each age are expressed as percentages of the total population, very small changes in per annum growth rate have a spectacular effect on the population age structure.

factors. Density-dependent factors have a more severe effect as the population density increases. Food shortage is a density-dependent factor. But some factors, such as a sudden extreme drop in temperature or a hailstorm, will kill the same proportion of the population no matter what the density. This is a density-independent factor.

An important characteristic of a population is its *age structure*. In Fig. 1-1, the age structures of three different human populations are depicted. In a fast-growing population (2 percent per annum), young people are being added to the population much faster than old people die, and so young people constitute an extremely large proportion of the total population: in population A, 52.4 percent of the population is under 20. Populations growing very slowly or not at all have much more equal numbers of people in different ages: in population C, only 31.0 percent of the population is under 20. The great importance of population age structure is that many important attributes of individuals depend on their age: birth rate, death rate, growth rate, tendency to migrate, feeding rate, and economic cost are a few examples. Thus, any change in population age structure alters the average of the population with respect to any of these attributes.

Two particular kinds of population age structure are sufficiently important to be assigned names: the *stationary population* and the *stable population*. For any given year, say 1970, we could calculate for each age of the people in a population the probability that they would survive to the next year, 1971. These probabilities could then be used to construct the stationary population, the number of people that would be alive at each age, given this set of probabilities, and a particular starting population at age zero, such as 100,000.

Thus, the stationary population based on 1967 mortality rates in the United States would be derived as in the following table (mortality rates from *Vital Statistics of the United States*):

age	mortality rate during preceding year per 100,000 living at specified age	survival per 100,000 during preceding year of life	number surviving to specified age
0	. . .	. . .	100,000
1	2,239	97,761	97,761
2	135	99,865	97,629
3	87	99,913	97,544
4	66	99,934	97,480
		etc.	

The stationary population, therefore, is the age distribution resulting from the application of survival rates for a particular year.

The stable population, on the other hand, is the age distribution that would gradually be produced in a population if a particular set of both age-specific birth rates and age-specific death rates were to continue indefinitely.

Four terms are used in this book in connection with the harvesting of populations by man: *biomass, standing crop, productivity,* and *yield*.

The biomass of a population is its total weight: the number of individuals in the population multiplied by the average weight per individual. The standing crop refers to the number of individuals in a population at a particular time, or their biomass. The productivity is the amount of living tissue produced per unit time by the population. The yield is what is removed in harvesting by man.

The next higher level of organization dealt with in ecology is the *community*, the total collection of species populations of plants and animals living in an environment. Since certain sets of species characteristically occur together, we can speak of the southern Wisconsin hardwood community, for example, or a tropical rain forest community, or a sagebrush community, or a tropical coral reef community.

Each species in a community is repeatedly found in certain types of habitats, performing certain types of functions. The combination of habitat and function constitutes the *niche* of the species (its place and its job). The niche concept is very real to the naturalist or the collector. For example, in visiting the same holes in the same reefs, year after year, he knows what to expect, not only about the identity of the species that will be present, but also about their relative abundance.

The concept of relative abundance leads directly to one of the most central concepts in community ecology: *diversity*. This term is applied to several characteristics of communities (species diversity, pattern diversity, etc.), but the same thing is always being measured: variety. The student can best grasp the meaning by seeing how it is measured. The diversity in any assemblage of items (individual animals, for example) is given by the formula:

$$\text{Diversity} = \log_2 \left(\frac{N!}{N_1! N_2! N_3! \cdots N_n!} \right)$$

or

$$\text{Diversity} = 1.443 \ln \left(\frac{N!}{N_1! N_2! N_3! \cdots N_n!} \right)$$

N represents the total number of individuals in a collection, N_1 represents the number of individuals in species 1, N_2 represents the number of individuals in species 2, and N_n represents the number of individuals in the nth and last species. When comparing diversity from community to community, the two samples compared must have the same N.

To illustrate the application and meaning of this formula, suppose we have collected six individuals from a community (in actual practice, of course, the collection size would be very much larger, but it is easier to explain the calculation with small numbers). We know, without considering anything about the assignment of these individuals to species, that the number of ways we can arrange six individuals is given by 6!, which is $6 \times 5 \times 4 \times 3 \times 2 = 720$. Now suppose we find that these individuals are in three species, three individuals in one species, two in the second, and one in the third. The denominator of the ratio is then $3 \times 2 \times 1 \times 2 \times 1 \times 1 = 12$, and a measure of the diversity in the set of individuals is $\frac{720}{12}$, or 60. The natural logarithm of this number is 4.1, the diversity is about 5.9, and the diversity per individual is about 0.98. When is the diversity in an assemblage of individuals the lowest possible and the highest possible? The greatest possible diversity per individual occurs when each individual is in a different species. The lowest possible diversity per individual occurs when all individuals are in the same species.

As Margalef (1958) has pointed out, in order to characterize the diversity of a community, we must include the diversity involved in the following aspects of community organization:

1 Allocation of individuals to species
2 Location of species in space
3 Location of species in space on the basis of their relative abundance
4 Location of each individual in space

Ecologists attach immense importance to measures of community diversity, because they tell us about the degree of organization in a community, more diverse communities being more highly organized. This high degree of organization in turn is characterized by communities that have developed beyond a juvenile stage and have a considerable degree of population stability.

Communities undergo a cycle of birth, growth, and senescence, like that in an individual organism. The difference is that communities do not die; rather one group of plants and animals replaces another. A community comes into being when some barren habitat, such as a lava flow, sand dune, bare rock, mud flat, or volcanic island is first invaded by a group of pioneer species. These species modify the habitat so that other species can invade and persist. Finally, the site is occupied by species that dominate the community and replace themselves, rather than being replaced by other species. Now we have a *climax community.* This whole process of replacement of groups of species, one by another, through habitat modification is called *succession.*

An even higher level of organization than the community is the *ecosystem.* Now we consider not only the total array of plant and animal species in an environment, but also the matter which cycles through the system and the energy which is used to power the system. Sunlight and warmth provide plants with energy to manufacture tissue out of carbon dioxide, water, and a host of minerals obtained from the soil. Some of this tissue is then eaten by animals, which then may be eaten by other animals. Ultimately, all living tissue is broken down by an immense variety of decomposers, and once again it becomes soil, available to be used again by the ecosystem.

We may visualize isolated sets of relations of the form:

$$\text{Plant species} \xrightarrow{\text{eaten by}} \text{herbivore species} \xrightarrow{\text{eaten by}} \text{carnivore species} \xrightarrow{\text{eaten by}} \text{top carnivore species}$$

as *trophic chains.* Well-known examples of trophic chains are:

$$\text{Vegetation} \longrightarrow \underset{Microtus}{\text{the field mouse}} \longrightarrow \underset{Mustela}{\text{the least weasel}}$$

$$\text{Detritus} \longrightarrow \text{crayfish} \longrightarrow \text{bass} \longrightarrow \text{muskellunge}$$

$$\underset{\text{(algae)}}{\text{Oceanic phytoplankton}} \longrightarrow \underset{\text{(tiny shrimp)}}{\text{krill}} \longrightarrow \text{blue whale}$$

Vegetation $\longrightarrow$ antelope $\longrightarrow$ lion

Carrot tops $\longrightarrow$ caterpillar of $\longrightarrow$ ichneumon fly
black swallowtail   parasite of
butterfly   caterpillar

When all these chains that constitute an ecosystem are linked together in a single diagram, what we have depicted is a *trophic web*. A chart of the biomass at each of the functional levels (plants, herbivores, carnivores) in an ecosystem typically takes the form of a pyramid (Fig. 1-2), if plants have the bottom level, and hence is called a *trophic pyramid*. However, not all known trophic pyramids have such diagrams; in some marine ecosystems the biomass at the phytoplankton level is less than that at the first herbivore level (zooplankton plus bottom fauna).

The light from the sun is converted by all plants in an ecosystem into an amount of organic matter which is expressed as the *gross photosynthetic production* per unit area. Only a fraction of the plant tissue eaten by a herbivore is converted to biomass; some of the remainder is defecated, some is unassimilated, and the rest is lost in respiration. The production efficiency at each trophic level is expressed in terms of the rate of biomass production at that level divided by the rate of biomass production at the level eaten. The rate of production in ecological systems is measured in kilocalories per square meter per year.

MATHEMATICS   As in physics, chemistry, and other sciences, environmental scientists are concerned with *rates of change in numbers*. Thus, if some variable N has been measured, we are concerned with the rate at which N changes through time. N may be the number of people in the world, the United States, or India, the density of spruce budworms in a spruce-balsam forest, the size of the tuna stock in the Pacific Ocean, or the number of people per 100,000 infected with influenza or dying because of lung cancer.

The concept of rate of change is illustrated in Fig. 1-3. One of the simplest rate of change formulas used in environmental science is that for the *exponential growth law*. This law describes the growth pattern in which the rate of change of N with respect to time, dN/dt, is proportional to the quantity present at any time, N. Suppose that r represents the dif-

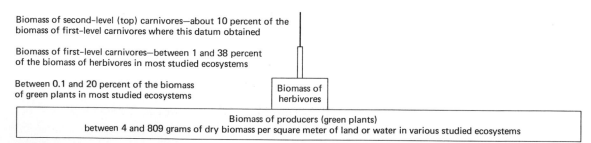

Biomass of second-level (top) carnivores—about 10 percent of the biomass of first-level carnivores where this datum obtained

Biomass of first-level carnivores—between 1 and 38 percent of the biomass of herbivores in most studied ecosystems

Between 0.1 and 20 percent of the biomass of green plants in most studied ecosystems

Biomass of herbivores

Biomass of producers (green plants)
between 4 and 809 grams of dry biomass per square meter of land or water in various studied ecosystems

FIGURE 1-2 The typical appearance of a trophic pyramid. Widths of the three top levels are each 10 percent of the width of the level immediately below. This width reduction is proportional to the typical biomass reduction from level to level in a trophic pyramid. The numbers mentioned were collated by H. T. and E. P. Odum (1959).

FIGURE 1-3 Rates of change of some variable N through time. Here the S-shaped curve depicts the typical curve for population growth with time. The rate of change at any point on the curve, such as A, B, or C, can be thought of in terms of a tangent to the curve at that point. Constructing a triangle about each of the three tangents, we see that when N is just beginning to grow, at A, the increment in N, dN, for a vanishingly small increment in t, dt, is small. The ratio dN/dt expresses the rate of change of N with respect to t at the point A. At B, growth has speeded up, and dN/dt at the point is larger. At C, growth has slowed down, and dN/dt is much smaller than at A or B. However, even though the rates of change at the three points are different, they can each be expressed in terms of a simple law:

$$\frac{dN}{dt} = rN \frac{N_{max} - N}{N_{max}}$$

where N_{max} is the maximum average size the population will attain and r is the instantaneous birth rate minus the instantaneous death rate. Because rates of change can often be expressed in terms of a simple formula, whereas the quantity (N in this case) cannot, formulas for rates of change figure prominently in environmental science as in all other sciences.

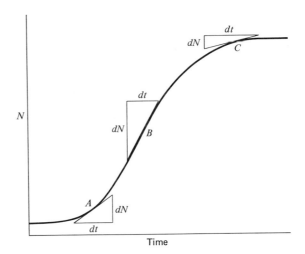

N Time

ference between the instantaneous rate of birth per capita and the instantaneous death rate per capita. Then r is the instantaneous net rate of population increase per capita. Then our exponential growth law is simply

$$\frac{dN}{dt} = rN$$

It can be shown, using integral calculus, that the value for N at any point in time is obtained from

$$\int \frac{dN}{N} = r \int dt$$

or

$$\ln N_t = \ln N_0 + rt \tag{1-1}$$

or

$$N_t = N_0 e^{rt}$$

where e stands for epsilon, the base of natural logarithms. The significance of Eq. (1-1) is that because populations grow at a rate proportional to the numbers present, population numbers should be converted to logarithms before subjecting them to any type of analysis.

By expressing variables in terms of logarithms rather than the raw data, we obtain a new way of looking at things that at first seems in violation of common sense, but actually yields more insight into the nature of the phenomenon.

To illustrate, suppose we wish to compare the fluctuations in two populations, which we shall call population 1 and population 2. For each of these, we have the following measures of density over a 10-year period:

year	population 1	population 2
1	3	302
2	18	516
3	70	721
4	25	613
5	7	347
6	2	234
7	4	372
8	21	523
9	90	876
10	13	440

Now if we plot trends in these two populations on ordinary graph paper, as in Fig. 1-4A, the fluctuations in population 1 seem insignificant in comparison to the fluctuations in population 2. However, if we plot the trends on semilogarithmic graph paper, as in Fig. 1-4B, it is population 1 that appears to have the most wide-amplitude fluctuations. The reason for the different effect of the two types of graph paper is that the logarithmic graph paper deemphasizes the effect of very different average densities of the two populations, and focuses attention on rates of change relative to population size, which is more biologically meaningful, in accordance with Eq. (1-1).

In nature, we might notice fluctuations in population 2 more than fluctuations in population 1, because the former is much more abundant. However, the latter is more *unstable*. This is the type of pattern we expect where diversity is low.

Subsequently, we will be concerned with cumulative effects, threshold effects, and interaction effects.

A *cumulative effect* is produced by some variable operating over a long period of time. Thus, the day-after-day exposure to air pollutants over two decades may have a lethal effect that would not be produced after brief exposure. A *threshold effect* is produced when the intensity of some causative agent rises above a certain threshold. It has been discovered, for example, that some animals do not begin looking for food until after their hunger level has surpassed a threshold. An *interaction effect* results when two or more variables interact so as to produce a combined result lesser or greater than what would be expected from the sum of each of them.

Throughout this book, we will be concerned with various measures of the variability in a set of observations. Two such measures are the *variance* of a *sample* and the *standard error* of a *population*. The concept of variability is illustrated in Fig. 1-5. The variance is a measure of the variability in a sample drawn from a population. The standard error, computed from the variance, is an estimate of the variability in the population itself.

In subsequent chapters there will be reference to *correlation* and *regression*. *Correlation coefficients* measure the extent to which variation in some one variable is associated with variation in another. *Regression coefficients* allow one to compute the value of a dependent variable Y, given corresponding values of an independent variable X. The meaning of these terms is illustrated in Fig. 1-6.

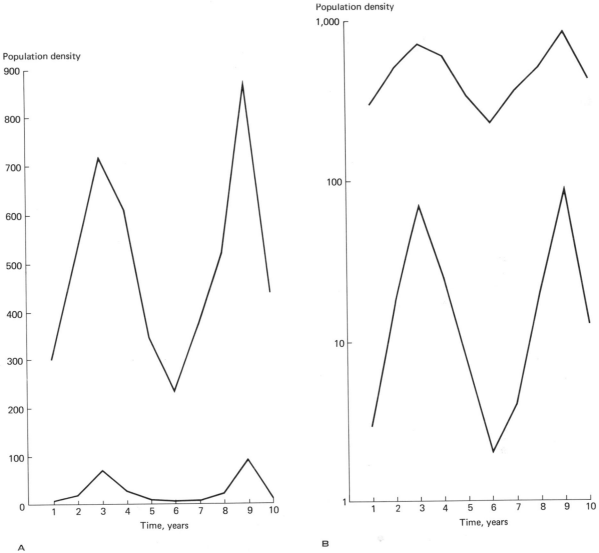

FIGURE 1-4 A Trends in densities of two hypothetical populations plotted on arithmetic graph paper. **B** Trends in densities of the same two hypothetical populations plotted on semilogarithmic graph paper.

Suppose we obtain pairs of measurements on two variables Y and X. If we are postulating that Y is somehow affected by X, then Y will be treated as the dependent variable and X as the independent variable. One of the simplest relations we could postulate is described by the straight line:

$$Y = a + bX$$

In this equation, b is a regression coefficient which can be calculated

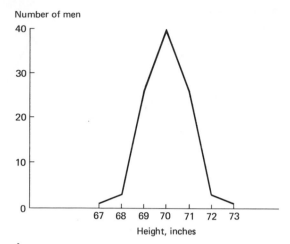

Number of men

A

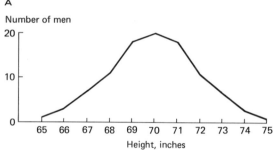

Number of men

B

FIGURE 1-5 The number of men at each height in two hypothetical samples of 100 men. There is very little variability in the height of the men depicted in A; there is a great deal of height variability in the men depicted in B.

from a set of measurements of Y and X. It measures the change in Y per unit change in X.

The correlation coefficient r expresses the degree of association between Y and X. The higher the value of b, the higher r is; and it becomes lower according to the greater scatter of the observed data points about the best line that can be fitted to the data using the statistical technique of least squares. This is an algebraic computation which computes the value of a and b for that line which minimizes the sums of squares of deviations of the data points about the line.

It is extremely important to point out that *correlation does not necessarily imply a causal relation between observed variables*. High correlation coefficients can occur by chance alone, or because the two variables are both being influenced by some third variable which has not been measured.

Regression coefficients can take any value. Correlation coefficients take only values from -1 to $+1$.

SCIENTIFIC METHODOLOGY   The chief theme of this book is that a wide range of phenomena can be understood in terms of a small number of *principles*. What is a principle? In this book, principle is a term used to describe broad generalizations which, to this point in the development of environmental science, seem

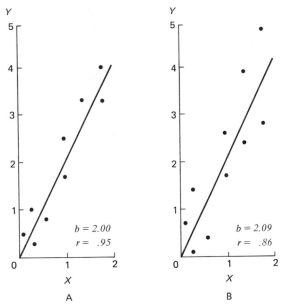
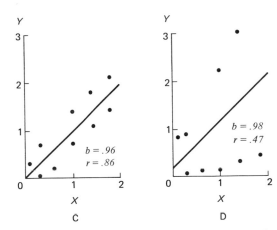

FIGURE 1-6 Four possible combinations of high and low regression coefficients (*b*) and high and low correlation coefficients (*r*). A and B have slopes about twice those of C and D, and the calculated regression coefficients are about twice as great. Also, the correlation coefficients are greater in A and B. B and D have greater scatter in the observed data, and consequently their correlation coefficients are lower than in A and C.

Even though the scatter in C is less than in B, the correlation coefficient is no greater, because the slope is less.

Thus, these four panels illustrate the dependence of the correlation coefficient on two factors evidently different in the four panels: slope and scatter.

to be useful in describing various specific situations. However, the principles have quite different origins.

For example, the first and second principles are thoroughly tested and accepted scientific laws, which have had a long history of demonstrated validity in many fields of science. Some of the other principles, however, are merely empirical generalizations, which seem to describe certain phenomena to the satisfaction of various workers. A few of the principles are accepted by only a few scientists, and they are still the objects of intense controversy. In subsequent texts in this field, they may be treated in more detail than here, and with more documentation. That will mean that they have withstood more tests of their validity, and are gradually being accepted as laws, rather than useful working generalizations, which are still widely regarded as *hypotheses*, requiring further tests to see whether they are generally applicable.

Other principles itemized in this book may not be remembered 50 years from now. It may be realized then that they applied to only a few special cases.

The student should not be upset by this. Science is a dynamic, cultural process expressing the evolution of human thought, and consequently it grows and changes, just as the living world changes.

The fact that so much is still to be discovered about the world, and so much of our present understanding is deficient or defective, should provide the student with a challenge to his imagination, initiative, curiosity, and creativity!

Four different methods of conducting research will be referred to in this book. Three of these can be thought of as variants of experimentation; the fourth is the comparative method which was so powerful in the hands of Charles Darwin.

The first method of testing hypotheses is the classical experimental approach (the method of "strong inference") which has been so prominent in the history of experimental physics, for example. To illustrate, suppose we were testing the hypothesis that reproductive rate of a beetle species was dependent on the population densities of beetles. Different vials, or cages, of beetles would be set up, each set of cages or vials containing different densities, so that one set would contain 4 beetles per cage, one set would contain 8, and others would contain 16, 32, 64, and 128. If the reproductive rate per beetle, on the average, in all cages is the same, then the hypothesis must be rejected.

A variant of this method is deductive, rather than inductive. That is, the hypothesis is tested, not by setting up an experiment to obtain data, but by using mathematics to deduce the consequences of an underlying assumption. To illustrate, let us assume that a population is growing in accordance with the logistic, or sigmoid, equation of population growth, as in Fig. 1-3. Where b represents the average birth rate and d the average death rate in the population, N is the population density at any point in time, N_{max} is the maximum population density that the environment can support on the average, and t represents time, the rate of change of population density with respect to time is given by

$$\frac{dN}{dt} = (b - d)N \left(\frac{N_{max} - N}{N_{max}} \right)$$

Now we might put forth the hypothesis that for any population growing in accordance with this law, there is some particular population density at which the growth rate of the population is at a maximum. We do not need to test this hypothesis experimentally, because given the preceding equation, which incorporates a set of assumptions, we can deduce whether or not the hypothesis follows from the assumptions expressed in the equation. From calculus, we can discover whether the preceding equation for rate of growth implies that there is a particular population density at which rate of growth is a maximum, proceeding as follows. We make use of the principle that when a rate, such as dN/dt is a maximum, the corresponding acceleration d^2N/dt^2 is equal to zero. We then compute the value d^2N/dt^2, equate it to zero, and solve for the value of N when the acceleration is zero (or the rate is at a maximum).

$$\frac{d^2N}{dt^2} = \frac{b - d}{N_{max}} (N_{max} - 2N)$$

Setting the preceding equation equal to zero, and solving for N, we obtain

$$2N = N_{max} \qquad \text{or} \qquad N = \frac{N_{max}}{2}$$

Thus, there is a particular value of N at which population growth rate is a maximum, and this value is half the maximum population density the environment can support. Note, however, that this generalization applies only under the specific assumptions we began with, namely, that the underlying population growth law was that given by the first equation. Thus, the deductive method also can be used to test hypotheses, but the conclusions are valid only for a given set of assumptions on which the deduction is based.

In the last few years, another method of testing hypotheses has become available: computer simulation. For example, we could program a computer with a mathematical model which mimics the behavior of a forest. Then we could test the hypothesis that of five alternative strategies for managing a forest, over a 100-year period, strategy 5 maximized the long-term productivity of pulpwood from the forest. The hypothesis would be tested by using the computer to simulate, or mimic, the behavior of the forest over the 100-year period, using each of the five alternative strategies. Clearly, this is a type of test that would not be feasible in nature because it would take too long and be too expensive, but using traditional mathematical deduction would not be possible either, because of the great complexity of the system of equations required to describe the behavior of the forest.

The fourth method of testing hypotheses is the comparative method of Darwin. Suppose we have the following hypothesis: Animal populations fluctuate less in response to a 10° change in temperature in a place where the temperature always fluctuates a great deal than in places where the temperature typically fluctuates very little. This hypothesis could be tested by collecting data on a great many populations in a great many places, and using a computer to compare the responses of each of the populations to a given amount of temperature change. No experiment has been done, but the hypothesis has been tested by comparing data on the same fundamental phenomenon in different situations.

It should be pointed out that some very powerful and ingenious methods of testing hypotheses can be devised by using combinations of the four methods just described. For example, we might use deduction (method 2) to obtain a new way of stating our hypothesis, then experimentation (method 1 or 3) or the comparative method (method 4) to test this new formulation of the hypothesis. Alternatively, we could use method 4, the comparative method, to obtain data, by computer analysis, which could then serve as the input to a deductive process which changed an equation so as to test a hypothesis.

We have now given a very brief introduction to the nature of the scientific method. Although it may sound simple, in practice a great deal of skill and experience are required to apply it.

REFERENCES Margalef, D. R.: Information Theory in Ecology, *Gen. Systems*, **3**:36–71 (1958).
Odum, H. T., and E. P. Odum: Principles and Concepts Pertaining to Energy in Ecological Systems, in E. P. Odum, "Fundamentals of Ecology," 2d ed., pp. 43–87, Saunders, Philadelphia, 1959.

2 the principles of environmental science

This chapter presents fourteen core principles of environmental science. This is done because in all sciences such a staggering number of data, theories, and models is accumulating in the literature that it is difficult to comprehend the significance of the material without recourse to a powerful underlying set of principles. This is particularly true in environmental science, an interdisciplinary field whose boundaries of knowledge are vague. Consequently, the order of presentation in this book is so designed that the student begins with a theoretical framework that will enable him to organize and grasp the detailed factual material presented later. Unfortunately, the full significance of these principles will not be clear until the student has read the rest of the book. Therefore, a suggested plan is to read the entire book and then reread this chapter in order to gain a deeper insight into the significance of the principles.

The set of principles should be stated in such a way that four desiderata are met.

First, the mode of presentation should expose logical relations between different principles in the set. It should be demonstrated that the data in this field are expressions of a powerful underlying general theory which has an internal logical structure of its own, as in chemistry, physics, Euclidean geometry, or metaphysics. A flow chart of the logical relations between principles appears later in the chapter for this purpose.

Second, the body of theory should be both a deductive and an inductive system. In addition to being deducible from preceding principles in the set, or from theory in other branches of science, each principle should be supportable by data and useful in accounting for data.

Third, the principles should be stated in a sufficiently fundamental, general fashion so that many principles already in the literature can be derived from them as corollaries, or special cases. In other words, it should be possible to reduce the total number of generalizations by stating principles so as to expose basic analogies between apparently different types of situations. This point is particularly important, because environmental science is now becoming a diffuse and fragmented field. The same principle is being rediscovered independently in several different subdisciplines without the authors' being aware that they are talking about some phenomenon which is already well understood, but in a quite different context.

Finally, the body of theory should be deliberately designed to be practical. It should be clearly indicated by the theory how one should proceed in building specific models to rationalize any given set of data.

Each principle is explained in accordance with the following scheme.

The principle is identified by a number, by which it will be referred to subsequently in the book. The principle is then stated.

The logical derivation of the principle from preceding principles is then given, if such derivation exists.

The evidence on which the principle is based is then briefly sketched.

Implications of the principle, special cases, extensions, and corollaries are then itemized.

Brief comments on the special implications of the principle for man, or the extent to which modern technological society operates in accordance with the principle or ignores it, are then presented.

PRINCIPLE 1   All energy entering an organism, population, or ecosystem can be accounted for as energy which is stored or leaves. Energy can be transformed from one form to another, but it cannot disappear or be destroyed, or be created.

No logical derivation for this principle is necessary, since it is merely a restatement of the first law of thermodynamics, one of the fundamental principles in the physical sciences.

This principle is also known as the law of conservation of energy. It can be expressed as a set of mathematical equations showing the equivalence of different types of energy. For example, $W = JH$ is a statement of the equivalence of mechanical work W and heat H, indicating that one unit of work is equal to J units of heat, where J is Joule's equivalent.

It follows from this principle that it must be possible to account for all the energy entering an organism, population, or ecosystem as energy stored, or energy leaving. The significance of this point is that living systems can be thought of as energy converters, and this suggests the possibility of identifying and characterizing different strategies for transforming energy. Consequently, it is just as useful and revealing for us to "keep books on calories" as it is for a business to keep books on the money paid into the business, and paid out. This calorie bookkeeping has become a major focus of research in ecology.

For example, we are concerned with how the incoming stream of calories, in the form of food, is partitioned by an organism into growth, reproduction, metabolism, and waste. In the case of animals, some of the wasted energy is lost in the feces, and some is lost after being eaten by exploiters (parasites or predators). Animal metabolism is further subdivided into the component that just keeps the basic machinery of the body functioning (basal metabolism) and that component which supports activity.

Figure 2-1 is a flow chart showing how the incoming stream of energy is partitioned into components for different purposes. The first partition eliminates a proportion of the incoming energy stream as unassimilated waste. The second partition divides the assimilated food into a component that will be used as fuel and a component that will be used as living matter. The third partition indicates the proportion of the energy lost by the average animal in this population to exploiters. The fourth partition splits the energy incorporated as matter into a portion that will

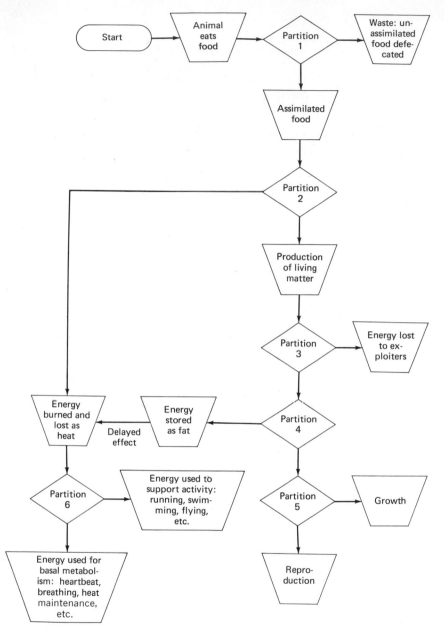

FIGURE 2-1 Illustration of the operation of the first law of thermodynamics in a hypothetical generalized animal. This flow chart is designed to show how the entire incoming stream of available energy can be partitioned into ultimate functions, by a series of partitions.

The complete disposition of all the energy entering the body is accounted for, using a flow chart of the type that would be used to describe a computer program. Each of the two geometrical shapes in the flow chart has a particular significance.

Diamond. This indicates partitioning of the incoming stream of energy into two components.

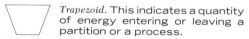
Trapezoid. This indicates a quantity of energy entering or leaving a partition or a process.

Note that this flow chart is not designed to mimic the order in which processes occur in the body; it is designed only to indicate how the ultimate disposition of food can be accounted for in terms of a small number of switch mechanisms, or partitions. In fact, of course, some conversion of food to fuel, fat, and protein would have occurred before all the waste was defecated.

be used immediately for growth and reproduction (protein) and a portion that is stored as fat, to be used later, as necessary for fuel. The fifth partition divides the growth and reproduction component into those two parts. The sixth partition separates the fuel into that used for basal metabolism and that available for activity over and above basic body functions.

Each of these partitions consists of a pair of proportions indicating how much of the energy that enters that diamond goes into each of the two outputs. We may think of each species of living organisms as playing a game against nature, in which success is measured by the change in biomass of the species from one generation to another. The game is being played successfully if the species maintains its biomass from one generation to the next, on the average (North Sea cod, painted lady butterflies, ants), and highly successfully if the species is increasing in numbers from generation to generation (man, white-tailed deer, opossum). It is not being played successfully if the species goes extinct (dinosaurs, passenger pigeons, saber-toothed tigers).

Different species, however, use very different types of games to keep their biomass up, from generation to generation, and these different games are largely defined by the values taken by the six pairs of coefficients in the partitions of Fig. 2-1. For example, partition 5 can be used to distinguish the "blue whale game" from the "herring game." In the blue whale game, most of the output goes into growth (blue whale females have only one young every second year, and it is only about 5 percent the weight of the mother). Thus, per annum, the blue whale partition 5 coefficients are:

Growth   .97
Reproduction   .03

In the herring, as in many fish and in insects and other invertebrates, the adult female reproduces every season she is alive (often only one season for invertebrates), and the proportion going into reproduction can be over 50 percent. Even in the case of a vertebrate, the field mouse, *Microtus arvalis*, the litter can weigh 53 percent of the mother's weight. In such cases, the partition 5 coefficients are:

Growth   .47 to .70
Reproduction   .30 to .53

What do these very different pairs of coefficients tell us about the fundamental difference between these two types of species? In the case of the blue whale, it is possible for the proportion of each generation going into reproduction to be small. This is because the initial size of the young will be enormous, and consequently the offspring will shortly be very large in comparison with most other organisms. Thus, it will have a fairly low per annum mortality rate due to predators, and it will be strong enough to swim against ocean currents that might otherwise take it away from areas of high food concentration.

Herrings, sardines, anchovies, smelts, and invertebrates are typically playing a very different game against nature. Their young are tiny and delicate, and very incapable of coping with air or water currents that blow them away from food or toward predators or parasites. However, although there is consequently a very low rate of survival per offspring, the initial number of offspring is so large (from scores of eggs to tens of thousands) that on the average exactly two survive to replace their parents.

Similar divisions which separate species on the basis of the games against nature they are playing can be found in the cases of the other five partitions. For example, if partition 6 puts a great deal of energy into activity, the species pays a price for this, in that it needs a lot of food for fuel, but in exchange for this payment it buys a lot of freedom from bad conditions at any particular place where it finds itself. A clam, a slow loris, or a lizard does not need to consume enormous amounts of food every day, but if conditions at its living site degenerate, then it must stay and perhaps die. A tiger, a man, and a hummingbird consume large amounts of food in proportion to their weight, but in exchange for the energy they burn off, they can exert a great deal of energy in flying, walking, or running to seek better conditions, including a richer food supply.

PRINCIPLE 2 No energy conversion system is ever completely efficient.

This is one way of stating the second law of thermodynamics, another of the fundamental principles in the physical sciences.

This means that while energy is never lost in the universe, it is constantly being converted into less useful forms. For example, the energy used by an animal is obtained in a useful, compact form, its food. However, the heat generated by an animal in running, flying, or swimming is lost, and of no further use. Indeed, there is a universal tendency for all energy on this planet to be irreversibly degraded into heat, which is lost by radiation out to space.

All biological processes are inefficient, in the sense that only a fraction of the useful energy input to an organism, population, or ecosystem is subsequently available for use by other organisms, populations, or ecosystems. Since energy is important to organisms, but energy conversion is inefficient, it follows that means for increasing the efficiency of energy use may often be important for organisms.

Table 2-1 illustrates the enormous energy loss down through a trophic pyramid. The term *trophic pyramid* derives from the fact that because of the great loss in usable energy down through the levels from green stuff to carnivores, each level can usually support only a much smaller biomass than the preceding level. Trophic pyramids are inverted, however, if we visualize the data as in Table 2-1, because the last line of the table represents the last trophic level to receive a unit of energy entering the ecosystem, and this level receives the smallest amount of energy.

Now, referring to partition 6, it will be noted in Table 2-1, column 4, that respiration as a proportion of consumption increases as we go up the pyramid. This is because plants just sit and herbivores are often surrounded by their food, needing to expend little energy to get it, whereas predators must expend a great deal of energy searching for and chasing their prey. The first column of Table 2-1 shows the tremendous loss of energy occurring down through the pyramid.

TABLE 2-1 Energy flow* in a terrestrial community

	flux-incident solar radiation, gross photosynthetic production, or consumption	respiration		efficiency
	λ_i	R_i	R_i/λ_i	λ_i/λ_{i-1}
Sunlight	47.1×10^8			
Vegetation	58.3×10^6	8.76×10^6	0.150	0.012
Microtus **(field mouse, herbivore)**	250×10^3	170×10^3	0.680	0.004
Mustela **(least weasel, carnivore)**	5824	5434	0.933	0.023

SOURCE: Golley (1960).
* All measurements in calories per hectare per year.

This type of table can be used for an interesting calculation. We can determine whether it is possible in principle for a given size of organism to live in a given environment. For example, suppose it takes a minimum population of 1,000 individuals to constitute a viable population of species X. Then from a table like Table 2-1, we can see how much incident solar radiation would be required to support this number of individuals; if a given environment does not have enough acreage to receive this much sunlight, species X cannot survive there. This is why very large species of fish do not live in small ponds or lakes, even if the ponds or lakes have physical characteristics the species could tolerate.

Another interesting problem on which Table 2-1 sheds some light is harvesting the sea. Many people feel that the oceans are an inexhaustible source of food. This is quite false, for two reasons. First, much of the world's oceans is very poor in mineral resources. The only very productive parts of the oceans are shallow; there minerals cannot sink to such great depths that they are lost to the system. Otherwise there must be strong upwelling currents to bring minerals back to the surface, where the phytoplankton use them to manufacture plant tissue. Shallow areas illustrating the first situation are Georges Bank, the Grand Banks, and the North Sea; one of the most spectacular instances of upwelling currents producing great biological productivity occurs off the coast of Peru, supporting an immense anchovy fishery.

The second reason for the low productivity of the oceans is the way we harvest oceanic trophic pyramids. Our harvesting strategy is analogous to harvesting lions and tigers on land, although Table 2-1 indicates that this would involve a tremendous energy loss. There is considerable inefficiency involved in converting energy from sun to plants, plants to herbivores, herbivores to carnivores, and small carnivores to large carnivores. When we harvest some of the ultimate or penultimate carnivores, such as tuna or salmon, we are using only a tiny proportion of the incoming solar radiation to the system because of all the losses due to conversion inefficiency that occurred at each of the steps just named. Thus, while we may be harvesting the top carnivores from a great expanse of open sea, the total amount of energy available to us is very small, compared with that being converted at plant or herbivore levels of the trophic pyramid.

definition of "resource"

A central concept in this book, whether we are talking about animals and plants in nature or the urban systems of man, is the notion of "resource." A moment's reflection reminds us that a resource is something useful, but how can this usefulness be measured, and how should "resource," therefore, be defined? One important biological measure of the usefulness of a resource is its effect on the reproductive rate. That is, the greater the availability of a particular resource that is not currently available in sufficient quantity, the greater the reproductive rate of that plant or animal population using the resource. However, many events can intervene in the life of a plant or animal between birth and the time an organism is ready to reproduce. Consequently, the beneficial effect of increasing one resource early in life might be overridden or concealed by the detrimental effect on reproduction of reducing availability of another resource later in life. Thus, the most useful and generally applicable measure of the effect of a resource is in terms of impact on energy conversion, not reproduction.

In practice, the rate of energy conversion in systems can be measured in many different ways. The energy content of the food eaten by an animal, or of its feces, can be measured in an oxygen bomb calorimeter, which measures the caloric content of materials by burning them. The rate of energy consumption of active animals is measured by measuring the oxygen consumption rate due to respiration. The rate of energy conversion in biochemical processes, such as photosynthesis, can be measured by radioactive tracer techniques. Thus, the photosynthetic rate in plants can be measured in terms of the rate at which the radioactive carbon isotope C^{14} is incorporated into plant tissue.

A resource will not have a positive effect throughout the possible range of availability levels. The positive effect occurs only while availability is being increased up to some optimal or sufficient level. Thus, we have the following definition:

A resource is anything needed by an organism, population, or ecosystem which, by its increasing availability up to an optimal or sufficient level, allows an increasing rate of energy conversion.

PRINCIPLE 3   Matter, energy, space, time, and diversity are all categories of resources.

It is clear from physical and chemical principles that energy conversion by a biological system must proceed at a rate that increases with the rate at which matter and energy are made available.

Space must operate in the same way. Consider the analogy between the concentration of fluorides and the distance from an animal to its nearest neighbor of the opposite sex in the same species. If the concentration of fluorides is higher or lower than the optimal concentration, this is harmful. If the distance to the nearest neighbor is too small, this indicates that the density of the population is likely to be so high that there will be interference with the mating process by other members of the population, owing to fighting between males, jostling, and other disturbances. If the distance to the nearest neighbor of the opposite sex is too great, the probability of finding a prospective mate will be too low, and the probability of fertilization actually occurring will be depressed from the highest level which can occur. Thus, the effect of space is fundamentally analogous to that of matter and energy, and space is a resource under the definition.

Time also is a resource, not just an independent variable. Consider, for example, desert mammals which depend for their existence on water holes. During a drying period, as the number of water holes decreases, the mammals must migrate from a water hole that has dried up to another that still has water. Whether or not this migration can be accomplished depends on whether the mammal has sufficient energy to keep moving long enough at mean cruising velocity to cover the minimum distance to another favorable site.

Many such instances could be cited, in which the probability of occurrence of an event increases with the amount of time available in which the event can occur. This is particularly apparent in the predation process. Predators are apparently efficient engines of destruction. One thing that prevents them from annihilating their prey and starving to death themselves is that time is a great enemy of a predator: it takes a surprisingly long time for a hunting pride of lions to stalk and run down a prey after they have found it. Three adult female lions have been clocked spending 45 minutes from the beginning of the stalk to the time they ran down a buffalo (Trimmer, 1962).

Diversity can sometimes be a resource. For example, the greater the variety of food eaten by a species, the less vulnerable that species is to an environmental change eliminating its food. That is, if one species eats only a single food species, it is vulnerable to local extinction if anything happens to that food item. For another species, that has 100 different food species, the chance of an environmental change that eliminates all 100 is remote. Consequently, more abundant and widespread species tend to be generalists, rather than specialists, simply because a higher proportion of the resources in the environment are available to them. Men, starlings, rats, grasshoppers, locusts, ants, tent caterpillars, and raccoons are generalists and abundant. The monarch butterfly, which eats only milkweed, the giant silkworm moths, each of which eats only a few species of shrubs and trees, and the koala bear, which eats only eucalyptus, are examples of organisms whose abundance is limited by a limited diet. This principle is only an empirical generalization which is valid on the average. An important exception is a species that eats only one or two plant species, provided these species are dominant items in the community over vast tracts. For example, the spruce budworm eats only spruce and balsam fir; but since these two species dominate the community over vast tracts of North America, the spruce budworm can become exceedingly abundant.

Figure 2-2 illustrates the effect of matter on biological productivity. This is the typical response of crop yield to chemical fertilizer; it is called Mitscherlich's law. It enabled the discoverer of the law to calculate how much fertilizer had to be applied in the field on the basis of pot experi-

Yield of crop, weight per acre per year

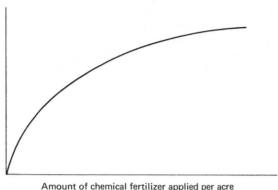

Amount of chemical fertilizer applied per acre

FIGURE 2-2 The effect of matter on biological productivity. The typical response of crop yield to increased application of a particular fertilizer is described by this kind of graph. This yield response curve is referred to as Mitscherlich's law.

ments in greenhouses. Figure 2-3 illustrates the effect of temperature on the rate of biological reactions. Here heat, one form of energy, is the resource. Figure 2-4 shows how populations grow through time, illustrating the role of time as a resource. Figure 2-5 shows the effect of space per car on the mean speed of traffic. Figure 2-6 shows the effect of the number of tree host species eaten by each lepidopteran species (butterflies and moths) on the relative abundance of different butterfly and moth species. This figure is intended to exemplify the effect of food species diversity on population productivity, since relative abundance is a measure of the amount of energy being converted by a species.

A great many specific implications follow from this general principle. A local concentration of energy flux will change the spatial distribution of matter by increasing the rate of energy conversion per population at that site. This is what a city is.

Matter is constantly being lost from a community of plants and animals when individuals die. Since constant cycling of matter through

Speed of development
(percent of total development completed in 1 hour)

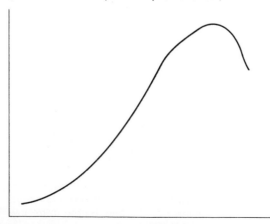

Temperature, °C

Figure 2-3 Typical example of the effect of energy on the rate energy conversion in biological systems. This is an idealized version of the effect of temperature on the speed of development of insect eggs. A large number of sets of experimental data conform to this pattern, including those of Powsner (1935), Birch (1944), and Browning (1952).

FIGURE 2-4 An example of the way in which time can be viewed as a resource. When a population is first introduced into a new environment, the population size builds up through the subsequent time period in a characteristic fashion until the population has grown to the maximum size the environment can support. Until this maximum size has been reached, the population size attained will be determined by the time that has been available for growth to occur, and in this sense, in such situations time is the limiting resource. The growth pattern for a wide variety of situations is described by this S-shaped curve, referred to as the sigmoid, or logistic, growth pattern. To demonstrate that many types of populations can be described by this curve, at least approximately, we can plot the population size of different populations on the same graph, by expressing all population sizes and times in dimensionless units. This is done by expressing the population size, for each population, as a percentage of the maximum size finally reached, and the times as percentages of the total time required to reach that maximum. To illustrate, the circles represent growth of the sheep population in South Australia from 1840 to 1890, and the squares represent growth of a laboratory population of the fruit fly *Drosophila melanogaster* over 21 days. (*Calculated from data of Davidson, 1938, and Pearl, 1930, respectively.*) The maximum size each population reaches is determined by the carrying capacity of the environment for that type of population.

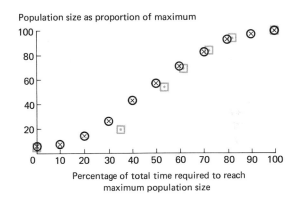

ecosystems is necessary for the production of biomass, this means that decomposer saprophagous organisms are critical to the maintenance of the ecosystem. Decomposers break down dead plants and animals so that the constituent chemical elements are again available to be converted into living tissue by plants.

Since not all resources will be equally scarce in relation to the needs of a biological system (individual, population, or ecosystem), the rate of energy flow through the system will be determined only by the

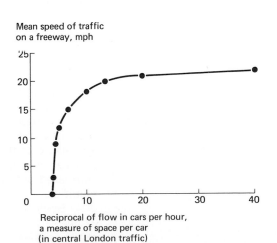

FIGURE 2-5 An example of the role of space as a resource: the effect of space per car on roadways in central London on mean journey speed of traffic. (*Calculated from data of Smeed, 1964.*) The technique of calculating the data points for this graph is illustrated by the following sample of three items:

flow in cars per hour	journey speed, mph	reciprocal of flow in cars per hour (space per car)
250	22	0.004
500	21	0.002
2,000	12	0.0005

Index of species abundance for forest Lepidoptera

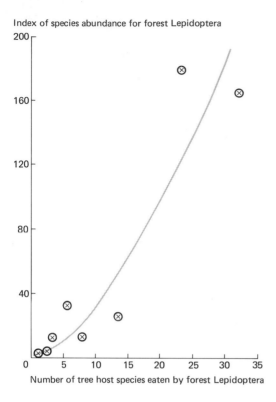

Number of tree host species eaten by forest Lepidoptera

FIGURE 2-6 Diversity as a resource. This figure shows the effect of the number of different tree host species eaten by species of forest Lepidoptera on the abundance of the lepidopteran species. The number of different species of trees eaten by an insect species is not a measure of the amount of food available, but rather the variety of food. Hence, it is one index of environmental complexity or diversity. The abundance of the different insect species is a measure of the rate at which the different species are processing energy per unit area of the habitat. (*Data from Watt,* 1965.)

resource or resources that are most scarce in relation to needs. Putting it differently, the rate of energy conversion will not be limited by resources that are superabundant in relation to biological needs. Of course, once the limiting resource deficiency is made up, some other resource becomes limiting. This is a generalization of Liebig's law of the minimum, which was based on observations on the effect of plant nutrients on crop yield. It typically applies to deficiencies of matter, but it could apply also to energy, space, time, or diversity.

A pollutant is a resource which is available in concentrations far in excess of a sufficient or optimal level. Since two of the components of matter and energy most likely to be limiting to terrestrial ecosystems are, respectively, water and temperature, it is reasonable to expect that a combined expression for these, actual evapotranspiration, would be highly correlated with production in different communities. Actual evapotranspiration is the amount of water entering the atmosphere from the soil (evaporation) and vegetation (transpiration) and thus may be thought of as the reverse of rain (Rosenzweig, 1968b). It is therefore a measure of the simultaneous availability of water and energy (heat). Rosenzweig has shown that actual evapotranspiration is in fact highly correlated with net aboveground productivity by plants in different communities.

Because at least one resource or pollutant will always be limiting in a finite environment, there will be a fixed upper limit to the total biomass of any species that can be supported in a given habitat. This fixed upper limit is referred to as the carrying capacity of that habitat for a particular species. It has been shown by Harper and White (1971), in comparing the carrying capacity of the environment for a variety of plant species populations, that

$$C = \overline{W}p^{3/2}$$

where C = carrying capacity

$\overline{W}$ = mean weight of individuals in the population

p = population density

This means that since C is fixed, we can increase the mean weight of the plants in a population or the population density, but not both. Increasing the mean density of the plants means that there will be a compensating decrease in the mean size of the plants. The same principle (lowered mean weight at higher density) is true of experimental animal populations, fish in ponds and reservoirs, livestock, and game. Consequently it is foolish to increase the stocking density in agriculture, forestry, or fisheries above that which allows for the desired mean weight.

Space itself is a resource, independently of any other resource in the space. That is, increasing the space per individual may have beneficial effects, even where other resources are held constant. An enormous variety of phenomena are explained at least in part by this simple point. In any population where egg-laying sites are scarce, the probability that a female can find an unoccupied oviposition site will depend on the competition by other females for that site. Here the critical variable is the space available per individual. The probability that prospective mates will find each other depends on the average distance between them. The probability that a predator will find a prey or a host will be found by a parasite within a given time period depends on the average distance the predator or parasite must traverse in order to find its victim.

This simple principle has many important implications for man. For example, because energy is a resource, and the amount of energy available, along with other resources, determines the carrying capacity of the environment, man's conversion from primary dependence on the sun to increasingly important dependence on fossil fuels has built up the carrying capacity of this planet for people. This carrying capacity will decrease suddenly, with tragic consequences, when we run out of fossil fuel, unless there can be a sufficiently massive development of nuclear power generation to replace fossil fuels. It is not so certain that this can be accomplished (Chap. 15).

People typically fail to see the significance of time as a critically important resource: that the probability of an event depends on the time available for it to occur. For example, it is widely believed that breeder reactors and the controlled fusion reaction will replace crude oil and natural gas before they are completely depleted; yet this belief ignores the very long lead times involved in developing new technologies and applying them on a vast scale. Particularly difficult are the problems involved in massive development of new technologies in underdeveloped countries short of capital and technological expertise.

The notion that space is a resource which can be limiting is being recognized more and more in our problems with saturation of airspace near large airports and with freeway congestion.

PRINCIPLE 4   For all categories of resources, when resource availability is already high, the effect of a unit increase in resource availability often decreases with increasing availability, up to some maximum. Beyond this maximum there is no further beneficial effect. [For all categories of resources except time and diversity further increase in resource availability beyond the maximum may be harmful because of toxic effects. This is the saturation or depletion principle.] For many phenomena there will be an increased probability of system breakdown as the maximum level of resource availability is reached. ⌐

The existence of a maximum follows in part from the first two principles. However, it is primarily an empirical generalization. For any process, there will be a fixed upper limit to the rate at which energy can be made available to support that process. Also, there will be fixed upper limits to the energy conversion efficiencies of whichever processes make energy available to the process in question. Further, as the limiting level for any process is reached, by increasing resource availability, there is an increased probability that a further increase in the resource will only force the process to bump into its limit.

This can be illustrated by reconsidering Fig. 2-3. There is a fixed upper limit to the heat that can be used by any biological system. This will be determined by a variety of factors, such as the species, age, temperature history, and time of year. Further increase in available heat, beyond the optimum, produces a decrease in the rate of biological energy conversion, in this case measured by speed of development. Similarly, optimal or sufficient levels of availability occur in all five categories of resources.

Figures 2-2 to 2-5 all illustrate this principle. Figure 2-6 does not, because even at the highest number of free food species, an optimal or sufficient level of food species diversity has not been reached. Thus Principle 4 must still be regarded as an empirical generalization. We do not yet have enough evidence to know whether it applies in all instances.

A first important implication of this principle is that there is often, as in Fig. 2-3, an optimum level of resource availability. For levels of resource availability either below or above this optimum, appropriately selected measures of the effect of the resource on the biological system will show decreased values. This has a very important consequence. To say there is some optimal level of resource availability per individual in the population is another way of saying that with respect to any single process there is an optimum number of individuals in the population for given availability of resources in the environment. This is why over long periods of time populations of plants and animals in a constant environment tend to fluctuate about a certain value, rather than gradually increasing or decreasing in numbers.

Density-dependent

Stated differently, there will be an intensification of the struggle for existence as the supply of resources per individual declines, or a relaxation as the supply increases. Consequently, excessive population density leads to population decline, and unusually low population density leads to population increase. In other words, trends in population density are strongly influenced by the current density. This phenomenon is referred to as density-dependent population regulation. Density operates on trend in population numbers by affecting the rates of birth, death, growth, and migration of individual organisms.

It is important to note, however, that many factors besides the current density of a population can operate to modify trends in population numbers that would occur if only current density were important. For example, changes in the abundance of a competitor or predator, when the upper Great Lakes were invaded by the alewife and marine lamprey, had a profound effect on trends in numbers of most fish species in the Great Lakes. Changes in the availability of food, pollution, and the weather (Chap. 5) can all have tremendous influence on population trends.

To say that populations tend to fluctuate about some equilibrium value over long periods of time does not necessarily imply that they return to this density after short departures from it. Rather, in some populations it implies only that large upward departures in density sharply increase the probability of a subsequent decline, and that large downward departures sharply increase the probability of a subsequent increase.

This principle also explains the evolutionary advantage that derives from aggregating in space, in high-density rookeries, roosts, colonies, nests, or burrows. There is an optimal space between individuals for various reasons. There is an optimal space to the nearest neighbor for purposes of information communication, feeding efficiency, and decreasing per capita vulnerability to predators.

In all cases the process being maximized is reproduction. In many cases there are other optima which must be compromised, such as the safety of an individual animal (from the soldier ant sacrificed in defense of the nest to the weak lion cub that drowns while a hunting pride swims across a stream to find higher food densities).

The advantage to prey in high-density aggregations is that the attack capacity of predators can be saturated. If each attacker can generate only K attacks per unit time at the most, and there are N attackers, then per unit time any prey in excess of NK must escape attack.

The advantage to prey of high densities is illustrated by Fig. 2-7A. At a host density of 100, 50 percent of the hosts are attacked. At a host density of 300, only 60, or 20 percent of the hosts are attacked. Clearly, for a constant number of parasites or predators, the hosts or prey minimize their percentage losses by aggregating in the largest possible groups. This must create intense selection pressure for large prey aggregations. It is also the reason for making convoys of ships and waves of bombers as large as possible during wartime. Percentage losses are minimized by saturating the attack capacity of packs of submarines, or the combination or fighter planes plus antiaircraft guns. This evolutionary advantage to clustering is one of many driving forces which

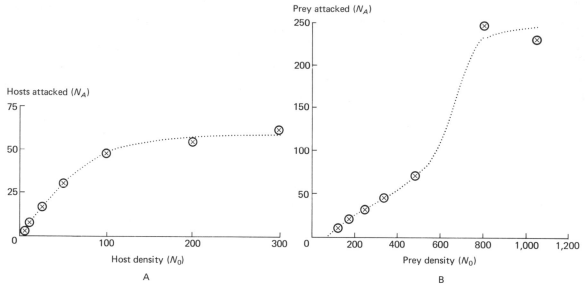

FIGURE 2-7 A illustrates the nonstimulatory relation between resource availability and resource utilization, and B illustrates the stimulatory relation. A is based on searching by the parasite *Nasonia vitripennis* for puparia of the housefly *Musca domestica*. Resource availability, plotted on the X axis, is density of the puparia vulnerable to attack by the parasites (N_0, for numbers of prospective hosts at the origin of the experimental time period). Resource utilization, plotted on the Y axis, is the number of hosts that the parasites were able to attack within the experimental time period (N_A). (*Data from De Bach and Smith*, 1941.) There is a nonstimulatory relation because the parasites do not learn how to increase their efficiency of resource utilization at intermediate availability levels. In B, resource availability is an index of the number of sawfly cocoons per acre which are vulnerable to attack by the deer mice (N_0). Resource utilization is the number of cocoons opened per mammal per day (*Holling*, 1959). The small mammals do learn how to increase their efficiency of resource utilization.

The second case points up a very interesting feature: The division of resources into stimulatory and nonstimulatory is *environment-specific*. In an environment without dog food, deer mice open cocoons in a nonstimulatory way.

creates organized colonies of men, ants, bees, tent caterpillars, baboons, and other organisms.

Of course, clustering has many other advantages. It aids in attack, as well as defense. In defense, it causes confusion of the attackers by numbers of targets, as well as saturating attack capacity.

The reason why greatly increased availability of time does not have a negative effect is as follows. The greater the length of time during which some event could possibly occur, the higher the probability that the event will in fact occur. Therefore, for those processes for which increased length of time is an advantage, there is no optimal time beyond which the magnitude of the advantage decreases.

Increasing density of any species, including man, has an increasing cost per individual in terms of all five categories of resources. In the case of time, the loss is due to interference (jostling, fighting, for example), waiting for a chance to get at a resource, such as a water hole, queuing, traffic congestion, or rerouting in the case of man. In the case of diversity, increasing density of one species if it is already abundant means that there will be, on average, less of the five categories of resources for other species that compete because of overlapping resource requirements. Consequently, the probability of local extinction of other species is increased, and this leads to a decrease in species and species-pattern diversity, which means that there is an increased potential for local environmental instability (see Principle 13). This instability works to the detriment of all species, including the one which caused it.

The reason why we had to specify that increases in density of *already abundant* species lead to a decrease in species diversity can be illustrated arithmetically. Suppose we have a tiny community of fifteen individuals, distributed in five species, with the population sizes being 5, 4, 3, 2, and 1. We may now ask what happens to the species diversity in two cases: in the first case there is an increase in the rarest species at the expense of the others, and in the second there is an increase in the most abundant at the expense of the others.

In the first case, the diversity changes from

$$\frac{15!}{5!4!3!2!1!} \quad \text{to} \quad \frac{15!}{4!3!2!1!5!}$$

so that there is no change in diversity. We could, indeed, adjust the numbers so that there is an increase.

In the second case, the diversity changes from

$$\frac{15!}{5!4!3!2!1!} \quad \text{to} \quad \frac{15!}{9!2!2!1!1!}$$

so that diversity decreases by a factor of 42.

The fact that the probability of systems breakdown rises exponentially as saturation or depletion is approached has important implications for man. Thus, for example, a 1 percent increase in automobile traffic on a freeway system has a vastly greater effect on the probability of system congestion when the system is already 90 percent saturated than when it is only 50 percent saturated.

This principle is violated or ignored in much long-term planning. For example, the market for airline transportation, or anything else, cannot continue to grow at the same growth rate every year, indefinitely, because ultimately the demand for the product is satisfied. Similarly, use rates for any substance cannot grow indefinitely at the same rate every year, because ultimately the resource will be depleted, whether it is silver, chromium, petroleum, or wilderness camping grounds.

PRINCIPLE 5   There are two fundamentally different types of resources: those whose increasing availability stimulates still further use of them and those from which this stimulatory effect is absent. *food problem*

There are two variants of Principle 5. On the one hand, we can imagine situations in which the availability of a resource has no effect on stimulating utilization of the resource. On the other hand, we can imagine at least two kinds of situations that will have such a stimulating effect. Figure 2-8 illustrates the difference. An example is the situation in which an animal is looking for various kinds of food. If it discovers that a particular type of food has suddenly become unusually abundant (ripe wild raspberries, for example), it will learn to focus its attention on that

Resource utilization

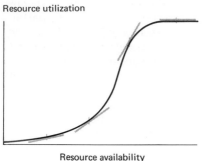

Resource availability

A

Resource utilization

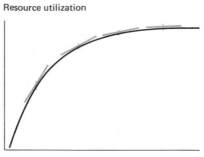

Resource availability

B

FIGURE 2-8 An idealized representation of the difference between a stimulatory situation (A) and a nonstimulatory resource utilization situation (B). In the case of B, each tangent drawn to the curve has lower slope than tangents drawn at lower levels of resource availability. That is, there is no range of values for resource availability within which increased resource availability increases the tangents to the curve (increases the rate of change to resource utilization with respect to resource availability). In the case of A, there is an intermediate range of values for resource availability in which increased resource availability increases the tangents to the curve. This happens because the resource utilization process is stimulated by the finding of more resource.

kind of food. Thus, increased resource utilization stimulates still more increase in resource utilization.

A and B of Fig. 2-7 illustrate nonstimulatory and stimulatory relations between resource availability and utilization, respectively. In both cases resource availability is the density of a food item which is being searched for. In A the resource utilization is numbers of puparia of the housefly attacked by a parasite, and in B resource utilization is numbers of sawfly cocoons opened per deer mouse per day.

A large number of commonly observed phenomena are illustrations of Principle 5. For example, where a population of organisms is growing in a new environment, the critically limiting available resource is time, and a measure of the extent to which the resource has been utilized is the size the population has attained. Thus we substitute population size for resource utilization in Fig. 2-8A, and time for resource availability. The reason why population growth constitutes a stimulatory resource utilization situation is as follows. The rate of population growth increases as population increases, simply because greater population implies more mothers to bear children. Thus, at intermediate levels of the critically limiting resource (time), growth rates are higher than when the population has just begun to grow. As the carrying capacity of the environment is approached, however, growth rates decline. Time is no longer the limiting resource, which is now food, space, shelter, or some other resource which operates to decrease birth rates and perhaps also increase death rates. As Davidson (1944) first pointed out, the ascending limb of the curve describing the relation between temperature and rate of egg development is perfectly described by the same curve that describes population growth with time. Here small increases in temperature have a stimulatory effect on rates of biological reactions, but as the optimal temperature is reached, deleterious effects begin to slow down the approach of reaction rate to the maximum possible reaction rate.

Figures 2-2 and 2-5 resemble Fig. 2-8B, and hence these involve nonstimulatory effects of increasing resource availability.

Now what does resource utilization really measure? For an individual plant or animal, it is the rate of conversion of energy per unit time. For a population, where there is no change in the rate of energy conversion per organism per unit time, it is simply N, the number of individuals in the population.

PRINCIPLES PERTAINING TO DIVERSITY

chemicals killing pests

PRINCIPLE 6   Those individuals and species which have greater numbers of offspring than their competitors tend to replace their competitors.

This is a statement of the theory of evolution of Darwin and Wallace. With the omniscience that invariably accompanies hindsight, it should always have been clear that if there is a heritable difference in the degree of adaptation to the physical or biological environment, and if population densities are high enough so that competition occurs, then it is the least-adapted that will be eliminated.

On the average, we expect the adapted to replace the nonadapted. This implies that the adapted leave a larger number of offspring than the nonadapted. The better-adapted individuals or species are those that have a more deleterious effect on others than others have on them (through fighting or interference) or are best-adapted to predation or the physical environment, or disease, or available food. In some cases, the better adapted individuals or species will be those that make most efficient use of resources.

On the average, then, species and communities characteristic of a particular environment are able to persist there because on balance they have a higher reproductive rate than any prospective invaders.

However, it follows that if conditions change, a different species or set of species may be better adapted. Thus, for example, the first wave of plant species to invade a sand dune or a cooled-off lava flow, by their actions in creating humus make the environment a suitable habitat for subsequent waves of species which then replace the pioneers. This process of successive replacement of waves of species one after another is called *succession*. Another interesting example of this replacement phenomenon occurs among the parasite complexes which attack caterpillars of species which fluctuate through a very wide range of densities. (Certain forest insect pests can attain population densities in peak-abundance years that are 10,000 or even 100,000 times as great as those of lowest-abundance years.) It appears that different parasite species have different degrees of tolerance of high density in their own species, that is, to available space in the sense of distance to the nearest neighbor. A parasite which performs well at low host densities because of high searching efficiency may be replaced at high host densities by other species for which the optimal distance to the nearest neighbor is smaller. This results in the rather odd situation that for such pests as the spruce budworm, which fluctuate through a very wide range of densities, we find one complex of parasite species at low densities of the budworm but another complex at peak budworm densities. Curiously, those parasite species which are commonly found attacking the budworm at outbreak densities may be very difficult to find at all at low budworm densities. Also, parasites which appear to be important controlling agents of budworm at low budworm densities may be very difficult to find at outbreak densities of the budworm.

Individuals, species, and ecosystems evolve mechanisms to make most efficient use of the energy in their habitat. Thus, fish which eat other fish have mouths at the front of their heads, those which eat surface insects have mouths at the top of their heads, and those which eat bottom fauna have mouths at the bottom of their heads. Fish which pick food particles out of corals have long, thin pointed snouts. Animals that live in very dry places have the necessary adaptations to water conservation, as cactids. Animals which live among tall grasses are adapted for bounding, like the antelope, deer, kangaroo, kangaroo rat, grasshopper, and rabbit.

A major implication for man is that we should think twice before replacing native game by imported livestock. Conversion of the American West from a buffalo grass system to a Shorthorn-Hereford system was an unwitting strategy decision that might be reevaluated. In Africa, it certainly appears clear that native game use the environment more effectively than imported livestock in terms of pounds of meat produced per acre per year (Chap. 10). Cattlemen are forced to be at odds with sheep herders if sheep are kept at high stocking densities because sheep are adapted to grazing grass so close to the ground that it is inaccessible to cattle.

PRINCIPLE 7   The steady-state diversity of communities is higher in predictable environments.

This follows from Principles 4 and 6. There is some fluctuation of environmental conditions in all habitats, but the predictability and magnitude of the fluctuations vary enormously from habitat to habitat (Chap. 5). Since there are optimal conditions, it is important to a species what proportion of the time a particular environment is optimal for that species. Further, by chance alone, in any environment, we expect a distribution of species, with respect to abundance, from very abundant (a few species) to very rare (a large number of species). If the physical conditions in an environment fluctuate so unpredictably and violently that all species are reduced in numbers thereby, there is created a situation intolerant of rare species, whose decline in absolute magnitude would be such as to militate against the probability of reestablishment.

In short, a stable physical environment is one in which a large number of species, common and rare, can evolve adjustments to conditions optimal for them. An unstable physical environment, on the other hand, is best suited for a smaller number of species, most of which are fairly abundant.

This principle has been described at length and in great detail by Valentine (1969) on the basis of fossil evidence, and it has been called the stability-time hypothesis by Sanders (1969). He found that in marine bottom fauna communities, the greatest species diversity (variety of species per sample of a fixed size) was found in habitats which had been stable over very long time periods. Slobodkin and Sanders (1969) interpreted this as the effect of environmental predictability on species diversity. That is, the longer the period over which there has been a high degree of environmental predictability in a particular habitat, the greater the species diversity that will have evolved there. This need not mean a warm, mild climate that fluctuates little from day to day within a year, or from year to year, as in Trinidad. It may mean also a cold, but very stable physical environment, as on the ocean bottom.

A stable physical climate over a long period of time produces not only great species diversity (variety of species); it produces also great pattern diversity (Pielou, 1966) with respect to the arrangement of entities in space. Thus, great physical stability in an environment allows for the accumulation of a higher degree of diversity in more than one sense.

What we mean here by predictability is pattern regularity. There can be a large temperature difference, for example, between night and day, or between the coldest and hottest days of the year. However, if the temperature (or light, or rainfall) is exactly the same at a particular time in one cycle and at the corresponding time in the next cycle, this is a perfectly predictable pattern, and the organism can evolve a response to it. Organisms cannot evolve a perfect response to the situation in which the coldest month one year has a mean monthly temperature 40°F lower than that for the coldest month in the same place the preceding year.

PRINCIPLE 8   Whether a habitat becomes saturated with diversity of species in a particular taxon depends on how niches are separated in the taxon.

It is reasonable to expect that if a particular taxonomic group of organisms is characterized by remarkably specialized environmental requirements, then each species has a narrow niche, and an environment can accommodate many such species without competition between them. If, on the other hand, a particular taxonomic group is characterized by species with very generalized feeding habits and tolerance to very wide ranges of environmental variables, then an environment can accommodate far fewer species.

Whittaker (1969) has noted that the niche responses of birds are to broad features of community structure and broad food categories, so that a community can readily become saturated with bird species. However, plants and insects have much more specialized environmental requirements. Many insect species, for example, eat only one or a small number of species of plant or animal foods. Plants and insects are more responsive than birds to microtype environmental differences between niches, such as those due to subtle differences in the biochemistry of the host or the environment. Thus, in the world of insects and plants, as the species themselves develop subtle biochemical differences and become genetically different, there is a self-augmentation of diversity through time, and much less likelihood that the community will become saturated with species in such taxa (taxonomic groups). That is, as species of plants and insects gradually evolve subtle biochemical differences from one another, the number of biochemically distinguishable niches in the environment increases, and it becomes possible for still more species of plants and insects to evolve.

PRINCIPLE 9   The diversity of any community is proportional to the biomass divided by the productivity.

Morowitz (1968) has argued that there must be a relation between the biomass, energy flow, and diversity of biological systems. Detailed presentation of his argument would digress too far from our main discussion here, but, in skeletal form, it may be put as follows. Suppose a system stores an amount of material B (for biomass) and has an energy flux through that material P (for productivity, a measure of net energy flow per unit time). Now suppose that the energy flow has associated with it a corresponding flow of matter. Suppose further that this matter is free to exchange with the stored matter. Then the average amount of time that material spends in this system will be given by

$$\bar{t} = K \frac{B}{P}$$

However, D, the diversity or organizational complexity of the system, is also proportional to $\bar{t}$ because D is a measure of the average amount of time that energy spends in the system on its way from the source to the sink (this follows from thermodynamic theory). Consequently, we have the proportionality

$$D = K \frac{B}{P}$$

Margalef (1969) has presented a graph showing that the primary productivity per unit of biomass is a decreasing function of the diversity (a measure of organizational complexity) in ecosystems. Carlander (1955) has shown that the standing crop of fish in different reservoirs is an increasing function of the number of species present. In this case, P is fixed by the incoming energy from the sun, and standing crop or biomass would be expected to increase in proportion to the species diversity, a measure of the organizational complexity in the community.

This principle means that the energetic efficiency of biological systems increases with increasing organizational complexity. That is, the energy cost of maintaining a unit of biomass decreases with increasing diversity.

PRINCIPLE 10   The ratio of biomass to productivity increases through time in a stable physical environment, up to an asymptote.

This follows from Principles 7 and 9. If D increases through time in a physically stable habitat, and D is proportional to B/P, then B/P must increase in a physically stable habitat.

This is an extremely important principle, because it means that biological systems evolve in the direction of increased efficiency of energy use, where the physical environment is stable enough to permit the accumulation of diversity. That is, if the maximum possible P is fixed by incoming solar radiation, but D and B increase through time, then a quantum of energy made available to a biological system is being used to support more biomass, because of a higher degree of organizational complexity.

Many types of examples illustrate evolution in the direction of maximizing the efficiency of energy use. Some such examples pertain to the morphology, physiology, or behavior of individual animals; some pertain to community energetics.

The size and shape of individual organisms appear to be determined in part by evolution in the direction of minimizing energy loss from the surface of the organism to the environment. This is a combined statement of Bergmann's rule and Allen's rule. Bergmann's rule states that homoiothermic animals from colder climates tend to be larger in size and hence to have a lower ratio of surface to weight than their relatives from warmer regions. Allen's rule states that there is a tendency toward shortening of extremities in colder climates, relative to the weight of the animal. Gates (1969) has developed a theory of climate spaces (air temperature, wind, and radiation) which any animal must occupy in order to survive thermodynamically, based on its intrinsic properties. His models show that increased body size has a dramatic effect in increasing the ability of animals to withstand cold temperatures.

Rosenzweig (1968a) has shown that while body size in eleven mammal species tends to be negatively correlated with temperature in their habitat, the positive correlation with latitude is about equally useful as a predictor. This may mean what while it is true that larger animals tend to occur in colder places, and maximizing efficiency of energy transfer may be the reason for the large size, the causal pathway is not only through the effect of the smaller surface/volume ratio in minimizing heat loss. In addition, Rosenzweig points out, larger body size may be useful for food storage in evening out the impact of year-to-year variations in the severity of the winter. A more complicated explanation is related to the theory of niche sizes. The greater long-term physical stability at lower latitudes has allowed for greater specialization, and hence niches there are smaller, on the average. The third hypothesis offered by Rosenzweig to explain the increase in body size with latitude springs from the observation that species diversity decreases with approach to the poles. Lowered numbers of prey species mean that a successful predator species with a constant degree of specialization will receive a larger and larger proportion of community production with increased proximity to the poles, and this would also operate to increase mean body size. Many possible explanations for increased mean body size near the poles, however, invoke the notion that evolution proceeds in such a fashion that species make the most efficient use of the energy available to them.

Royama (1971) has speculated that there may be selection for maximization of hunting efficiency in bird predators. We would expect to see in mature communities the employment of distribution and dispersal patterns which make the most efficient possible use of energy.

If this principle is correct, then we would expect to find that in a particular region of the world, the biomass/productivity ratio for communities is higher for later than for earlier successional stages. In fact, this ratio does increase for later successional stages, because of the increased number of species present, and the development of woody trunks which can support multistoried canopies of vegetation.

Several important implications follow from this principle. A trophic level or a community can be kept immature by wide-amplitude, irregular weather fluctuations, predation by man or other animals, or any perturbations with a low level of predictability, such as flooding.

Margalef (1968) has noted that the animals at the ends of major food chains are more efficient than those near the beginning. This is to be expected if this principle is true, because larger animals tend to be more immune to environmental fluctuations than smaller animals because of their greater size and mobility (range); also, environmental perturbations impacting on the community will have their influence dampened by passage through different levels of the trophic pyramid before they affect the top predators. Consequently, top predators in a trophic pyramid in effect live in a more stable environment than the herbivores, and have had a longer period of time with any given degree of stability; hence they should have a higher biomass/productivity ratio. Also, energy is scarcer at the ends of food chains, and thus efficiency is more important in determining fitness.

One would expect that natural populations and ecosystems would be organized in space so as to minimize the energy cost of movement (or transportation) relative to the energy gain from an activity. This point is the germ of a theory of distribution, dispersal, and food gathering and transportation, which ties together such diverse phenomena as the pattern of dispersal of starlings away from a starling roost every day, the distribution of sand crab burrows on a beach, and the role of space location in economic geography of urban dynamics.

The most important significance for humanity, regarding this principle, is that we are violating it. As human society becomes more developed, there is a decrease in the energy cost per unit of gross national product, but at the same time the gross national product per capita is rising so rapidly that there is an increase in the energy cost per person.

PRINCIPLE 11   Mature systems exploit immature systems.

That is, mature ecosystems, trophic levels, and populations remove energy, biomass, and diversity from their immature counterparts. Putting it differently, energy, matter, and diversity flow along a gradient of increasing diversity, or organizational complexity, or from subsystems of lower diversity to subsystems of higher diversity (Margalef, 1968).

This follows from Principle 5, which states that increasing availability of a resource (such as diversity) may stimulate more utilization of that resource. It also follows from Principle 9, which states that increasing diversity in a system implies increased efficiency of energy use. One way to increase efficiency of energy use is to exploit other parts of the system which expend their energy gathering up matter and energy. As Margalef has indicated (1963, 1968), this is a surprisingly widespread occurrence, which explains all kinds of specific situations. Energy in a plankton system flows along a gradient of increasing diversity.

But in addition, young people move from farms, villages, and towns to big cities, so that age distribution also changes along a diversity gradient. Poor countries, states, and cities train doctors, teachers, scientists, engineers, and other professionals at great expense, who then take their training to cities with more economic activity (greater diversity) that will be able to pay them more. Thus talent moves along a gradient of increasing diversity. In almost all economic transactions between developed and underdeveloped countries, even though they are supposed to benefit the underdeveloped country, there is greater net economic benefit to the developed country. Birds disperse out of a woodlot into the surrounding grassland, and by constantly cleaning it of seeds and other biomass and stored energy, they suppress its succession to maturity. Ants and bees constantly move material from the area which they exploit back to their nests, just as cities gather up material from the surrounding countryside.

This principle implies that more diverse systems or subsystems not only parasitize their less diverse neighbors, they actually suppress development of diversity in other parts of the system, thus preventing them from reaching maturity, with its associated high diversity. For example, smaller cities within 200 miles of New York, Chicago, San Francisco, or Los Angeles have notable difficulty in developing competitive variety of shops, entertainment or cultural offerings, or diversity of economic activity. The most talented young people, the most innovative people, and the capital constantly gravitate toward the largest city.

This principle also explains the curious phenomenon pointed out by Huffaker (1971) that wolves have a territory limited to one-tenth of a square mile, which keeps the wolf population too low to prevent periodic eruption of the prey population. As Huffaker notes, the size of the territory and its compressibility may be an evolved response to the typical mean food level, which may be far lower than the maximum food density which sometimes occurs. This may mean that the wolves have evolved a higher degree of organization (diversity) than their food, and constant predation by wolves and other predators keeps the prey trophic level immature. Hence it never grows rich in the kind of organization that would be reflected in genetic and other mechanisms preventing wide-amplitude population fluctuations.

PRINCIPLE 12   The perfection of adaptation of any attribute depends on its relative importance in a given environment.

This follows from Principles 6 and 7. If selection occurs, but diversity increases through time in a stable environment, then we would expect to find continuing improvement in the adaptation to a particular environment. Thus, it follows that in mature ecosystems that have evolved in a stable physical habitat, the need for homeostatic responses to wide-amplitude, unpredictable fluctuations in the physicochemical environment will not be necessary, and instead we will find sensitive behavioral and biochemical adaptations to the biological and social environment. These adaptations will often involve types of phenomena not seen in temperate zones, such as the brilliant colors of tropical insects and fish, and biochemical integration of mature communities frequently leading to relations mediated by complex communication systems of various kinds (Margalef, 1968).

Evolution in an unpredictable environment calls for maintenance of a plastic response on the part of the population. This implies a high degree of diversity. Evolution in a stable but biologically complex environment equates success with a perfect, complex, but rigid set of responses to an information-rich set of environmental cues.

A very important body of research bearing on the evolution of adaptations to the biological rather than the physical environment has been developed in connection with coevolution of butterflies and plants (Ehrlich and Raven, 1965). It happens that there are consistent patterns of association between particular taxonomic groups of butterfly species and particular taxonomic groups of plants. Further, a body of experimental work exists which explains these patterns of relations in terms of the chemical nature of the plants and the effects of those chemicals on the different species of butterfly larvae. For example, Thorsteinson (1953) found that larvae which normally ate Cruciferae would also eat other plants if certain mustard oil glucosides from the Cruciferae were smeared on the other, normally unacceptable leaves. In general, it appears that the choice of food plants by butterfly larvae is determined by a combination of chemical attractants in the plants of their choice, and repellents in plants they do not eat. It is noteworthy that butterflies are basically a tropical group, and consequently have had ample time to coevolve with plants in a stable physical environment. It is also noteworthy that the fantastically wide-amplitude fluctuations found in some moth larvae are all characteristic of cold and continental climates and high mountain habitats.

The same type of evidence for biochemical interactions between insects and plants has been demonstrated by Way (1971) for aphids.

Connell (1971) has reported that the unpredictable barnacle supply of Scotland has a generalist predator, whereas the more predictable barnacle supply of the state of Washington has a specialist predator. The highly adapted but rigidly specialized predator is the product of the more stable of the two environments. Connell also notes that around tropical tree species there are few of their own seedlings, but the young of other tree species can grow. This indicates the importance of biological adjustments within and between species in a mature community, and also supports Principle 7. The more diversity accumulated in a community, the more likely it is to accumulate, because of the development of mechanisms which actually promote the accumulation of community diversity.

An important implication of this principle is that there is no single best evolutionary strategy; which strategy is best depends on the stability of the physical environment.

A corollary of Principle 12 is that populations in immature systems respond less to a unit change in the physicochemical environment than populations in mature ecosystems. Populations in environments with physicochemical stability for a long time have little need for evolved ability to adjust to instability. The populations in mature ecosystems are vulnerable to being exterminated by catastrophic changes in the physical environment because of their genetic rigidity. Thus, an important consequence of living in a highly predictable environment is that it leads to irreversibility, because selection pressure places a premium on perfection, and hence rigidity of response. Since there is little free energy available in a mature community to adjust to great environmental perturbations, the latter may cause a totally unsuspected magnitude of response. This is particularly true because of the complex system of control mechanisms jointly linking all species in the community. Such mechanisms mean that a perturbation applied anywhere in the system will produce shock waves throughout the causal web. Because many of the control mechanisms are exponential, and have threshold effects, cumulative effects, and interaction effects, some consequences of the original perturbation may have much wider amplitude deviation from mean values than the perturbation itself. This is the amplifier effect. The crown-of-thorns starfish may be an excellent example. The increasing population of this starfish is threatening to destroy the corals that make barrier reefs, and hence the most complex communities known on earth.

Chapter 5 reports on a large study to determine how much populations fluctuate in stable, maritime climates, compared with extremely unstable continental climates. It turns out that the populations fluctuate more in the maritime climates, in relation to their mean levels of abundance, because they are much more sensitive to a unit change in weather than the populations in continental climates, where weather fluctuates a great deal more.

PRINCIPLES PERTAINING TO POPULATION FLUCTUATIONS

PRINCIPLE 13   Physically stable environments allow the accumulation of biological diversity in mature ecosystems which in turn promotes population stability.

Four lines of argument based on preceding principles support this principle.

First, Principle 7 states that organizational complexity accumulates in a physically stable environment. In other words, there will be a great increase in the number of species and the variety of trophic pathways in the community. This means that in a mature community, the number of pathways by which incoming energy flows through the system is increased, and if anything should go wrong with one pathway, the number of others which can replace it is higher than would be the case in a simpler community (a less mature community). Thus, in a mature community, there is more spreading of risk, and consequently an increased degree of stability within the ecosystem, provided there is a persistently high degree of stability in the physicochemical environment. This argument is well understood by field biologists, and it was stated formally in the context of information theory by MacArthur (1955).

Second, Principle 13 follows from Principle 7 for another reason: if stability of the physical environment is a prerequisite for the accumulation of organizational complexity and biological diversity, then that physical stability itself will promote population stability in a mature ecosystem.

The third argument in support of the validity of Principle 13 follows from Principle 12, which states that sensitive and complex adaptations and control systems will have evolved in response to the biological and social environment in a mature community. These adjustments will operate to ensure homeostasis and stability. The mature community is a tremendously complex feedback control system.

Finally, another argument for Principle 13 follows from Principle 9, concerning the relation between maturity (high diversity) and efficiency of energy use. High efficiency implies minimal waste, and wide-amplitude population fluctuations are made possible by a high rate of biomass turnover (birth rate less death rate), which is a measure of waste. Consequently, wide-amplitude fluctuations are characteristic of immature ecosystems, trophic levels, or populations.

Very large-amplitude instability in population fluctuations occurs either in very simple habitats in high latitudes or altitudes, or in tropical situations altered by man. The very large-amplitude population fluctuations described for Switzerland by Baltensweiler (1971) and the spruce budworm in Canada by Morris et al. (1963) and in Germany by Schwerdtfeger (1941) have no known counterpart in natural systems in the tropics.

Huffaker (1971) has presented an interesting example of a regulatory mechanism in a mature community. Predators can enrich the diversity of lower trophic levels by increased cropping of any prey species that begins to increase in numbers as compared with other prey species. Thus, the functional response of a predator (increase in prey eaten per predator with increasing prey density) is a mechanism for constantly maximizing the diversity at the prey trophic level (the diversity per individual). Thus, predators damp wide-amplitude fluctuations in mature communities. The known exceptions, as in lemmings, hares, locusts, and spruce budworm outbreaks, would appear to be characteristic of immature communities, where the species diversity and pattern diversity are too low; herbivore populations erupt and "escape" from regulation by their predators.

Principle 13 has been rediscovered independently by the architectural writer Jane Jacobs (1969). Such parallel discoveries argue for the emergence of a new field of environmental science which has considerably broader boundaries than the traditional plant and animal ecology and draws on a common body of theory. She contrasts the Manchester-Detroit type of city with the Birmingham (England)–Cambridge (Massachusetts)–Palo Alto type. The economies of the former are dominated by a small number of very large industries and are therefore very vulnerable to instabilities due to changes in the market conditions for those industries. The latter type have a very large number of small industries, and therefore what den Boer (1968) calls spreading of risks and, probably, long-term stability as a consequence. An interesting confirmation of Mrs. Jacobs's thinking is the massive unemployment now facing certain cities which have been affected by cutbacks in the aerospace industry, which was a mainstay of their economy, and Seattle, which is tremendously dependent on one very large employer, the Boeing Company.

However, the parallels with the natural world are not perfect. In man's world, the cities with small numbers of very large industries have great efficiency of energy use, because of economies of scale which are lacking in small industries. As in the plant-animal world, however, great efficiency is linked to rigidity and irreversibility, and vulnerability to catastrophic change. It is man's conversion from dependence on a flow energy source (the sun) to dependence on stock energy resources (fossil and nuclear fuels) that has broken down the natural relation between number of energy flow pathways and efficiency of energy use. In the world of man, efficiency of energy use is not regarded as being as important a desideratum as maximization of energy throughput. This type of thinking clearly supposes that unlimited energy resources will be available indefinitely, and that there are no thermodynamic limits to the rate at which we can burn energy. Clearly, such thinking is characteristic of a very brief growth phase in the history of mankind. The type of situation we observe in nature is the product of a long-term equilibrium state.

It is frightening, in view of the importance of this principle, how mankind is decreasing both the species diversity and the pattern diversity of the planet. The number of higher plant species is decreasing in many countries, and many species of large mammals and game birds are in danger of extinction over the next century or two if present trends are not reversed. Even the number of insect species is sharply reduced by the activities of man, and this is part of the explanation for the great pest outbreaks that occur in vast tracts of land under monoculture. The parasites and predators existed at much lower densities than the pests they regulated, and if agricultural practices reduce the densities of pests and their enemies by, say, 99 percent each, the absolute effect will sharply increase the probability of extinction for the enemies of the pest but not for the pest itself.

PRINCIPLE 14   The degree of pattern regularity in population fluctuations depends on the number of generations of prior population history which influence the population.

Different populations vary greatly with respect to the pattern regularity of their population fluctuations. The meaning of "pattern regularity" can best be grasped by comparing A and C of Fig. 5-6.

This principle follows from the converse of Principle 13: the lack of diversity in the food chains of immature communities allows for a high degree of population instability. Further, if the nature of this instability is such that a small number of species interact in such a way that there can be extended time lags in the system, then the possibility of highly regular fluctuations is created. Suppose, for example, that lynxes are highly dependent on hares as a principal food item, hares are dependent on a few plant species, and the plant species are dependent on certain soil nutrients.

Suppose hares become excessively abundant in year t, and there is a mass starvation and resultant low survival of hares to year $t + 1$ for this reason. This could lead to an increase in soil nutrients by year $t + 3$, increase in plant tissue production by $t + 4$, increase in hare populations by $t + 5$, and increase in lynx populations by $t + 6$ or $t + 7$. In other words, the lynx population density is under the influence of a system characterized by long time lags, or a high degree of momentum, or inertia.

We will first state the general principle, then illustrate its operation with a numerical example.

If the trend in numbers of a population from generation g to generation $g + 1$ is determined almost entirely by the state of the system at g, then population fluctuations can be quite erratic. This will be true, for example, if the trend from g to $g + 1$ is largely influenced by some type of environmental perturbation at g, such as the coldness of the water, or drought, or depth of snowfall. If, on the other hand, the state of the system at $g - 1$, $g - 2$, and earlier generations has a large effect on the trend from g to $g + 1$, relative to the effect of the state at g, then the pattern of population fluctuations will be very regular. That is, long time lags build inertia, or momentum, and hence pattern regularity into a system.

This is the same principle that is operating when business analysts and economists detect cycles in market behavior by smoothing out a time series of raw data on prices of stocks or some commodity by taking running sums, or averages.

Table 2-2 and Fig. 2-9 illustrate the effect on pattern regularity of increasing the number of generations of prior history which influence population fluctuations. The first column of Table 2-2 shows us the population densities that would occur each year if there were no random environmental perturbations, and if population fluctuations were totally

TABLE 2-2 Principle 14. The degree of pattern regularity in population fluctuations depends on the number of generations of prior population history which influence the population.

year number	population size as produced by 10-year cycle	deviations due to random perturbations	sum of columns 1 and 2	3-year running sum	3-year running mean	5-year running sum	5-year running mean
1	20	0	20				
2	26	−4	22				
3	29	8	37	79	26		
4	29	−12	17	76	25		
5	26	12	38	92	31	134	27
6	20	−8	12	67	22	126	25
7	14	−4	10	60	20	114	23
8	11	16	27	49	16	104	21
9	11	−8	3	40	13	90	18
10	14	0	14	44	15	66	13
11	20	4	24	41	14	78	16
12	26	−4	22	60	20	90	18
13	29	12	41	87	29	104	21
14	29	−16	13	76	25	114	23
15	26	4	30	74	25	128	26
16	20	4	24	67	22	130	26
17	14	0	14	68	23	122	24
18	11	8	19	57	19	100	20
19	11	0	11	44	15	98	20
20	14	−12	2	32	11	70	14

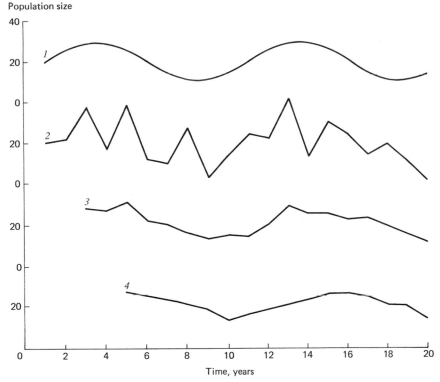

Population size

Time, years

FIGURE 2-9 Graph of columns 1, 3, 5, and 7 in Table 2-2. Line 1: Pure cyclical process. Line 2: Cyclical process strongly modified by random environmental perturbations. Line 3: Pattern of population fluctuations obtained when each point is the average of corresponding point and two previous points in line 2. Line 4: Pattern of population fluctuations obtained when each point is the average of corresponding point and four previous points in line 2.

controlled by a perfectly regular 10-year cyclical process. The second column introduces random perturbations of the type we would expect from weather fluctuations. The perturbations were generated by a random process such that the frequency of a perturbation is inversely proportional to its magnitude (the largest perturbations occur only one-third as frequently as the smallest perturbations). The third column illustrates the type of population trend we would expect in a system in which an underlying 10-year periodicity was largely masked by environmental perturbations. This column is intended to be the analog of Fig. 5-6C. Columns 5 and 7 show the pattern of population fluctuations that would be obtained if each point in the time series represented the average of 3 years or 5 years, respectively, of system history.

When columns 1, 3, 5, and 7 are plotted, in Fig. 2-9, three interesting points are made. First, the regularity of the top line which is marked by the random perturbations in the second line is progressively more visible again in the third and fourth lines. Thus, the degree of pattern regularity is in fact dependent on the number of years of prior population history which influence the system. Second, line 2 appears like the bottom line of Fig. 5-6, and line 4 appears like the top line of Fig. 5-6. Third, the

3-year running mean line lags the top line by 1 year, and the 5-year running mean line lags the top line by 2 years.

In vast land areas with homogeneous, optimal food supply and wildly fluctuating weather, some populations may increase in numbers so rapidly that they "escape" from regulation by their usual controlling factors. This is another way of saying that the factors which control them operate with long time lags, because of course all populations must ultimately be controlled. Where the controlling system has agents with time lags of two or more steps as important elements in the whole system of control, there is a resultant extraordinary pattern regularity in the fluctuating system. Predators which follow prey with such fairly regular fluctuations have an even larger number of time steps in the system which regulates them, and hence would be expected to fluctuate in an even more regular fashion than their prey. This is the case with the lynx of northern Canada, which has one of the most regular patterns of population fluctuation ever discovered.

Another mechanism which can give rise to such lags operates in the case of forest insect defoliators which strip their host trees to such an extent as to affect the quantity and quality of foliage that will be available for the following generation. If this in turn has an effect on the quality or quantity of eggs that the female larvae will have as a result of eating such foliage, then the degree of defoliation by generation g can have an effect on the number of larvae three generations later.

This concludes discussion of fourteen core principles of environmental science. There are many other principles of environmental science, but these pertain to more special topics and will be introduced in subsequent chapters. To conclude this chapter, we present Fig. 2-10, a flow chart that summarizes the logical relations among the fourteen core principles already presented.

This flow chart reveals that the set of principles constitute an interlocking body of theory, in which Principles 7, 9, 12, and 13 have a central and critical logical relation to the web of relations within the set. An important implication of the fact that these principles have this set of logical, as opposed to empirical, relations is that the whole body of theory is more convincing for this reason. That is, not enough data are yet available to provide a completely satisfactory empirical basis for certain of these principles. However, because the principles are interconnected by logic, each principle in the set is more credible than it would be simply on the basis of the evidence in support of it.

For example, ideally, in order to document Principle 12 fully, it would be desirable to know a great deal about the genetics and population dynamics of coral reef fish species. However, while such information is scarce, our inclination to believe in the validity of Principle 12 is increased by our knowledge of the data in support of Principles 6, 7, 9, and 10.

The student is encouraged to try to think of examples illustrating these fourteen principles. Moreover, to increase comprehension of all the material in this book, reference will be made to the principles in subsequent chapters, in which primary emphasis is placed, not on a logical system, but as it relates to the processes occurring in the real world.

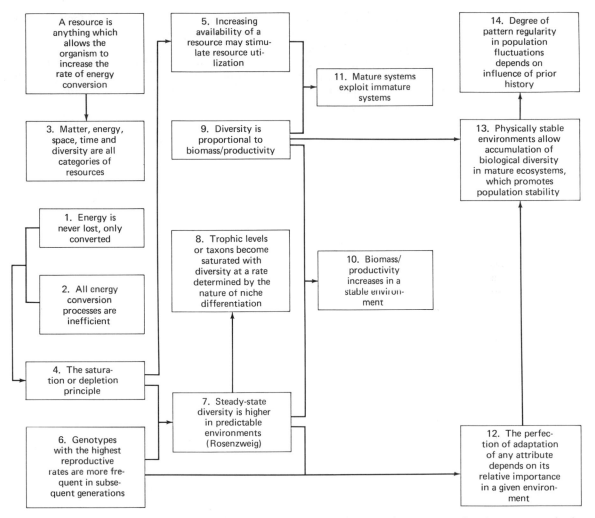

FIGURE 2-10 Flow chart of logical relations among fourteen core principles.

SUGGESTIONS FOR INDIVIDUAL AND GROUP PROJECTS

Conduct experiments and make observations to check on the applicability of Principle 3. For example, set up different sets of eggs and incubate them in different incubators, each at a different temperature. Do you find the pattern in Fig. 2-3? Find out how population size increases with increasing availability of time, for *Drosophila* in pint milk bottles or *Tribolium* in shell vials of flour. Certain comprehensive catalogues of biological data can be used to see whether generalists are in fact more abundant than species with only a few food species. For example, *United States National Museum Bulletin*, no. 216, part 2, by Henry and Marjorie Townes, is a catalogue of parasitic Hymenoptera that gives for each of many species an abundance index (the number of specimens in collections) and a food diversity index (the number of known food species).

Does the average abundance increase with the number of food species?

Conduct experiments to find the types of situations in which A and B of Fig. 2-7 are encountered.

Collect data bearing on the applicability of Principle 11 to the relation between cities and rural towns. (See, for example, chap. 8 in Colin Clark, "Population Growth and Land Use," Macmillan, London, 1967.) What types of processes explain these data?

Do all fourteen of the principles in this chapter seem important to you? Can you think of other instances where each of them applies? Can you think of any other principles which seem important but are omitted?

REFERENCES   Baltensweiler, W.: The Relevance of Changes in the Composition of Larch Bud Moth Populations for the Dynamics of Its Numbers, in P. J. den Boer and G. R. Gradwell (eds.). *Proc. Adv. Study Inst. Dyn. Numbers Population*, pp. 208–219, Center for Agricultural Publishing and Documentation, Wageningen, 1971.

Birch, L. C.: An Improved Method for Determining the Influence of Temperature on the Rate of Development of Insect Eggs, *Australian J. Exptl. Biol. Med. Sci.*, **22**:277–383 (1944).

Browning, T. O.: The Influence of Temperature on the Rate of Development of Insects, with Special Reference to the Eggs of *Geryllubos commodas* Walker, *Australian J. Sci. Res.*, (B)**5**:96–111 (1952).

Carlander, K. D.: The Standing Crop of Fish in Lakes, *J. Fisheries Res. Board Can.*, **12**:543–570 (1955).

Connell, J. H.: On the Role of Natural Enemies in Preventing Competitive Exclusion in Some Marine Animals and in Rain Forest Trees, in P. J. den Boer and G. R. Gradwell (eds.), *Proc. Adv. Study Inst. Dyn. Numbers Population*, pp. 41–63, Center for Agricultural Publishing and Documentation, Wageningen, 1971.

Darwin, C.: The Descent of Man and Selection in Relation to Sex, chaps. 1–12, J. Murray, London, 1871.

Davidson, J.: On the Ecology of the Growth of the Sheep Population in South Australia, *Trans. Roy. Soc. S. Australia*, **62**:141–148 (1938).

———: On the Relationship between Temperature and Rate of Development of Insects at Constant Temperatures, *J. Animal Ecol.*, **13**:26–38 (1944).

De Bach, P., and H. S. Smith: Are Population Oscillations Inherent in the Host-parasite Relation? *Ecology*, **22**:363–369 (1941).

den Boer, P. J.: Spreading of Risk and Stabilization of Animal Numbers, *Acta Biotheoret.*, **18**:165–194 (1968).

Ehrlich, P. R., and P. H. Raven: Butterflies and Plants: a Study in Coevolution, *Evolution*, **18**:586–608 (1965).

Gates, D. M.: Climate and Stability, *Brookhaven Symp. Biol.*, no. 22, pp. 115–127, 1969.

Golley, F. B.: Energy Dynamics of a Food Chain of an Old-field Community, *Ecol. Monographs*, **30**:187–206 (1960).

Harper, J. L., and J. White: The Dynamics of Plant Populations, in P. J. den Boer and G. R. Gradwell (eds.), *Proc. Adv. Study Inst. Dyn. Numbers Population*, pp. 41–63, Center for Agricultural Publishing and Documentation, Wageningen, 1971.

Holling, C. S.: The Components of Predation as Revealed by a Study of Small

Mammal Predation of the European Pine Sawfly, *Can. Entomologist*, **91**:293–320 (1959).

Huffaker, C. B.: The Phenomenon of Predation and Its Roles in Nature, in P. J. den Boer and G. R. Gradwell (eds.), *Proc. Adv. Study Inst. Dyn. Numbers Population*, pp. 327–343, Center for Agricultural Publishing and Documentation, Wageningen, 1971.

Jacobs, J.: "The Economy of Cities," Random House, New York, 1969.

Lack, D.: Natural Selection and Family Size in the Starling, *Evolution*, **2**:95–110 (1948).

MacArthur, R.: Fluctuations of Animal Populations and a Measure of Community Stability, *Ecology*, **36**:533–536 (1955).

Margalef, R.: On Certain Unifying Principles in Ecology, *Am. Naturalist*, **97**:357–374 (1963).

———: "Perspectives in Ecological Theory," University of Chicago Press, Chicago, 1968.

———: Diversity and Stability: A Practical Proposal and a Model of Interdependence, *Brookhaven Symp. Biol.*, no. 22, pp. 25–37, 1969.

Morowitz, H. J.: "Energy Flow in Biology," Academic, New York, 1968.

Morris, R. F. (ed.): The Dynamics of Epidemic Spruce Budworm Populations, *Mem. Entomol. Soc. Can.*, no. 31, pp. 1–332, 1963.

Pearl, R.: "The Biology of Population Growth," Knopf, New York, 1930.

Pielou, E. C.: Species-diversity and Pattern-diversity in the Study of Ecological Succession, *J. Theoret. Biol.*, **10**:370–383 (1966).

Powsner, L.: The Effects of Temperature on the Duration of the Developmental Stages of *Drosophila melanogaster*, *Physiol. Zool.*, **8**:474–520 (1935).

Rosenzweig, M. L.: The Strategy of Body Size in Mammalian Carnivores, *Am. Midland Naturalist*, **80**:299–315 (1968a).

———: Net Productivity of Terrestrial Communities: Prediction from Climatological Data, *Am. Naturalist*, **102**:67–74 (1968b).

Royama, T.: Evolutionary Significance of Predators' Response to Local Differences in Prey Density; A Theoretical Study, in P. J. den Boer and G. R. Gradwell (eds.), *Proc. Adv. Study Inst. Dyn. Numbers Population*, pp. 344–357, Center for Agricultural Publishing and Documentation, Wageningen, 1971.

Sanders, H. L.: Benthic Marine Diversity and the Stability-time Hypothesis, *Brookhaven Symp. Biol.*, no. 22, pp. 71–81, 1969.

Schwerdtfeger, F.: Uber die Ursachen des Massenwechsels der Insekten, *Z. Angew. Entomol.*, **28**:254–303 (1941).

Slobodkin, L. B., and H. L. Sanders: On the Contribution of Environmental Predictability to Species Diversity, *Brookhaven Symp. Biol.*, no. 22, pp. 82–95, 1969.

Smeed, R. J.: The Traffic Problem in Towns, *Town Planning Rev.*, **35**:133–158 (1964).

Thorsteinson, A. G.: The Chemotactic Responses That Determine Host Specificity in an Oligophagous Insect (*Plutella maculipennis* Curt. Lepidoptera), *Can. J. Zool*, **31**:52–72 (1953).

Trimmer, C. D.: "Uganda National Parks Report for the Quarter Ending 30 Sept. 1962," 1962.

Valentine, J. W.: Niche Diversity and Niche Size Patterns in Marine Fossils, *J. Paleontol.*, **43**:905–915 (1969).

Watt, K. E. F.: Community Stability and the Strategy of Biological Control, *Can. Entomologist*, **97**:887–895 (1965).

Way, M. J., and M. E. Cammell: Self Regulation in Aphid Populations, in P. J. den Boer and G. R. Gradwell (eds.), *Proc. Adv. Study Inst. Dyn. Numbers Population*, pp. 232–242, Center for Agricultural Publishing and Documentation, Wageningen, 1971.

Whittaker, R. H.: Evolution of Diversity in Plant Communities, *Brookhaven Symp. Biol.*, no. 22, pp. 178–196, 1969.

3 the fundamental ecological variables: matter, energy, space, time, and diversity

MOTIVES FOR USING THIS METHOD OF CLASSIFYING VARIABLES

There are many different schemes by which we could classify the variables that operate to govern ecological phenomena. For example, we might classify factors as animate or inanimate, or density-dependent or density-independent (Fig. 3-1). Density-dependent factors operate with an intensity that depends on the density of the population. Starvation is an example: the more dense the population is, the less food there is per individual, and the more harmful is the effect that starvation has on growth, survival, and reproduction. Density-independent factors kill a proportion of all the organisms present regardless of the density of the populations. For example, the percentage of tropical marine fish killed by freezing weather off the coast of Texas is density-independent. Indeed, there are a great many ways in which we could classify environmental factors.

However, by classifying variables as *matter, energy, space, time, and diversity*, we achieve a number of objectives.

1 By showing how all five categories of variables interact with each other in any given ecological process, we demonstrate a strongly systems-oriented way of looking at the world. That is, we are trained to bear in mind how each part of a system affects other parts. This point of view leads to very deep understanding about how complex phenomena operate, and the significance of component processes for a whole process.

2 If we can show how all five of these categories of variables interact in any given process, then we will have linked up all parts of ecology into a single unified body of theory. Up to now, there has been an unnatural gulf between five subjects: *physiological ecology, population ecology, community energetics, community organization,* and *speciation* (or

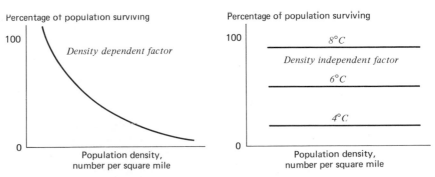

FIGURE 3-1 Examples of the difference between density-dependent and density-independent factors.

evolutionary ecology). This is just one unnatural way in which ecologists divide themselves. It is as illogical as any other. It should be pointed out that this unnatural division would not correspond to other possible ones (human ecology, wildlife ecology, etc.); the important fact is that all parts of ecology are intricately interwoven in the real world.

Population ecology has dealt with the factors that regulate the distribution, abundance, and stability of populations. Because this subject was concerned with distribution, it was concerned with space. Because it was concerned with the abundance of organisms, it dealt with the factors that regulated variation in abundance through time.

Community energetics was concerned with how energy passed through a community, from the plants which used the sun's energy for photosynthesis, to the animals which ate the plants (*herbivores*), to the animals which ate the herbivores (*carnivores*). Also, since energy is a source of power for allowing the plants to make green tissue out of matter, and for allowing animals to eat plants and catch prey, community energetics has also been concerned with matter.

Diversity has been of concern to two other branches of ecology.

Speciation was the science that dealt with the way in which species are formed in nature; it considered the role that the genetic structure of populations played in this process (genetic structure is a way of storing diversity in populations and individuals).

Community organization considered the relation between the structure and composition of communities and their stability. By composition of communities we refer to the relative number of individuals of each species. Structure refers to the spatial arrangement of living and nonliving objects in the environment and the structure of the food web (Fig. 3-2).

All these ways of describing the composition of communities deal with diversity in the community. Thus, we use the word *diversity* in two senses: *genetic diversity* and *community composition and structure diversity*. In this book we propose to show how matter, energy, space, time, and diversity are each involved in every ecological process and phenomenon.

3　Viewing ecological phenomena in terms of these five categories of factors, we are led to a strongly dynamic and process-oriented approach to ecology. This type of approach gives much more insight into the actual dynamics of phenomena than a descriptive approach, which describes the state of a system at different points in time.

4　Looking at the interaction of these five categories of factors from the evolutionary point of view suggests that we examine processes as the basis of strategies, balancing costs against benefits involved in various processes. The concept of strategies for obtaining energy will be dealt with at considerable length in this chapter.

5　A major advantage of looking at processes in terms of the interaction of matter, energy, space, time, and diversity is that this particular mode of classifying variables reveals underlying similarities among various phenomena which otherwise would appear to be quite unrelated. For example, a parasitic wasp seeking a caterpillar to lay her eggs in, a barra-

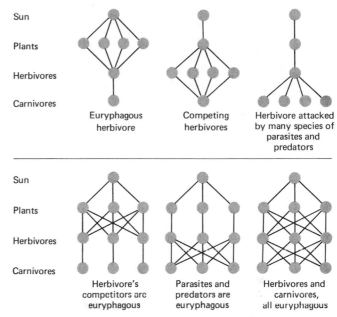

FIGURE 3-2 Examples of the different types of structural relations found between the different species populations in a food web. The circles represent species; the lines represent paths along which energy flows. A line from one species to another at a lower level in the diagram means that the species on the lower level eats the species on the higher level.

cuda hunting for prey, a predacious water bug seeking tadpoles, a man going to work in the morning, or a regional planner worrying about the optimum site for a new town, all have certain problems in common. In each case, a cost is being balanced against a benefit. The cost is the time and energy expended in movement; the benefit is the energy acquired as a consequence of making the movement. If the benefit is less than the cost, then the operation is a failure; if cost and benefit are equal, the operation breaks even; and if benefit exceeds cost, the operation is a success. In the various cases, costs and benefits are measured in different terms, but all can ultimately be expressed in terms of the accumulation of energy and diversity. If an individual animal repeatedly spends more energy searching for food than it obtains from the food it finds, then the animal will soon exhaust its energy stores (fat) and will starve. If a man spends more time and energy making trips to work than can be justified by what he receives from that work, then he faces economic ruin. For a wealthy family, such a course might not be critical, but for a poor family it might make the difference between being able to hold a job and not being able to.

Thus, there are many advantages in classifying the variables operating on ecological phenomena as these five categories. Not only does it allow us to make penetrating use of the method of strong inference, but the fifth justification suggests the possibility of using the comparative method of Darwin also.

AN EXAMPLE OF INTERACTION AMONG THE FIVE CATEGORIES OF ECOLOGICAL VARIABLES

The five fundamental categories of variables in any ecological process are matter, energy, time, diversity, and space; three coordinates are required to describe position with respect to the last of these. Any complex phenomenon can ultimately be understood in terms of these categories of variables, and, conversely, all of them must be involved in a complete explanation of any phenomenon. Their operation in a typical ecological situation can be explained in terms of the hunting down of a buffalo by lions.

The lions go through a behavioral cycle of digesting their last meal, searching for a new meal, pursuing it when they have found it, and finally eating it. The length of time required by the search will depend on the average distance between the lions and a prospective food item, and this, in turn, will depend on the density of their food (prey). When the lions finally locate their food, marked changes in their acceleration and velocity will be required if the pursuit is to be successful. Some simple algebra will indicate how time, diversity, energy, space, and matter are involved in this act of predation.

Let $E_m =$ the energy consumed in a meal, in caloric equivalents

$D_f =$ the average distance a predator must search in order to locate item of food (prey)

$E =$ the additional energy used per unit distance in searching for food

$E_p =$ the additional energy expenditure required by the actual pursuit of a prey

$E_r =$ the resting metabolic rate per unit time in the interval between obtaining one meal and obtaining the next

$T_e, T_d, T_s, T_p =$ the times spent in eating, digestive pause, searching, and pursuit, respectively

Then the net energy gain E_g from eating one meal is given by

$$E_g = E_m - [D_f(E) + E_p + E_r(T_e + T_d + T_s + T_p)]$$

The role of energy in this equation is obvious, since the equation is expressed in calories. This equation is critical for the predator because it expresses the amount of energy obtained from a meal which can go into activity, growth, and reproduction. If the quantity is negative, that is, if more energy is expended in getting the meal than is obtained by eating the meal, the organism has negative growth (loses weight). Time enters the equation because the length of time between meals ($T_d + T_s + T_p$) determines how much energy will be burned off as basal metabolism between meals. Space enters the equation as the distance the predator must move to find food (D_f), and matter is the amount of food eaten. This quantity will be some multiple of E_m. Of course, each of time, space, matter, and energy enters this equation in many ways. For example, time is involved in the energy output in the pursuit phase because increase in velocity per unit time is acceleration, and force equals mass times acceleration. Thus, the acceleration determines the energy consumption required by the actual attack.

A similar analysis can be presented for mating, because here also the distance between prospective mates has an effect on the energy consumed in searching for a mate, and hence in the total process of mating.

Note that in the preceding discussion it is important to distinguish between basal metabolism and the energy expenditure over and above this which supports activity.

Diversity is important throughout this process. Both the lion and the buffalo have behavior patterns and physiological capabilities which are the product of their genetic constitution (*inherited diversity*) and their learned experience (*acquired diversity*). Also, the distribution of each in space, the distribution of alternate prey and competing predators, and the distribution of plants or other cover which provides hiding places for prey or ambush sites for predators, all constitute diversity in this system. Thus diversity is being used here as an all-inclusive term to describe the amount of nonrandom structure in the universe, from genetic structure at a molecular level to structure in the sense of distribution of plants, animals, and topographic structure of the terrain.

Each of the five categories of variables is important in this situation, and each has implications for the other four and interacts with them. For example, the average distance between a lion and a prospective food item determines the average amount of energy lions must expend to obtain food, all other factors being equal (such as the distribution of the food). The amount of energy the lion has stored within it determines the amount of time the lion can exist without obtaining new energy. Similarly, the amount of matter represented by the lion (its weight) determines the size and number of prey the lion must obtain per unit time in order not to starve to death. The more one considers the true nature of predation, the more obvious it becomes that the essential feature of all categories of variables involved in the situation is their *interrelatedness*, or putting it differently, the complexity of the patterns of interaction. In fact, this feature describes all ecological phenomena, not just predation.

However, in order to set the stage for examining how these categories of factors interact in ecological phenomena, it is necessary to consider each of the five separately. Then we will be able to consider the different strategies involving the variables and their interactions that are being played by various species and assemblages of species (communities).

MATTER An enormous amount of detailed information can be presented about minerals, water, soil, and air, but much of it represents merely different manifestations of six fundamental principles. Therefore, before considering those four groups of factors, we will lay a theoretical foundation for the discussion by considering the six principles which apply to all types of effects of matter on living organisms. These principles are all related to each other and constitute a unified body of theory which facilitates comprehension of detailed information about the effect of matter on organisms. The principles can be understood readily in terms of the different panels of Fig. 3-3. Throughout this chapter, principles will be numbered by continuation of the numbering system used in Chap. 2. A principle that has already been introduced in Chap. 3 retains its original number.

principle 5 For any organism Y and any factor X, there is some availability of that factor which maximizes the rate of any given process in the organism. The factor may be the concentration in water of a chemical element, such as potassium, the diameter of soil particle sizes, or the concentration of a gas, such as carbon dioxide in the air. The process may be growth rate, reproduction rate, rate of movement, or any other biological process.

Different processes may have different optima for the same factor, and the same process (for example, respiration) may have different optima at different times of the year or at different stages in the life cycle of an organism. At availability levels of the factor either greater than or less than X_{max}, that which produces the maximum rate Y_{max}, the rate is lower. At concentrations of X considerably higher or lower than X_{max}, lethal

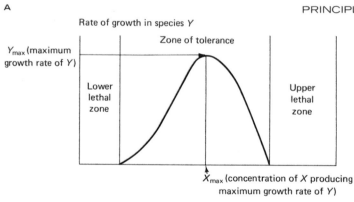

A

Rate of growth in species Y

Zone of tolerance

Y_{max} (maximum growth rate of Y)

Lower lethal zone

Upper lethal zone

X_{max} (concentration of X producing maximum growth rate of Y)

PRINCIPLE 5 The effect of factor X on species Y varies as a function of the concentration of X so as to produce an optimum concentration.

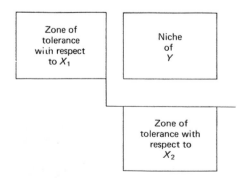

B

Zone of tolerance with respect to X_1

Niche of Y

Zone of tolerance with respect to X_2

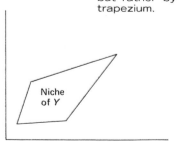

C

Niche of Y

PRINCIPLE 15 Zones of tolerance for factors X_1, X_2, etc., determine the type of habitat in which species Y can live. If there is interaction amongst X_1, X_2, etc., the habitat will not be defined by a square, cube, or n-dimensional variant, but rather by an n-dimensional trapezium.

FIGURE 3-3 The principles describing the effect of matter on living organisms.

zones for Y are encountered, and species Y dies. The zone inside the lethal zones might be labeled a zone of tolerance.

Several phenomena familiar to everyone may be understood immediately in terms of this principle. Suppose X represents the concentration of water in the soil, and we wish to maximize growth rate of a crop of species Y. If X exceeds X_{max}, we use drainage pipes, ditches, or canals to drain the soil and bring the water level back to X_{max}. If the water concentration is below X_{max}, we water, sprinkle, or irrigate the soil. If X is potassium and there is too little potassium in the soil, we use fertilizer to bring the concentration up.

Figure 3-3A makes an important point that is not widely understood. It is meaningless to say that a given substance is harmful or helpful in the environment unless a particular concentration is specified. Substances normally considered helpful, such as water and salt, are harmful if present in excessive quantities, and substances not normally considered helpful, such as trace elements like molybdenum, can be

D

Zone of tolerance with respect to X_1

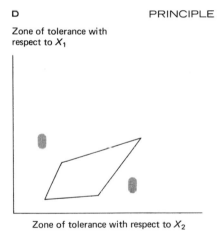

Zone of tolerance with respect to X_2

PRINCIPLE 16 The law of the minimum (Liebig's law). The organism survives in the range of conditions outlined in black, but dies in either of the ranges of conditions in the shaded sections, even though each of these ranges lies within the zone of tolerance for X_2. In each case, X_1 is not suitable.

E

Relative abundance of species

PRINCIPLE 17 A gradient in species types produced by an environmental gradient.

Species Y_1

Species Y_2

Species Y_3

Very low

Concentration of factor X

Very high

present in such insufficient concentrations that they inhibit growth of plants and animals. An example of misunderstanding concerning this point is the widely known controversy about the addition of fluorides to drinking water to prevent tooth decay. It is certainly true that fluorides are poisonous *if present at excessive concentrations.* However, *all* substances commonly considered helpful are also poisonous if present in excessive quantities. What public health authorities advocate is that the *optimal* concentration of fluorides be added to the water. Clearly, it is not sufficient to know that factor X is available; it must be available at the correct concentration, in the correct form, and at the correct time and place. If injected into the bloodstream, even air is a poison.

PRINCIPLE 15   The parts of the earth which can be inhabited by any given species at any given time are determined by the zones of tolerance of all the factors taken together.

Suppose there are three factors, availability of which determines whether a particular species can live in an area. These factors would define a three-dimensional space, analogous to the two-dimensional space under Principle 15 in Fig. 3-3. This space is the niche within which the species can live. Outside this, there are lethal zones in which the species is not able to replace itself by reproduction. If these factors do not interact, then the zone of tolerance for the species can be computed as in Fig. 3-3B. However, if there is interaction among the factors, then the zone will be as in Fig. 3-3C. All together, these zones define a particular type of habitat which the species can occupy. If, in addition to material requirements, the habitat is defined in terms of energy characteristics and food availability, we can produce a total picture of the environmental requirements for a species which is referred to as the niche for that species.

PRINCIPLE 16   For a long time ecology has had a principle called *Liebig's law of the minimum.* This law states that even if all but one of the substances necessary for the growth of a plant are present in appropriate concentrations, the inadequacy of the one that has too low a concentration will prevent growth.

This law can be generalized to cover excessive as well as insufficient concentrations, as illustrated in Fig. 3-3D. The quadrangle represents the niche of the species with respect to only two factors, X_1 and X_2. The circular areas outside the quadrangle represent environments in which the species would die, even though each environment fell within the range of X_2 values that could support life. Stated differently, growth is produced by the interaction of many nutrients. Inadequacy of even one can stop it.

PRINCIPLE 17   Because each species differs slightly in environmental requirements from other species, slight environmental differences through space or time produce series of species that replace each other in space (*coenoclines*) and time (*succession*). (This is illustrated in Fig. 3-3E.)

One of the most striking examples of this is in the Hawaiian Islands, on all of which there are very high mountains and all of which are subjected to prevailing winds from the northeast. These winds hit the mountains laden with moisture, and then drop a great deal of moisture on the windward side of the mountains but very little on the lee, or southwest. On some of the islands, such as Oahu, there are striking rainfall gradients over such short distances that they can be seen by an observer standing at an appropriate viewing position. Plate 1 shows a rainfall map of Hawaii, a panoramic view with color infrared film taken from the angle at which the prevailing winds hit Mauna Kea, 13,796 feet high, and views of the vegetation at the sites indicated on the map and the photograph.

PRINCIPLE 18   This is an extension of Principle 5. If it is necessary that various categories of matter be available at the correct time and place and in the correct form, then it follows that living organisms are dependent on the *mechanisms that allow for the availability of necessary substances.*

For example, upwelling currents in the ocean are of critical importance in the production of living stuff in the ocean because they bring up to the surface minerals that have drifted down below levels at which they can be useful to green plants. For example, the great anchovy fishery off the coast of Peru depends on such an upwelling. Similar situations are found off the west coasts of Africa and North America. Another example concerns the availability of water in the soil. Different types of soil are characterized by differences in the extent to which they make their water available to plants, and hence the physical properties of the soil which determine this characteristic are profoundly important for plant growth.

PRINCIPLE 19   Matter and energy differ from each other in that, whereas a quantum of energy passes through a food chain only once, matter cycles through repeatedly (the *law of conservation of matter*) and hence the productivity of the food chain depends on those processes which determine the rate at which matter is recycled through the system.

We will now examine the various kinds of matter in some detail to determine how they illustrate the six principles just enumerated.

minerals   Table 3-1 gives the requirements of typical plants and animals for various chemical elements and the availability of those elements in the environment. This table makes it clear that certain elements are much more likely to be limiting in the environment than others. For example, silicon is a tremendously abundant element on the earth's surface, and yet it is not an important constituent of plant or animal tissue; therefore, it is only rarely limiting. Phosphorus, however, may be rare and is an important constituent of plant and animal tissue, and therefore it is likely to be limiting frequently. Because the three substances most likely to be limiting to plant growth are nitrogen, potassium, and phosphorus, the principle function of fertilizers is to make up for soil deficiencies in these elements. Perhaps the most commonly found deficiencies in plants are in nitrogen, phosphorus, potassium, iron, magnesium, and calcium. Each of these deficiencies is represented by a highly characteristic set of symptoms in the plant. In fact, it is very unusual to find a completely even distribution of necessary substances for plant growth over the surface of the earth. This is why uneven growth of trees in orchards or crops in fields is such a common sight, and this is why fertilization is such a common practice.

Unfortunately, because there is an *optimal* concentration of each substance for each biological process (Principle 5), continued use of fertilizers to meet a need for one type of process at one place may lead to a buildup of the fertilizer components that is excessive for the physiological tolerances for other processes at other places (Principle 17). One ex-

TABLE 3-1 Requirements of plants and animals for chemical elements and the availability of those elements in the environment*

chemical element	average composition of terrestrial matter, in percent, including atmosphere, hydrosphere, and lithosphere (the solid mass)	average composition of a corn plant, in percent (dry weight)	average composition of the human body, in percent (dry weight)
Oxygen	46.68	44.57	29.13
Silicon	27.60	1.17	0.00
Aluminum	8.05	0.11	0.00
Iron	5.03	0.88	0.01
Calcium	3.63	0.20	3.75
Sodium	2.72	0.10	0.40
Potassium	2.56	0.92	0.90
Magnesium	2.07	0.18	0.10
Hydrogen	0.145	6.26	8.40
Titanium	0.696	0.00	0.00
Chlorine	0.095	0.14	0.40
Phosphorus	0.152	0.18	2.50
Carbon	0.149	43.70	45.00
Manganese	0.116	0.00	0.00
Sulfur	0.100	0.17	0.60
Fluorine	0.030	. . .	0.14

SOURCE: A. J. Lotka, "Elements of Physical Biology, © 1925, The Williams & Wilkins Co., Baltimore.
* Not all organisms have the same requirements, of course, but this table does make several interesting points. First, some elements which many people consider to be poisons are actually part of the body (e.g., fluorine). This fact reinforces the point that being poisonous is a matter of concentration and may depend on the compound in which the element occurs. Also, the human and the corn plant, like all living species, make little use of some very common elements (silicon and aluminum), and are dependent on obtaining adequate supplies of some rare elements (phosphorus and chlorine). However, the situation is even more unbalanced than this table reveals, because some extremely rare elements, such as molybdenum, which is less than 0.001 of 1 percent of terrestrial matter, may be the limiting factor for an entire community of plants and animals (Goldman, 1960).

ample of this is related to the current fear of the effect of excessive fertilizer-produced concentrations in the soil on the well-being of babies, expressed by Barry Commoner (1970) and others. Also, it is possible that excessive soil concentrations of nitrates may lead to vigorous competition from weeds that require such concentrations.

soil Five groups of factors determine the characteristics of the soil at any location: the geological material, or rock from which the soil was, in part, formed; the climate, particularly the rainfall; the topography; the vegetation; and the age of the soil. All these factors interact. Thus, the degree of similarity of the soil to its parent rock will depend on the age of the soil. Very young soils will be quite similar to the rock; an old soil in an area where climate and vegetation are constantly acting on it may be quite dissimilar to its parent rock.

These five groups of factors are important in determining the productivity of soil because they determine the chemical characteristics of the soil, the physical characteristics, and the way in which the soil is structured in horizontal layers. These resultant characteristics, in turn, are critical for soil productivity because they determine whether adequate concentrations of mineral nutrients will be available for plants, whether they will be accessible to plants, and the response of the soil to water. We now need to consider how the character of soil is determined by the five groups of factors, and how the resultant characteristics in turn determine the productivity of the soil.

If a soil is young—for example, a recent lava flow—or if climate and vegetation have not been able to age it much, as in a desert, then the parent rock will be the principal determinant of the physical and chemical character of the soil. Igneous rocks are primarily silica (about 59 percent) and alumina (15 percent). Shales or cemented clays have somewhat similar composition. Sandstones are about 78 percent silica and 5 percent alumina. Limestones are quite different from the preceding three rock types, having about 42 percent calcium oxide and 42 percent carbon dioxide. However, it is rare for a soil to be primarily the product of the parental rock type. In typical soils silica and alumina are by far the most important constituents.

For most soils, the interactions of climate, vegetation, sufficient age, and appropriate topography determine their physical and chemical character. Appropriate topography means that the soil must be located on a reasonably flat site, or it will never stay in one place long enough to age under the influence of climate and vegetation to attain the properties of a mature soil.

One of the methods of classifying soils is based on the preeminent importance of climate as a determinant of soil type. This method classifies many of the world's soils in two broad categories, the pedocals of arid or semiarid environments and the pedalfers of more rainy climatic zones. The pedocals have an accumulation of calcium carbonate and sufficient lime (calcium oxide), and they are generally basic. The pedalfers have an excess of hydroxides of iron and aluminum, the lime has been washed out, and the soil is generally acidic. The reason for the acidity of the pedalfers is that rainfall has caused the decomposition and washing out of many of the elements originally in the soil, such as calcium, magnesium, sodium, and potassium. Thus, the highly productive appearance of the tropical rain forest is deceptive. If the forest is cut away and replaced by conventional agriculture, there is a danger that the tropical latosols, lacking the just-mentioned elements, will form cemented layers, called laterite, which are so stonelike that they can be used in construction.

Vegetation and soil form a reciprocal circular causal system: each is an important factor operating on the other. The higher plants are absolutely dependent on the water and minerals contained in the soil, but they are also important contributors to the physical and chemical material of the soil when they die and are decomposed by microorganisms within the soil. In order to comprehend fully how important plants are to cumulative improvement in soil productivity, it is necessary to consider the fundamental nature of soil physics and chemistry.

The particles which make up the soil can be divided into different size classes. By convention, the boundaries between classes have been chosen because certain physical changes of great significance in soil dynamics occur at approximately these boundaries. Elementary soil particles fall into the following four size classes:

diameter of particle, millimeters	name of particle
Over 2	Gravel
0.02 to 2	Sand
0.002 to 0.02	Silt
Less than 0.002	Clay

The significance of the size class boundaries is that collections of particles of more than 2 millimeters in diameter do not hold water; at about 0.02 millimeter, particles are small enough to aggregate; and at 0.002 millimeter, the particles are just twice the diameter at which they can be considered as colloidal particles.

A soil is almost never all composed of one size of particle: the best soils, called loams, are in fact a mixture of many different sizes, with roughly 40 percent sand, 40 percent silt, and 20 percent clay. Wide variations of about 10 percent more or less than all these figures will still yield a reasonably good loam soil. It is of interest to consider why such a loam soil is desirable. On the one hand, if a soil is too sandy, it will hold very little water, and it will be an inhospitable microhabitat for the soil microorganisms so necessary to convert dead organic matter to the inorganic salts needed by plants (Principle 19). On the other hand, if a soil is all clay, it has the consistency of glue when wet, and it has several major drawbacks if used for agriculture. First, it is almost impossible to work, because of its viscosity. Second, the particles are so small, and the interstices between them are so fine, that water and dissolved salts will cling to the particles and will not be sufficiently available to the plant roots. The best soil, however, must contain all sizes of particles. It needs enough of the very fine particles to provide sufficient organic matter, and enough coarse particles to give the soil a good crumb structure that allows adequate movement of air and water so that a variety of organisms can live in the soil (Principle 18).

Vegetation is important in soil formation because dead vegetation breaks down into humus, which becomes the very fine particles that intermix with coarser inorganic materials from the parent rock.

A soil which is largely one size of particle lacks stability and is vulnerable to bad management. A soil too rich in clay can be packed into an unworkable slab by excessive trampling in wet weather; a soil that is too sandy is vulnerable to erosion because it lacks viscosity that colloidal particles give a soil.

water   Besides all the obvious points one might make about water, there are two especially important facts of ecological significance.

First, water is immensely important as an agent in energy transfer and utilization. It takes a great deal of heat to melt ice, and a great deal of heat to convert water to a gas. Specifically it takes 80 calories of heat to convert 1 gram of ice to water when both are at 0.0°C, and it takes 536 calories to convert 1 gram of water to vapor at 100°C. Further, it takes a lot of heat even to raise the temperature of water. It takes 1 calorie to raise the temperature of water 1°C. The three heats we have mentioned are called the latent heat of fusion (the melting heat), the latent heat of vaporization, and the specific heat. The significance of these high heats for water is that water is a tremendous force for the amelioration of wide-amplitude swings in incoming solar radiation in the environment. Trees in a large oasis make it the warmest part of a desert at night, and the coolest place in the daytime, because of the water stored in the leaves and the liquid to vapor conversion at the surface of the leaves. It is interesting to note how cool a park is in comparison with the rest of the area in a hot downtown part of a city on a summer afternoon. The pleasant climate of coastal cities is due to the nearby ocean. Vancouver is more pleasant than Calgary, Los Angeles than Las Vegas, and Copenhagen than Moscow, in each case because of the ameliorating effects on seasonal temperature fluctuations by the ocean.

The second major significance of water is that it contains a great variety of chemicals as dissolved salts. Many of these occur in water at vanishingly small concentrations, but even so, they may have extremely important biological effects. Trace quantities of a wide variety of substances are deadly in water. For example, there is now some evidence that DDT may be inimical to photosynthesis of oceanic phytoplankton (green algae) at fractions of a part per billion. Certain heavy metals can be dangerous at very small concentrations (mercury, lead, cadmium). (Principle 17 implies that community stability will be sensitive to fluctuations in concentration of such substances.)

air   Air is important not only for respiration but for protecting us from small and large particles hurtling through space. It is the medium which determines the amount and quality of the energy that will reach us from the sun, and like water, air can be made toxic by very small concentrations of many substances.

The air around the planet is divided into two layers that have biological significance: the troposphere, or lower air mass, and the stratosphere, or upper air mass. The boundary between the two, the tropopause, occurs at an altitude of 10 or 12 kilometers. The difference between the two layers is that the troposphere is an area of constant vertical mixing of air masses owing to convection, whereas the stratosphere is isothermal (about −55 or −60°C, from top to bottom). In the Troposphere, the temperature drops steadily from about 20°C at ground level to about −55 or −60° at the tropopause.

The significance of these two layers is that because the bottom one has vertical mixing, whereas the top one is isothermal and does not, any fine particles in air that impede incoming solar radiation will have effects that last different lengths of time, depending on how far up they get. If particles, such as smoke from forest fires or air pollution, get no higher than the tropopause, they will be brought down out of the atmosphere in a few weeks. Thus, although they may be distributed all over the planet, their effect will not be permanent. On the other hand, if enormous masses of fine particles happen to ascend to great heights, then the particles will drift down until they hit the troposphere. If the particles go high enough in the first place, and are fine enough and light enough, it may take them several years to drift down. Thus, they can be diminishing the proportion of the incident solar radiation that hits the earth's surface the entire time they drift down. This is why volcanoes have such a tremendous effect on weather. They emit particles as small as 2 microns wide, that are very light, and may blast them to a height of up to 80 kilometers. Thus major volcanic eruptions which blasted vast quantities of fine particles into the upper atmosphere have had an effect on the global temperature which was profound enough to chill the whole world for between 5 and 10 years, and affect all living processes on the planet. It has been estimated that Mount Tambora, which erupted in Sumbawa in 1815, blasted 150 cubic kilometers of fine ash into the atmosphere. The following year, 1816, central England had the coldest July in the instrumental record, and there was scarcely any summer in North America.

Air pollution can of course have similar effects if the particle concentrations build up to a sufficient level, and evidence now available suggests that this is happening.

ENERGY Energy figures in the dynamics of living systems in many ways, some of which are obvious, some not so obvious. Plants must get enough energy from the sun as light to power photosynthesis, so that they can use the minerals available to them to build living stuff. Animals need energy in their food to support basal metabolism, activity, growth, and reproduction. Energy is important to plants and animals also because it determines the rate at which all processes will occur. Warming increases the rate of reaction of any particular system or process up to a maximizing temperature, after which further increase in warming will diminish the reaction rate. Very low and very high temperatures are lethal. However, there are other causal pathways by which energy is important for living things. It has already been mentioned that water has a high latent heat of vaporization. This means that there will be a great loss of heat at a surface at which water is being converted to water vapor. This can have a positive effect, as when a man sweats in a desert. However, if a plant or animal is in a climatic region short of water, it must battle the effect of heat on conversion of water to water vapor by developing a tough skin through which little water passes. It will be noticed in Plate 1 that the plants above the 6,700-foot cloud line on Mauna Loa have thick, tough leaves in contrast to the broad, thin leaves of the plants at lower altitudes where there is torrential rainfall much of the year.

Similarly, animals need adaptations to protect themselves from energy dissipation through chilling. Various mechanisms have evolved for keeping body heat up, from feathers, fur, and contraction of skin capillaries to prevent loss of blood heat, to shivering, which uses muscle energy without performing work and thus releases heat. Another type of energy transformation occurs when animals bask in the sun and supplement their energy intake from food with this direct method of obtaining energy from the sun.

However, by all odds the most important forms of energy acquisition are eating by animals and photosynthesis by plants. We will now consider these types of transformations.

An immense amount of information exists about the energy metabolism of different kinds of organisms. We shall consider a few examples, to illustrate the vast differences in strategies for using incoming food energy, and then we shall examine data on groups of species comparatively, to see whether any rules can be discovered to account for these differences between species.

One of the most extreme energy disposition strategies is found in hummingbirds, as reported by Greenewalt.[1] A hovering hummingbird has an energy output per unit weight about ten times that of a man running 9 miles per hour. A man puts out 3500 kilocalories per day (basal metabolism: 1700 kilocalories; activity: 1300 to 1800 kilocalories). If a hummingbird weighed 170 pounds, it would put out 155,000 kilocalories of energy every day. A man eats about $2\frac{1}{2}$ pounds of food every day, but a hummingbird the size of a man would eat 285 pounds of hamburger. Whereas the man eats roughly 1.4 percent of his weight every day, the hummingbird eats 50 percent of its weight in sugar daily. However, it eats only a tenth of 1 percent of its weight in protein every day, which is about the same for a chicken, a dog, or a man. Thus, the big difference between these different kinds of animals is not the percentage of their weight in protein eaten every day, but their caloric intake. This is why the discussion in this book emphasizes the energetic content of food rather than its mass or mineral composition. It should be noted, however, that in many known cases mineral or vitamin concentration, and not calories, is the critical item for the fate of a population.

[1] C. H. Greenewalt, "Hummingbirds," ©1960, Doubleday & Company, Inc., New York.

We have already established, in comparing the blue whale and the herring, that different species have very different strategies with respect to partition 5 in Fig. 2-1. The comparison of a man and a hummingbird indicates that similar large differences can be found with respect to partition 2.

Partition 4 is an expression of the extent to which a species must tide itself over cold parts of the year when food is not available. If a species lives in a tropical region where food is always available, then the proportion of the living matter stored as fat for future use will be small. However, a bear, for example, will store a great deal of fat as it approaches the winter in northern Canada or the high-altitude country in the United States.

Enough examples have now been given to make the point that there are large differences between species with respect to the energy strategies being used. We now need to discover those rules which allow us to perceive some kind of pattern in this great mass of detailed information. We will do this by considering each of the switches in turn with respect to the underlying laws that appear to govern species-to-species differences in the switch settings, and differences within a species from time to time and place to place. This discussion should give us an overview of the energy strategies being played by different organisms, insofar as this is possible without considering time, space, matter, and diversity.

The first switch determines the proportion of the incoming food that is wasted and the proportion that is assimilated. Paloheimo and Dickie (1966) have reviewed the literature and performed statistical analyses on the data from many fish growth and metabolism studies. It is clear from their findings that in fish, growth efficiency (unit weight added per unit time per unit of food eaten) is lower when the available food is more abundant, or when the fish is larger. However, we must be careful in interpreting this finding, because gross growth efficiency does not include energy expenditure, and it could be argued that in large fish the gross growth efficiency is lower because a higher percentage of the incoming food goes into energy, rather than into waste. However, the available information concerning the second switch indicates that larger organisms put less energy per unit weight into basal metabolism, not more. Hence, the most reasonable interpretation of all the evidence is that larger amounts of available food and larger size produce greater waste of the food that is eaten.

Another factor that seems to operate on switch 1 is temperature. For example, Brooks (1968) has studied the metabolism of small birds, the redpolls, which live in the arctic and subarctic; they are only 12 to 14 grams in weight. Figure 3-4 shows the effect of temperature on the coefficient of metabolic utilization (percent assimilated of the total calories ingested) in two species of redpolls, which were tested at 7 hours of light and 24 hours of light. For *Acanthis hornemanni* there was a statistically significant increase in efficiency of food utilization at the extremes of temperature after a 24-hour trial. The general conclusion which can be drawn from this observation is that the body makes more efficient use of available energy when it is stressed. (This illustrates Principle 12, that

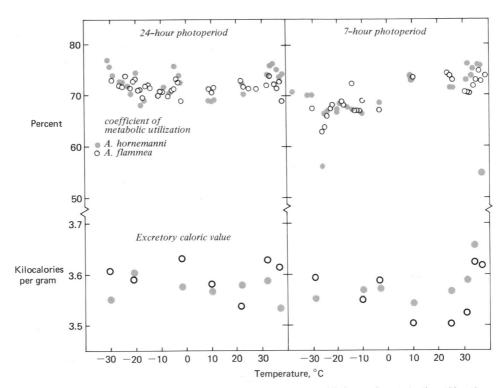

FIGURE 3-4 Excretory caloric value and coefficient of metabolic utilization (digestive efficiency) in relation to temperature for redpolls at 24- and 7-hour photoperiods. (*Brooks*, 1968.)

the perfection of an adaptation depends on its importance in a particular environment.)

Paloheimo and Dickie noted effects of environmental factors on efficiency of energy utilization in fish, but it is difficult to reduce all available observations to a simple generalization. The reason is probably that the effects of lethal or sublethal conditions become important at extreme ends of the zone of tolerance, whether the factor is temperature, salinity, or some other one. There is a deeper generalization in all the findings, however, and that is that the efficiency of food utilization is a mechanism which allows the organism to obtain the energy necessary to support a physiological equilibrium within the tolerable range of conditions.

Switch 2 determines the proportion of the assimilated food going into immediate energy expenditure and the proportion going into tissue manufacture. This switch distinguishes the hummingbird strategy, in which a great deal of energy is being burned up, from the bear strategy, in which a much smaller proportion of the incoming energy stream is allocated to energy and a higher proportion goes to tissue manufacture.

In general, small animals burn up more energy per unit mass per day than large animals. If we make the additional assumption that a major energy cost for animals is the heat exchange per unit surface of their bodies, then the cost per unit mass of this loss declines with increasing mass. An assumption that we might make is that this is because for a sphere the ratio of surface area to volume is given by

$$\frac{4\pi r^2}{\frac{4}{3}\pi r^3}$$

Since the surface is a more rapidly rising function of radius than volume, this means that with increasing mass, the surface per unit mass declines.

This line of reasoning can be put to an empirical test as follows. The mass of an animal, assuming a spherical shape, is proportional to the cube of the radius. The surface is proportional to the square of the radius. Therefore, the surface is proportional to the mass$^{2/3}$. More precisely, where M represents the standard metabolism in kilocalories per day and W represents the body weight, we would expect to find that M and W are related by the equation

$$M = aW^{0.667}$$

Zar (1968) used a very precise statistical routine for computing the values of the coefficients in the equation relating M and W for nearly 100 species. He found that instead of 0.667 for the index in the above equation, he obtained the value 0.737. However, this is surprisingly close to the value obtained by an exceedingly simple line of reasoning, assuming a spherical body shape. Consequently, this is a strong indication that the energy cost of surviving for an animal is weakly related to body surface, rather than mass.

Partition 4 is affected by the time of year in mammals: more energy goes into fat than into other tissue prior to hibernation. Partition 5 distinguishes the blue whale and herring games. Partition 6 distinguishes the slow loris from the hummingbird strategy, and also describes the effect of temperature on organisms (there is a cessation of activity in many species when the temperature goes below a threshold value).

The discussion concerning the six partitions is summarized in Table 3-2. In effect, we can think of each individual or species as playing a game of strategy against nature, in which the game is played with calories, the object of the game is survival of the species, and the six switches are set so as to define a strategy. The particular strategy that evolves in each species under the influence of natural selection depends on the average values and the fluctuations in environmental variables.

If there is an enormous and guaranteed supply of food available, as in the case of baleen whales that eat krill, a very high proportion of the incoming calories goes into maintenance of an extraordinarily high growth rate per individual, rather than a large number of individuals. If the animal lives in a very unstable environment, or has a very unstable food supply, the optimal strategy is to have a very large number of tiny offspring as a hedge against catastrophic deterioration in the probability of survival owing to environmental deterioration (herring, sardine, smelt). If the species lives in a very stable environment, or is able to make it stable because of intelligence, behavior, or self-regulation of body temperature (homoiothermy), there is a very small number of offspring. Desert

TABLE 3-2 Summary of information about factors determining partitioning of energy eaten by an animal

partition (switch)

number	function	factors determining switch setting	fundamental signif-icance of this switch
1	Determines propor-tions of food wasted and used	Food availability Age Temperature	More waste when animal is less stressed
2	Determines propor-tions of assimi-lated food going into energy and tissue manufacture	Size Activity level (hunter or cropper)	More food goes into activity when animal is small or when its food is rare or active, re-quiring extensive search or pursuit
3	Determines propor-tions lost to para-sites and predators	Size Speed Protective mech-anisms (skunk, hedgehog) Use of protective mimicry or warn-ing coloration	High reproductive rate associated with large turnover allows for wide-amplitude popula-tion fluctuation
4	Determines propor-tions of tissue used for energy storage (fat) and for growth and reproduction (protein)	Need for diapause or hibernation, because of food-less and rigorous winter season	Energy storage is needed if there are rigorous and foodless periods during year
5	Determines propor-tions of protein going to growth and to reproduction	Probability of survival for each individual Homoiothermy	Whether species maintains its bio-mass from genera-tion to generation by reproductive volume or high survival
6	Determines alloca-tion of energy to basal metabolism or activity	Activity level (hunter or cropper)	More energy needed at carnivore levels of trophic pyramid

reptiles do not need a high level of active metabolism or basal metabo-lism because they need little energy to maintain their body temperature in the hot deserts where they live (lizards avoid heat loss when the sun is not shining by hiding in burrows or crevices where heat loss by radiation will be minimal).

Another fact which has a bearing on the energy strategies of orga-nisms is that different foods have different energy contents per unit mass.

PRINCIPLE 20 The energy content per unit weight varies greatly between different foods: this affects the food acquisition strategies of organisms.

To illustrate this principle, examples of the energy content of different categories of foods are tabulated in Table 3-3. We see in this table the reason why rodents choose to store nuts as food for the winter: nuts have an extraordinarily high caloric content. What an overwintering animal needs is fuel to keep its temperature up and maintain vital organ function (calories), not protein with which to grow. The table reveals the great importance of cereal grains as a source of energy, and the even greater importance of meat or fish as a source of protein. Also, where fish and meat are available to an animal, meat is more attractive because although the protein content of the two foods is similar, meat is a better source of energy.

Energy has a positive effect on organisms because it supplies the light to plants for photosynthesis, and the heat and food to animals which support their activity, growth, and reproduction. However, like matter, it has negative effects when energy variables approach or enter the lower or upper lethal zones of organisms. Excessive amounts of light, heat, and nuclear rays are all damaging.

PRINCIPLE 21 The species filling a particular niche can typically make more efficient use of resources in that niche than foreign species which might attempt to invade and replace them.

This principle relating to the efficiency of energy use has far-reaching practical implications. For example, it has been shown by Dasmann (1964) and Matthews (1962) that the maximum possible stocking density of cattle on a Southern Rhodesian ranch yielded a net profit only 78 percent as great as the net profit from a sustained yield of thirteen species of native game on the same acreage (see Chap. 10 for further discussion). This should not be too surprising, however, because if other organisms could make better use of the energy and other environmental materials (food, site factors) at a particular place than the animals there originally, the original animals would have been replaced long ago. The only exceptions occur where a potentially superior competitor has been kept away by some geographical barrier. In such cases, removal of the barrier or accidental introduction of the superior competitor can have spectacular results (Elton, 1958).

TABLE 3-3 Nutrient content of typical representatives of different food categories

category	example	grams of nutrient constituents per 100 grams of edible portion		
		calories	protein	carbohydrates
Leaves of plants	Lettuce, headed	15	1.2	2.9
Fruits	Black raspberry, raw	74	1.5	15.6
Nuts	Almonds, dry	578	18.6	19.6
Cereal grains	Whole wheat meal	344	12.7	75.3
Meat	Hamburger, cooked	364	22.0	0.0
Fish	Herring, raw	191	18.3	0.0

SOURCE: Data from "Heinz Nutritional Data," H. J. Heinz Company, 1964.

Another principle related to this has been stated by Margalef (1968) as follows.

PRINCIPLE 22   Structures that endure through time are those most able to influence the future per unit of available energy.

In order to understand this abstract principle, we must consider the history of a community of plants or animals. E. P. Odum (1969) has pointed out that as communities develop, the biomass supported per unit of energy flow increases. Now since the filling up of the space in a region with living stuff clearly has an influence on the future biological history of that region, three trends are related: community development, increased B/E ratios (increased energy efficiency) resulting from community development, and increasing influence of the system on the future of the system.

The fundamental phenomenon being described here is succession, the process by which the organisms in an environment transform that environment, by their activity, so that it becomes more suitable for a new wave of organisms which will replace them. This phenomenon is often illustrated in textbooks by the gradual conversion of a lake to a swamp as the lake fills up with organic matter, or the gradual conversion of a beach to a forest as wave after wave of plants occupies the beach and stabilizes the sand dunes. However, there is another particularly graphic example of succession found near active volcanoes. Succession in this instance, as elsewhere, illustrates the role of time as a resource, as well as matter and energy.

Figure 3-5 illustrates four different stages in the development of plant communities on the big island of Hawaii after all life has been eliminated by a lava flow. A of Fig. 3-5 gives some notion of the devastating effect of a lava flow by illustrating the contrast between the area the lava did not touch (the trees in the background) and the strip of ground over which it flowed to the ocean (the area in the foreground in which there are no trees, but only lichens, ferns, and herbs). This picture was taken several years after a lava flow. Immediately after a lava flow the ground would be absolutely bare.

The next three pictures illustrate different steps in the process by which plants recolonize the bare lava after it cools. The first invaders are lichens (symbiotic union of algae and fungi) which grow on the bare lava and, as they die, gradually build up soil, particularly in the cracks in the lava flow. C illustrates how ferns and grasses then make use of this soil in the cracks, largely a product of the lichens, to become established. In D, the lava is now covered with a fairly continuous growth of grasses and other herbs, and this process of recolonization is continually building up soil so that invading shrubs can get enough for adequate root development.

There are several important principles to note in connection with this process. First, it illustrates the importance of time itself as an important limiting resource: because of weathering of the lava surface, a long time is needed for enough soil to be built up by plants in order that progressively higher plants can invade. Second, the fundamental idea in succession is that each wave of plant species invaders modifies the envi-

A

B

C

D

FIGURE 3-5 Different stages in the succession of plant communities after all life has been removed from an area by lava flow. All pictures taken on the island of Hawaii.

ronment by its own activities with the result that as modifications in the environment become sufficiently suitable, succeeding waves of species can invade and outcompete their predecessors. Third, each wave of species invaders tends to include larger plants than earlier waves, and also tends to fill the air space over the area more completely than earlier waves. Thus, as succession proceeds, the canopy of leaves more completely covers the area and more completely utilizes the incoming solar radiation. The air space over the area fills up with biomass, so that the ratio of biomass present per unit of incoming solar radiation rises. In the final stage of succession, the climax community, the process of succession comes to an end, because the climax association of plant species replaces itself, rather than paving the way for still another wave of species to invade. At this point equilibrium has been reached. The association of plant species that has the highest energy efficiency (highest B/E) of any wave occurs somewhat before the climax stage. However, because the climax replaces itself, it has the highest degree of control over the future. Also, the plant species in this wave tends to include very large individuals whose size has a great influence on the landscape, through their effect on microclimate, and soil formation and maintenance. Thus, we have Margalef's general principle. This principle is of immense significance for mankind, because an essential characteristic of our civilization is that it burns up energy at a fantastic rate. If natural ecosystems are any guide, this makes one wonder whether our civilization can survive (Chaps. 6 and 15).

An important characteristic of radiation emission from the sun to the earth is that the annual pattern of variation in energy received shows much more season-to-season difference as the poles are approached. Figure 3-6 shows how the incoming energy per day per unit area varies every 10° in latitude from the equator to 80°N. It is particularly noteworthy that the summer energy input per day at the highest latitudes is actually greater than at the equator. This intense peaking of the annual energy input into a short period in the summer means that plants and animals in the arctic must be evolved so as to make extremely efficient use of the energy when it is available. Also, as previously mentioned, it is critical that they have mechanisms for withstanding the environmental rigors of the remainder of the year. This intense peaking at the pole is the reason why, in general, density-independent factors are more important for animals and plants near the poles, but density-dependent factors are more important near the equator.

There are other issues relating to ecological energy metabolism which concern the entire planet. The planet continues to support life because various biogeochemical cycles maintain themselves in an approximate state of equilibrium. Specifically, we may speak of a carbon

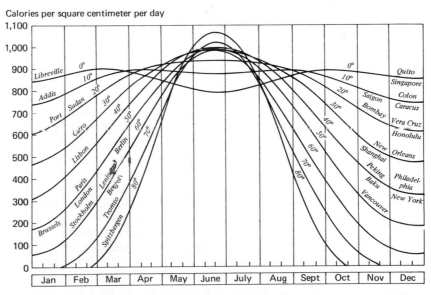

FIGURE 3-6 Seasonal and latitudinal variation in incident solar radiation. (*From Gates, 1962.*)

cycle, which includes carbon dioxide, a nitrogen cycle, an oxygen cycle, a water cycle, a potassium cycle, or a phosphorus cycle. There are many others. If, for any of these cycles, the substance in question is being lost to the biosphere faster than it is being produced, then large-scale problems may follow. For example, potassium must be brought to the surface of the ocean from the depths, or added to the ocean from rivers, about as fast as it sinks to the ocean bottom, or the productivity of the ocean will diminish. Oxygen must be produced by green plants about as fast as it is consumed by respiration and combustion, or the biosphere will suffer from anoxia, and everything living will die. Carbon dioxide must be used up by plants in photosynthesis as fast as it is produced by respiration and combustion, or it will accumulate in the atmosphere and cause planetary heating. This latter phenomenon occurs because carbon dioxide, though nearly transparent to visible light, absorbs infrared radiation (heat). Thus, increase in the carbon dioxide concentration of the earth's atmosphere would not diminish incoming light rays from the sun, but it would diminish radiation of heat from the earth out to space. Consequently, increasing carbon dioxide concentrations in the lower atmosphere has the same effect as the glass in the roof of a greenhouse: heat is kept in and the temperature is kept up. Doubling the carbon dioxide content of the lower atmosphere would increase the temperature at the surface of the earth about 2 to 3°C (Manabe and Wetherald, 1967).

Alteration of biogeochemical cycles could affect energy utilization by plants and animals in a number of ways. First, living organisms have rates of reactions which depend on temperature and, in the case of plants, on light. Any process which has marked effects on the availability of heat or light consequently has important effects on the rate of energy utilization by organisms. Second, plants cannot produce living stuff at the maximum possible rate unless appropriate concentrations of the requisite minerals are available; on the other hand, excessive concentrations of minerals can be toxic to plants and animals.

Are there any grounds for worry about these conditions? The point has been made that modern civilization is both increasing combustion (cars, airplanes, factories) and cutting down forests. Could we choke the planet? At first inspection, it would appear that we are not burning fossil fuel fast enough to overpower the ability of plants to convert the carbon dioxide in photosynthesis. The carbon dioxide annually consumed in photosynthesis throughout the world is 11×10^{16} grams, and the worldwide burning of fossil fuel releases only about 1.5×10^{16} grams of carbon dioxide per year (Johnson, 1970). However, two complications suggest that this problem may become more significant in the future. The rate of use of crude oil in the world, a reasonably good index of industrial and transportation activity, is increasing about 6.9 percent per annum, or doubling every 10 years. Thus, in another 30 years, if this trend continues, worldwide burning of fossil fuel will release close to eight times 1.5×10^{16}, or 12×10^{16}, grams of carbon dioxide to the atmosphere. More important, we are gradually cutting down on the ability of plants to photosynthesize on a worldwide basis, both by cutting down forests and by poisoning marine plankton with DDT (Wurster, 1970).

Fortunately, however, biogeochemical cycles are characterized by homeostatic mechanisms, such that a fairly large perturbation in one part of the system can be compensated for by an adjustment elsewhere. Thus, in the case of the carbon dioxide cycle, increase in carbon dioxide production by man is compensated for by the buffering effect of oceanic phytoplankton which absorb about two-thirds of the extra carbon dioxide we produce. Also, the global biomass of plants (as in the Amazonian forest) may have increased as a result of our increased carbon dioxide production. Reducing the plant life in some parts of the planet would therefore have little if any effect on the carbon dioxide cycle, because of these buffering mechanisms. However, it would eliminate the food for many animals.

Nitrate fertilizer is being applied to soil in the United States faster than plant roots can take it up. The excess washes out of the soil into rivers, to become part of the national water cycle, and some of it turns up in our food. This has a deleterious effect on the health of humans and animals (Wright and Davison, 1964).

SPACE   Space itself has many different kinds of implication for organisms, apart from any resource that might be within the space. These implications can be considered within three categories: the amount of space per individual, the structure of available space, and the spatial interspersion of different species.

the amount of space per individual   A number of experiments have been conducted to show the importance of space as an important resource for organisms. Ishida (1952) studied experimental populations of bean weevils, in which the number of available oviposition sites per female and the volume of the container per female were the two factors that varied from one experiment to another. He was able to show that an increase in the density of females either per oviposition site or per volume of the container depressed the fecundity rate of female weevils.

PRINCIPLE 23   The distribution of individual organisms in space determines in part the efficiency with which energy is utilized by the individual and its population.

This principle is very important for human society as well as for plants and animals. If the amount of space per individual is large, then the mean distance between individuals will be great, provided that all individuals are evenly distributed. If the density of food items is low, then the average distance to the nearest food item will be great. The former statement implies greater distances that must be traversed to find a mate; the latter implies greater distances that must be traversed to find food. A major result of increasing mean distance is greater energy cost of traversing space. In order to move, the organism must obtain energy over and above that required to maintain basal metabolism. Either this energy cost of movement may be met by wind or water current energy which the organism is able to exploit, or the cost may have to be met by the organism itself, in the form of additional food it must eat. The relations between this energy cost of movement and the distribution of organisms in space are brought out in three graphs. Figure 3-7 depicts hypothetical curves for gross energy intake from feeding per unit time, energy loss from moving, and net energy intake, all plotted against the distance outwards from an animal's burrow, nest, or roost. These curves assume that the animal searches outwards from its nest daily looking for food, and that its nest has been in the same location for some time. Consequently, the food discovered per unit time close to the nest will be low, because this area will already have been well searched over. Thus, there is an increase in energy intake per unit time as the animal moves farther from its nest, because the area will have been less thoroughly worked over in earlier searches. However, as the distance from the nest increases beyond an optimum distance (the heavy vertical line), there is an increasing amount of time lost in food gathering because of time spent fighting with the nearest neighbor, whose territory has been invaded. B in Fig. 3-7 depicts the energy cost of moving outwards from the nest, which is a linear function of the distance. C is obtained by subtracting the energy cost of movement in B from the energy gained in movement in A. C shows that there is a rather sharply peaked curve for the energy gained as a function of the distance out from the nest. The heavy vertical line is the optimal outer edge of the territory in which the animal searches.

The implications of Fig. 3-7 for the frequency distribution of territory sizes in an animal's ranges are brought out in Fig. 3-8. Three different patterns of territory size distribution are depicted. Which would we expect to be most common in nature? Are territories approximately equal in size, or is there a wide variation in size, or a small variation? An interesting form of evidence comes from the burrows of sand crabs on sand beaches. Early in the morning, before people and other animals have been walking over the beaches, the tide has smoothed out all evidence of the previous day's activity, and the only marks on the beach are the piles of sand made by the sand crabs digging out their holes overnight. A of Fig. 3-9 shows that the size distribution of territories, as evidenced by the distribution of distances between crab burrows, is closer to Fig. 3-8A than to Fig. 3-8B. This suggests that there is very

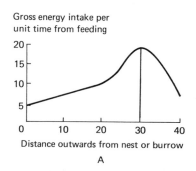

Gross energy intake per unit time from feeding

Distance outwards from nest or burrow

A

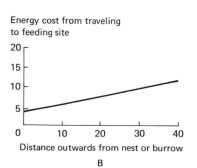

Energy cost from traveling to feeding site

Distance outwards from nest or burrow

B

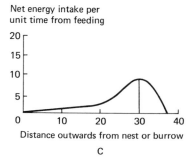

Net energy intake per unit time from feeding

Distance outwards from nest or burrow

C

FIGURE 3-7 Gross energy intake from feeding, energy cost due to movement, and net energy intake as a function of distance outwards from an animal's nest or burrow. These are hypothetical figures, and all are understood to be in calories per unit time. C is obtained by subtracting the figures in B from those in A for each distance traveled outwards. The heavy vertical lines represent the optimal distance to travel outwards from the nest or burrow.

Top view of
animals'
territories

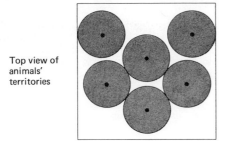
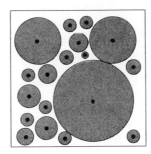
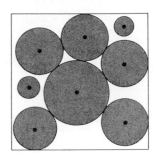

Relative
abundance of
territories

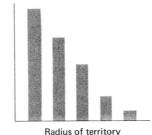
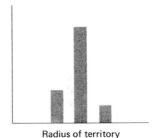

Radius of territory

All territories the same size

Radius of territory

Many different sizes of territories
with abundance of each size
inversely proportional to size

Radius of territory

Most territories have mean radius
but a few are larger or smaller

FIGURE 3-8 Frequency distributions (histograms) of territory sizes of animals for three different patterns of territory size distribution.

strong selection operating within a food-gathering system for approximately equal territory sizes.

Other kinds of evidence bearing on the relation between distribution of organisms and energy utilization are found in the other parts of Fig. 3-9. B illustrates the very tight packing of fish in a school of juvenile fish. This packing may result from the need to minimize energy loss to predators. Therefore, the prospective prey stay together in a close group so as to saturate the attack capacity of any predator that finds them. C and D of this figure compare two cities surrounded by physical barriers: mountains and water. In one case, efficiency of energy use has not been an object, and subdivisions are so interspersed with farmland that commuting costs will be tremendous (Santa Clara County, south of San Jose). In the other case, the society makes much less use of gasoline per capita per day, and the town is much more compact (a small town near Lake Amatitlan in Guatemala).

The mean distance between an organism and any entity it is searching for, whether food or a mate, is also significant because it determines the probability that an event will occur. If the distances are too great because prospective mates or food are present at very low densities, then the organism may not find a mate or food before it runs out

A

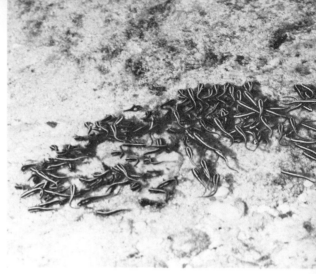

B

C

D

FIGURE 3-9 Examples of the implications of spatial distribution patterns for the efficiency of energy utilization in a species. **A** The distribution of sand crab burrows on Poipu Beach (Kauai, the Hawaiian Islands). **B** A school of tightly packed small fish close to the shore near Fiji. **C** and **D** Two small urban areas surrounded by mountains: Santa Clara County, south of San Jose, and a small Guatemalan town. The degree of compression of the urban areas is related to the availability of money with which to buy gasoline for cars (**C** *Courtesy of Air Photo Company.*)

of stored energy with which to pay for the cost of movement. In that case, the organism will perish or will not reproduce; if this happens to enough individuals in a species, the entire species will perish.

Another implication of the amount of space per individual works the opposite way: not only may there be too much space per individual to search over, but there may be too little space per individual. This occurs if there is so much crowding of individuals that competition for space required for vital functions becomes excessive. An example is plants competing with each other for space in which to obtain light for photosynthesis. White and Harper (1970) have shown for a wide variety of plant populations, from wheat through vegetables to trees, that the mean dry weight per plant w is related to the density of surviving plants p by the equation $\log w = a - b \log p$, where b is close to 1.5 for a very wide variety of plants and situations. The theoretical basis for this empirical relation is as follows. The space taken up by a plant is proportional to the weight raised to the $\frac{2}{3}$ power. (Space is proportional to the square of the length; the weight is proportional to the volume, which is proportional to the cube of the length.) Since the population density is inversely proportional to the space occupied by one plant, the relation between population density and weight is given by

$$Cp = w^{-2/3}$$

where C is a constant or, raising both sides to the $-\frac{3}{2}$ power,

$$w = C'p^{-3/2}$$

where C' is a new proportionality constant.

Too small an amount of space for animals will occur if there is so much crowding that it interferes with feeding, mating, or seeking a place to oviposit, nest, or rear young. It should be noted that different types of space may be required for different processes.

Since too little space is deleterious and too much can be also, then the curve relating the relevant measure of success (productivity, survival, reproduction, or growth) to density will have an optimum density. Whether this curve has an ascending limb as well as a descending limb will depend on whether finding a mate as well as food is a relevant process for the species. In the case of plants, they do not need to search for mates, and so typically only the descending limb of the curve is present. The curve for needle-leaved evergreens, however, might well have an ascending limb too because they inhibit growth of broad-leaved deciduous plants by shedding acidic leaves.

the structure of available space
The structure of space, as well as the amount of space per individual, is important to organisms in several different ways.

Certain kinds of organisms are not able to live in a habitat at all unless a portion of the space in the habitat has a particular kind of structure. For example, the young of many aquatic animals depend on very shallow spaces at the edge of bays into which they cannot be pursued by deep-bodied aquatic predators. The voracity of predacious fish for tadpoles is so great that but for this mechanical exclusion of fish predators from tadpole habitats, the large fish predators would quickly eliminate the entire tadpole population.

Another space factor which has many important implications for the dynamics of all plant and animal populations is the degree of patchiness (heterogeneity) as opposed to uniformity (homogeneity) of the habitat. A great deal of spatial heterogeneity, like so many other factors, has both good and bad effects on any given species. In general, if we consider only species x, it is to the advantage of x to live in a very homogeneous environment, in which large contiguous tracts of space are all in the same climatic zone and have the same soil, topography, and biota. Considering only x, there are real disadvantages to environmental heterogeneity. Heterogeneity increases the probability that an individual will be inadvertently moved to an inhospitable part of the environment, by wind or water currents, for example. This can happen to seeds, small insects, fish, or invertebrate larvae. To illustrate, a small caterpillar might be blown a considerable distance from a small patch of suitable trees, to an adjacent patch of bare rock, water, or trees of a species inedible by that species. Or a population of larval sardines might be washed off their nursery grounds out to the open ocean by strong ocean currents. Clearly, considering *only* species x, spatial heterogeneity seems to be a bad thing.

However, when an environment is heterogeneous, because it is chopped up into small dissimilar elements in a mosaic pattern, difficulties are created not only for x, but also for the species which eat x. The net effect of all these effects might be to yield benefits to x that more than compensate for the losses to x caused by environmental patchiness. For example, suppose that x is balsam fir trees. It has been shown by Morris et al. (1963) that where balsam fir grows in partially isolated stands, because of cutting operations, spruce budworm populations do not build up to the same level as where there are large contiguous forests. This is because of greater spruce budworm larval mortality due to dispersal onto inhospitable habitat surrounding small isolated stands. The phenomenon is diagramed in Fig. 3-10. The lowered pest survival in isolated stands of trees means that the amount of pest damage per tree is much lower on the average than in the large unbroken forest

Stand is part of a large
contiguous forest

Stand consists of isolated
patches of trees due to
patchy forestry cutting pattern

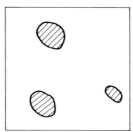

Pest
population
densities

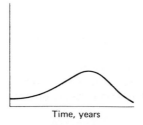

Time, years   Time, years

FIGURE 3-10 An idealized representation of the relation between the structure of forest stands and the fluctuation in numbers of an insect pest. Where the stand is part of a large contiguous forest, dispersal mortality to insect pest larvae is low, and the pest populations build up to very great densities at the peak of the outbreak. Where the stand consists of small isolated clumps of trees, dispersal mortality is much higher, and peak pest population densities are much lower. (*The drawing is an idealized version of the information on p. 224 of Morris et al., 1963.*)

stand. This is important for forest management. Optimal cutting strategy is to cut down the forest in a mosaic pattern, rather than clear-cutting (cutting very large tracts all at once).

Just as the heterogeneity of distribution of plant populations can influence the effects of herbivores on the plants, so the heterogeneity of herbivore distribution patterns can influence the effects of predators or parasites on the herbivores. Also, variations in the structure of the environment can create niches in which the prey can hide from the predators, thus preventing extinction of the prey and, subsequently, the predators. Huffaker and Kennett (1956) found that in a predator-prey system of prey mites and predator mites on strawberry plants, the fluctuations in numbers of both prey and predators were small, since the prey were spared severe overexploitation by the predators because of protective microareas where the prey could hide.

Another important feature of the structure of space is uneven topography, where the sun has different kinds of temperature effects on hilltops and valleys because of their different characteristics as heat sinks. Figure 3-11 shows an additional effect: rainy and cloudy days are

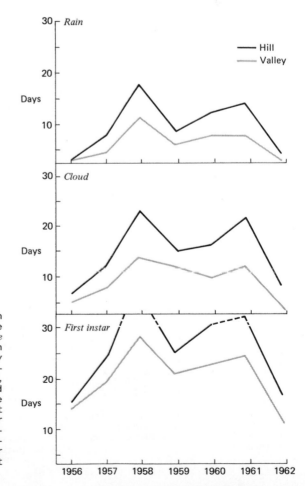

FIGURE 3-11 Cloudy and rainy days recorded in valley and on adjacent hill during the emergence and development of first-instar *Malacosoma pluviale* larvae near Victoria, British Columbia. Broken lines represent years in which there was virtually no survival of first-instar larvae. Period of first-instar development always included part of April, but occasionally it began late in March or ended early in May. Length of period varied in response to weather (bottom panel). Note that lines are not parallel. If we were to express weather effect or biological effect at either location as a linear function of weather at a nearby meteorological station, the constant b in the equation $Y = a + bX$ for the hill would differ from that for the adjacent valley. (*Wellington*, 1964.)

more common on a hill than in a valley, and so the number of days required to complete the first-instar larval period is greater for caterpillars on the hill than in the valley. Consequently, survival of insect populations is greater in the valley than on the hill, and at times of very low population densities because of bad weather, caterpillar populations may disappear from the hill altogether (Wellington, 1964).

the spatial interspersion of different species

The concept of interspersion is illustrated by Fig. 3-12. Extending the argument applied to the relation between balsam tree distribution patterns and pest population buildup, we can see the relation between interspersion of many plant species and the probability that their insect defoliators will build up to very high densities. The extreme example of

+	+	+	+	X	X	X	X	X
+	+	+	X	X	X	X	X	O
+	+	X	X	X	X	O	O	O
+	+	X	X	X	O	O	O	O
X	X	X	X	O	O	O	O	O
X	X	X	O	O	O	O	O	O

Plant species separated spatially by environmental gradients

A

O	+	+	O	X	X	O	X	X
X	X	O	X	X	X	O	O	+
X	O	X	O	X	X	O	O	X
O	+	+	O	O	X	X	X	O
O	X	+	O	X	+	+	X	X
O	X	O	+	X	X	+	O	+

Random degree of plant species interspersion

B

O	+	X	O	+	X	O	+	X
X	X	O	X	O	X	X	O	+
X	O	X	O	+	O	+	X	O
O	+	O	X	O	X	X	O	X
X	O	X	O	+	X	+	X	+
O	X	O	+	X	O	X	X	X

Complete interspersion of plant species produced by use of plant exohormones to suppress growth by each species of seedlings of its own species adjacent to parent plant

C

FIGURE 3-12 Different degrees of interspersion of plant species. Each symbol represents an individual plant. The figure was constructed as follows. B was constructed first, by taking the first nine columns and first six rows of a random-number table, and using + for all numbers beginning with 0, 1, and 2, × for all numbers beginning with 3, 4, 5, and 6, and 0 for all numbers beginning with 7, 8, and 9. Then A and C were created, using the same number of symbols and each kind as in B. A principal point of this figure is to illustrate the three kinds of environmental patterns that can be observed in nature: random distributions (B), more contagious, or underdispersed, than random patterns (A), and more regular than random, or overdispersed, patterns (C).

this is the tropical rain forest, where a tremendous number of tree species are completely interspersed. One may have to walk a considerable distance down a forest trail to get from one individual of a particular species to the next individual of the same species. In the case of the rain forest, we would not expect any one insect species to build up to high population densities, because the food preference of most insect species is confined to a small number of plant host species, and no plant host species is present in large numbers. Therefore, the tropical rain forest is characterized by a large number of insect species, no one of which is very common. On the other hand, large tracts of contiguous forest with a small number of tree species are characterized by small numbers of insect species which, on the average, have very high average population densities.

The degree of interspersion of plants depends on the presence of environmental gradients and the extent to which a plant species uses chemical effluents to suppress growth of seedlings of its own species adjacent to it. In Fig. 3-12, A shows a low level of interspersion, where the different plant species are separated spatially by environmental gradients. B illustrates a random degree of interspersion, and C illustrates complete interspersion. C illustrates a distribution pattern with more spatial mixing of species than we would expect from chance alone, as in the case where each species is suppressing growth of its own seedlings.

TIME Much of life is a race against time to obtain energy while it is available, before energy resources are unobtainable, because of nightfall, the end of a season, or exhaustion of an available energy supply. Even for man, life is a race to develop the technology that will allow us to use progressively more exotic sources of energy before existing sources of energy are depleted. For example, suppose the human species is not able to develop a massive breeder and fusion nuclear power technology before fossil fuels are exhausted and before uranium ores are burned up in nonbreeder reactors. The human population would have to decline until the population density reached a level at which it could be sustained by the annual rate of influx of solar radiation and the rates at which this could be converted to food, fuel, and building materials by plants.

Whether there will be adequate time available to sustain life for any given plant or animal species depends on two sets of factors with respect to time: those that characterize a place and those that characterize species that are prospective inhabitants of that place.

available time as a characteristic of a place Any place on earth can be characterized by the length of time in a day, a year, or a sequence of years that is suitable for operation of a particular biological process. The length of time during which conditions are within the zone of tolerance so that growth can continue uninterrupted determines the total amount of biomass in any species that can accumulate prior to the termination of this suitable time period. Unless the species can take in enough energy to sustain it until the next suitable time period, it will not be able to persist, at this particular site. The length of time available in which conditions allow for the accumulation of biomass determines the maximum possible biomass that can be accumulated. In a marginal habitat, it is the average length of the unbroken series of suitable time intervals that determines the mean population density of a particular species at that site. For example, these are the estimates of 4-year-old smallmouth bass in South Bay, Lake Huron, a marginal habitat for this species:

1947	12,200	1950	9,000	1953	36,800
1948	7,500	1951	44,900	1954	1,000
1949	2,700	1952	29,000	1955	1,000

Almost all the year-to-year variation is attributable to the water temperature in the year in which the year class is spawned (Watt, 1959). In only three of the years between 1943 and 1951 the water temperatures were high enough to allow the population to attain any size. Development of this species is constantly being curtailed by a break in the sequence of consecutive years with favorable weather conditions. Only those species which can build up biomass fast enough to take advantage of the short growing season at high latitudes will have a sufficiently large population to keep the species going the following season. Rarity in a population increases the probability of extinction too much for population survival where the environmental conditions fluctuate markedly.

characteristics of species which determine how they are affected by time

Shortage of time is one of the critical resource limitations for all living organisms. Time is required to search for, and if necessary pursue, food, to find a mate, to choose a place to rest or breed, or to hide from enemies. It is needed also for growth and reproduction. Even the largest, most powerful animals must perish if they cannot combat this enemy and win. If a lion is incapable, for some reason, of replacing energy as fast as it burns it up, on the average, it will eventually die. Time is critical for predators and parasites. Because they are higher up in the trophic pyramid than herbivores, because of the great energy wastage down through the pyramid, their food is less dense than that of the herbivores. If a parasite or predator is handicapped by a low probability of encountering a host or prey per unit time, it must deal with this problem by expending a great deal of energy in searching, by depending on acute senses for detecting prey, or by producing a very large number of eggs, as invertebrate parasites do. Extending this idea, there is a critical interaction between the length of time available for searching and fluctuations in the density of food. For a predator, a combination of reduction in searching time with environmental fluctuations that reduce prey density can be catastrophic. Once a predator begins to starve, both its energy and the length of time it can search are diminished; a vicious circle is reinforced: more starvation means less searching time, which means still more starvation. In general, because the absolute numbers of predators are typically smaller than those of their prey, any environmental change which causes a similar proportional reduction in predators and prey has a greater effect on the dynamics of the predator populations. Suppose insect prey and insect predators have densities of 100,000 and 10,000 per unit area, and pesticides reduce the densities of each to 1 percent. The prey now number only 1,000, but the predators are only 100, and for the latter, problems such as the time required to find a mate become far more critical. This is one way of explaining the observation that continued pesticide use tends to eliminate the insect enemies of insect pests proportionately more than the pests themselves.

However, time is critical also for herbivores. They may need to eat during a large part of each day in order to obtain sufficient food. If, then, predation pressure should enforce hiding a large part of each day, the herbivore would have inadequate time for food collection and would starve.

Several attributes of a species determine how it is affected by time. For example, the velocity with which an organism can move determines the distance it can travel in a unit of time. This will be critical in determining whether an animal can move rapidly enough to find resources it needs for continued existence during days or seasons that are suitable for activity. The reaction rates of various processes within organisms will determine the number of events which occur per unit time, and this means the difference between survival and death of an individual, or a population. For example, the most successful species in a particular environment may be those for which the time between egg laying and the development of larval stages is shortened the most by increased temperature. Some species, the homoiotherms, are adapted to maintain a high rate of reaction at a wide variety of habitats, but others, the ectotherms, are more likely to be physiologically adapted to perform at their best only within a limited range of temperatures. In order to complete the life cycle in minimal time and with minimal risk in the short arctic season, insects have a number of special adaptations (Downes, 1965). Stages may be dropped out of the life cycle, or prerequisites for egg development may be dropped, as in the case of blood meals for blackflies, and to cut risk of failure to find a mate out of the hazards facing a species, males may be dropped and parthenogenesis adopted (reproduction without mating).

Another point to consider in determining how species are affected by time is whether they have a small number of large meals or a large number of small meals. In the former case, it is important for the species in a marginal habitat to be adapted to ensure obtaining the appropriate number of meals before it runs out of suitable time for pursuit.

Time is important also because of Principle 7, concerning the accumulation of diversity through time. The longer the time during which an environment has been sufficiently salubrious to allow for the accumulation of large numbers of species, the larger will be the number of species that has accumulated: the greater will be the diversity of the biota. This is the time-stability hypothesis. Figure 3-5 depicted the role of time in the gradual redevelopment of communities on lava flows as well as in succession. It has been demonstrated experimentally that time is a critical resource in determining the number of species that colonize a depopulated island (Wilson, 1969).

DIVERSITY   The diversity of any set of items is measured by the number of possible ways the items can be combined. There are many additional kinds of information that we might collect to characterize the nature of a community, in addition to those mentioned in Chap. 1. The genetic characterization of populations is a measure of their information content (diversity). So is the organization of species into food webs. The degree of order expressed by temporal trends in a species is a measure of information content. Other writers, such as Pielou (1969), have written extensively about such measures of community organization as the patchiness and the nature of spatial distribution patterns.

The significance of all these measures is that they are a means of describing the relation between the structure and the dynamics of communities, as indicated by the discussions related to Figs. 3-10 and 3-12.

the origins of high diversity

It is possible to perform statistical analyses on the origins of high diversity by comparing the biotas of different islands. Johnson, Mason, and Raven (1968) have done this, and by combining their results with those of other workers, such as Sanders (1969), we arrive at the following principle.

PRINCIPLE 24   The species diversity in an environment depends on the area, geographic isolation, richness of the environment, and ecological diversity. Ecological diversity, in turn, depends on the length of time there has been uninterrupted climatic stability in a habitat.

Thus, for example, while there is a general trend toward increasing species diversity as one moves from the poles to the equator, there is also very high diversity in communities at the bottom of the ocean wherever the physical environment is very predictable.

Johnson, Mason, and Raven arrived at their conclusions on the basis of a study of plant species diversity in fourteen islands, four island groups off the coast of California and Mexico, and ten areas on the coastal mainland of California and Mexico. They found that the area of an island or a mainland region is the best single predictor of species diversity. Elevation is another good predictor of it. Both these variables measure ecological diversity, an important factor increasing the probability of speciation, the process by which existing species split up into two or more new species. Environmental richness was measured by the amount of rainfall, increasing rainfall causing increasing environmental richness, and aridity leading to desert conditions. Increasing richness of the environment increases species diversity. Increasing distance from the mainland also increases species diversity, by increasing the impact of isolating mechanisms, which also promote speciation.

The effect of environmental gradients on species diversity depends on which environmental factors are most important for a particular group of organisms. For example, in general, there is a gradient of decreasing rainfall from eastern to western North America, at the latitude of Utah and Nevada. Associated with this gradient of increasing drought from east to west is an increase in lizard species diversity, but there is a decrease in the diversity of parasitic Hymenoptera (small wasps apparently sensitive to too much drought).

What makes possible very large species diversity? How is it that within certain environments on earth, very large numbers of species can maintain themselves as distinct entities over long periods of time? The key to the answer lies in specialization. One way to illustrate this is to examine the structure of the mouths in communities that have great species diversity, because great variations in animals' mouths will indicate whether a great diversity of functions is being fulfilled by the various species. A particularly interesting place to do this is the shallow coral seas off the coast of tropical oceanic islands. At least 584 species of fish are believed to live off the coasts of the Hawaiian Islands (Gosline and Brock, 1965). Plate 2 shows the mouth structure of six of these. Clearly, the large number of different fish species found in this habitat is made possible by adaptations to a wide variety of different niches. Consequently, the different species are not in direct competition with each other and are able to coexist indefinitely.

the effects of trophic diversity

Several authors have considered the relation between the structure of trophic pyramids and the stability of species populations in the ecosystem. This discussion is based on the treatment by Southwood and Way (1970). Different patterns of trophic web structure are diagrammed in Fig. 3-13. In that figure, dots represent species and lines represent paths

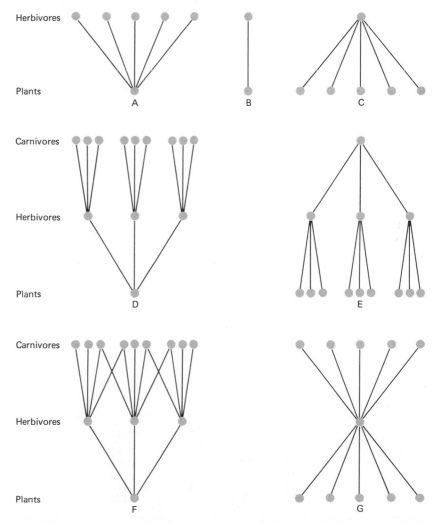

FIGURE 3-13 Seven different patterns of trophic web structure. (*Redrawn from Southwood and Way, 1970.*)

along which energy flows, from the species on one trophic level to the species on higher trophic levels which eat it. The seven panels represent seven different types of web structure. Southwood and Way argue that four factors must be considered in determining the effect of trophic web structure on species population stability. First, how much species diversity is there at each level in the web? The situations in A and C have more diversity than in B. Second, how is this diversity related to the direction of energy flow? In A, energy flows from low diversity to high diversity; in C, it flows from high diversity to low diversity. Third, how much cross-linking is there between parts of the pyramid? F has cross-linking, whereas D does not. Finally, what is the exact nature of the trophic links: how precise, rapidly responsive, and elastic is the response of one species to changes in availability of another? This last subject is the topic of the next chapter.

The best evidence now available suggests that the following rules describe the relation between web link structure and stability:

1   The least stable system of all is that depicted in B: one species eats one species.

2   The situations in A and D promote stability: a large number of species compete at the trophic level to which energy flows. Data in support of this notion are given in Table 3-4.

3   The situations in C and E operate against stability: the larger the number of food species available to an organism, the less stable it will be. Table 3-5 illustrates the evidence.

TABLE 3-4 The effect of number of insect species eating the same host plants on the stability and abundance of forest Macrolepidoptera (data on all regions of Canada pooled)

mean number of insect species eating same host plants	number of insect species	mean arithmetic mean count	mean number of insect species eating same host plants	number of insect species	mean standard error of logs of counts
Gregarious species					
17	5	546	21	4	0.44
52	2	3,024	52	2	0.66
71	13	2,895	71	13	0.57
85	5	2,734	85	5	0.50
109	3	1,475	110	3	0.36
141	1	41	142	1	0.61
242	3	11	228	2	0.32
Solitary species					
23	51	43	21	24	0.37
52	85	122	53	71	0.32
69	116	116	69	90	0.35
91	67	55	90	55	0.31
119	91	11	119	72	0.26
157	45	6	158	36	0.27
232	65	4	232	36	0.22

SOURCE: Watt (1965).

TABLE 3-5 The effect of number of tree host species on the stability and abundance of forest Macrolepidoptera (data on all regions of Canada pooled)

number of tree host species	number of insect species (divisor for means)	mean arithmetic mean count	number of insect species (divisor for standard errors)	mean standard error of logs of counts
Gregarious species				
1.0	4	16	2	0.50
2.0	3	18	3	0.46
3.7	7	335	7	0.41
5.0	1	43	1	0.66
7.6	3	1,646	3	0.40
16.2	8	1,957	8	0.49
26.0	5	7,206	5	0.70
42.0	1	5,465	1	0.78
Solitary species				
1.0	98	4	37	0.24
2.0	81	7	43	0.23
3.3	105	38	75	0.28
5.5	68	62	62	0.32
7.9	62	30	61	0.31
13.5	72	71	72	0.36
23.4	29	453	29	0.37
32.2	5	377	5	0.39

SOURCE: Watt (1965).

4   Situations as in F are highly stable because of the cross-links.

5   Situations as in G are highly unstable, because the effect of many food species seems to override the effect of many enemies: locusts and mice would appear to be examples. In each case the species can build up to tremendous numbers quickly in the face of many species of enemies.

INTERACTIONS OF THE FIVE FUNDAMENTAL CATEGORIES OF VARIABLES

It will be clear already that all five of these categories of variables are so inextricably interwoven with each other that it is impossible to consider any process without considering all five. Thus, the availability of matter determines the density of living stuff, and hence the space available per individual. This in turn determines the time that will be required to find a mate, and the time that will be required to find food. The amount of energy coming into an environment per unit time determines the extent to which living organisms can make use of the matter that is present in the environment.

However, merely stating that these interactions exist is of little help in understanding how the world operates, or in managing any process in the world which we wish to optimize in any sense. Thus, we are led to quite elaborate conceptual models which allow us to perceive the role of each factor in determining the behavior of the whole system. Such models can then be used in computer simulation studies to attempt to learn how to manage the system in the real world that corresponds to the mathematical model in the computer.

It is impossible, on the level of earlier mathematical training expected of students reading this book, to demonstrate actual models that can be used in computers to bring all five categories of variables together in a single system of equations. What we can do, however, is to use a system of simple, stylized equations to indicate the kind of approach that can be adopted in a real situation on a much larger scale.

The approach is a simplification and modification of that used by Hamilton and Watt (1970); it is fundamentally reductionist, in that it begins by attempting to look at a whole process and then splits each part down into subcomponents, until the model becomes a quite realistic mimic of an actual process.

Whether we consider an animal or a man, we can begin by reducing the energy flow into its components:

$$E_N = E_C - E_W - E_B - E_G - E_R - E_A - E_L \qquad (3\text{-}1)$$

This equation merely states that E_N, net energy gain, in calories, can be computed from these constituents of energy flow:

E_C = energy consumed

E_W = energy wasted (defected, unused)

E_B = energy cost of basal metabolism

E_A = energy cost of activity, above basal metabolism

E_G = energy cost of growth

E_R = energy cost of reproduction

E_L = energy lost by exploitation of other species

Each of these constituents can then be expressed in terms of all the factors that can operate on it, and in this way the four other categories of fundamental variables are included in a system of equations.

For example, the amount of food consumed can be expressed as a product of T_S, the time spent searching for food; N_C, the number of food items consumed per unit time spent searching for food; and E_M, the average energy content of one food item.

$$E_C = T_S N_C E_M \qquad (3\text{-}2)$$

Then each of T_S and N_C can be broken down into the component variables that determine them. The time spent searching is dependent on the average distance between food items and between the animals searching for food, and the number of food items consumed per unit time will depend on the same variables. Proceeding in this stepwise, reductionist fashion, we can gradually develop a system of equations which can be used, for example, to show how to increase the net food gain by selecting that area in which the average distance between competing searchers and between food items maximized the probability of obtaining food per unit time. That is, increased distance between searchers is an advantage for any individual searcher, because it minimizes time lost to direct competition with other searchers. On the other hand, too much distance between food items is a disadvantage, because too much time is spent trying to find them. Since one variable represents a loss and the other a gain, there is a strategy problem involved in selecting the optimum trade-off point. However, when there are many different kinds of gains and losses (costs and benefits), the problem of finding optimal strategies is really significant, and it is at this point that computers have great importance.

In general, it would appear that the energy strategies of natural communities and the energy strategies of man have been quite different, chiefly because of the availability of fossil and nuclear fuels to man. Natural communities in effect are maximizing energy efficiency, whereas

TABLE 3-6 A tabular model of ecological succession: trends to be expected in the development of ecosystems

ecosystem attributes	developmental stages	mature stages
Community energetics		
1. **Gross production/community respiration (P/R ratio)**	Greater or less than 1	Approaches 1
2. **Gross production/standing crop biomass (P/B ratio)**	High	Low
3. **Biomass supported/unit energy flow (B/E ratio)**	Low	High
4. **Net community production (yield)**	High	Low
5. **Food chains**	Linear, predominantly grazing	Web-like, predominantly detritus
Community structure		
6. **Total organic matter**	Small	Large
7. **Inorganic nutrients**	Extrabiotic	Intrabiotic
8. **Species diversity – variety component**	Low	High
9. **Species diversity – equitability component**	Low	High
10. **Biochemical diversity**	Low	High
11. **Stratification and spatial heterogeneity (pattern diversity)**	Poorly organized	Well-organized
Life history		
12. **Niche specialization**	Broad	Narrow
13. **Size of organism**	Small	Large
14. **Life cycles**	Short, simple	Long, complex
Nutrient cycling		
15. **Mineral cycles**	Open	Closed
16. **Nutrient exchange rate, between organisms and environment**	Rapid	Slow
17. **Role of detritus in nutrient regeneration**	Unimportant	Important
Selection pressure		
18. **Growth form**	For rapid growth ("r-selection")	For feedback control ("K-selection")
19. **Production**	Quantity	Quality
Overall homeostasis		
20. **Internal symbiosis**	Undeveloped	Developed
21. **Nutrient conservation**	Poor	Good
22. **Stability (resistance to external perturbations)**	Poor	Good
23. **Entropy**	High	Low
24. **Information**	Low	High

SOURCE: E. P. Odum, copyright 1969 by the American Association for the Advancement of Science.

man is maximizing power throughout, without regard to efficiency of energy use.

Another approach to integrating all five categories of variables is to consider how relations among matter, energy, space, and diversity change together through the successional process as a community develops (Table 3-6). This table, like this chapter, demonstrates the linkages between all five categories of the basic variables in all ecological phenomena. In subsequent chapters, this point will be illustrated repeatedly with examples from various ecosystems, including that operated by man.

SUGGESTIONS FOR INDIVIDUAL AND GROUP PROJECTS

Take pictures of the distribution of ants around an ant nest with a wide-angle lens. Note the type of food the ants are collecting and where they find it. From the picture, make a drawing, showing the position of each ant. Superimpose on the drawing a series of concentric circles an equal distance apart, with the nest center the center for each circle. How many ants are in each of the rings? How do you explain this in terms of energy consumption and use?

Select particular groups of plants or animals for collecting or observation (birds, mammals, insects, fish, trees, shrubs, or wild flowers). Make diversity measurements for the group you selected in several equal-sized plots. Select a number of different plots in each of several different types of environment. Do the diversity measurements between types vary more than the plot-to-plot measurements within a type? Do you see any analogous contrasts between diversity of building types from area to area in your city?

Are there any areas near your home with startling changes in habitat over a short distance (steep mountains, sea or lake coast, swamps, salinity, aridity, or nutrient soil gradients)? What changes in the species composition of the plant and animal community do you notice, going from one end of the gradient to the other? What does this tell you about zones of tolerance?

What do you notice about successional sequences in the country near your home? What types of trees are maturing under the tree species that make up the top canopy of vegetation now? The same types as those now dominating the community? Is there a difference in diversity between the shrubs in different woodlots near your home? Why? How does this relate to the species composition of the trees making up the tallest vegetational level in each woodlot?

REFERENCES

Brooks, W. S.: Comparative Adaptations of the Alaskan Redpolls to the Arctic Environment, *Wilson Bull.*, **80**:253–280 (1968).

Commoner, B.: Threats to the Integrity of the Nitrogen Cycle: Nitrogen Compounds in Soil, Water, Atmosphere and Precipitation, in S. F. Singer (ed.), "Global Effects of Environmental Pollution," D. Reidel, Dordrecht, Holland, 1970.

Dasmann, R. F.: "African Game Ranching," Macmillan, New York, 1964.

Downes, J. A.: Adaptations of Insects in the Arctic, *Ann. Rev. Entomol.*, **10**:257–274 (1965).

Elton, C. S.: "The Ecology of Invasion by Animals and Plants," Methuen, London, 1958.

Gates, D. M.: "Energy Exchange in the Biosphere," Harper & Row, New York, 1962.

Goldman, C. R.: Molybdenum as a Factor Limiting Primary Productivity in Castle Lake, California, *Science*, **132**:1016–1017 (1960).

Golley, F. B.: Energy Dynamics of a Food Chain of an Old-field Community, *Ecol. Monographs*, **30**:187–206 (1960).

Gosline, W. A., and V. E. Brock: "Handbook of Hawaiian Fishes," University of Hawaii Press, Honolulu, 1965.

Greenewalt, C. H.: "Hummingbirds," Doubleday, New York, 1960.

Hamilton, W. J., III, and K. E. F. Watt: Refuging, *Ann. Rev. Ecol. Systematics*, **1**:263–286 (1970).

Huffaker, C. B., and C. E. Kennett: Experimental Studies on Predation and Cyclamen-mite Populations on Strawberries in California, *Hilgardia*, **26**:191–222 (1956).

Ishida, H.: Studies on the Density Effect and the Extent of Available Space in the Experimental Population of the Azuki Bean Weevil, *Res. Population Ecol.*, **1**:25–35 (1952).

Johnson, F. S.: The Oxygen and Carbon Dioxide Balance in the Earth's Atmosphere, in S. F. Singer (ed.), "Global Effects of Environmental Pollution," pp. 4–11, D. Reidel, Dordrecht, Holland, 1970.

Johnson, M. P., L. G. Mason, and P. H. Raven: Ecological Parameters and Plant Species Diversity, *Am. Naturalist*, **102**:297–306 (1968).

Lotka, A. J.: "Elements of Physical Biology," Williams & Wilkins, Baltimore, 1925.

Manabe, S., and R. T. Wetherald: Thermal Equilibrium of the Atmosphere with a Given Distribution of Relative Humidity, *J. Atmospheric Sci.*, **24**:241–259 (1967).

Margalef, R.: Information Theory in Biology, *Memorias de la Real Academia de Ciencas y Artes de Barcelona*, **23**:373–449 (1957) (in Spanish). Translated in *Gen. Systems*, **3**:36–71 (1958).

——: "Perspectives in Ecological Theory," University of Chicago Press, Chicago, 1968.

——: Diversity and Stability: A Practical Proposal and a Model of Interdependence, *Brookhaven Symp. Biol.*, no. 22, "Diversity and Stability in Ecological Systems," pp. 25–37, 1969.

Matthews, L. H.: A New Development in the Conservation of African Animals, *Advan. Sci.*, **18**:581–585 (1962).

Morris, R. F. (ed.): The Dynamics of Epidemic Spruce Budworm Populations, *Mem. Entomol. Soc. Can.*, no. 31, 1963.

"Nutritional Data," H. J. Heinz Company, Pittsburgh, 1964.

Odum, E. P.: The Strategy of Ecosystem Development, *Science*, **164**:262–270 (1969).

Paloheimo, J. E., and L. M. Dickie: Food and Growth of Fishes. III. Relations among Food, Body Size, and Growth Efficiency, *J. Fisheries Res. Board Can.*, **23**:1209–1248 (1966).

Pielou, E. C.: "An Introduction to Mathematical Ecology," Wiley, New York, 1969.

Sanders, H. L.: Benthic Marine Diversity and the Stability-time Hypothesis, *Brookhaven Symp. Biol.*, no. 22, "Diversity and Stability in Ecological Systems," pp. 71–81, 1969.

Southwood, T. R. E., and M. J. Way: Ecological Background to Pest Management, in R. L. Rabb and F. E. Guthrie (eds.), "Concepts of Pest Management," pp. 6–29, North Carolina State University, Raleigh, 1970.

Watt, K. E. F.: Studies on Population Productivity. II. Factors Governing Productivity in a Population of Smallmouth Bass, *Ecol. Monographs*, **29:**367–392 (1959).

——: Community Stability and the Strategy of Biological Control, *Can. Entomologist*, **97:**887–895 (1965).

Wellington, W. G.: Qualitative Changes in Populations in Unstable Environment, *Can. Entomologist*, **96:**436–451 (1964).

White, J., and J. L. Harper: Correlated Changes in Plant Size and Number in Plant Populations, *J. Ecol.*, **58:**467–485 (1970).

Wilson, E. O.: The Species Equilibrium, *Brookhaven Symp. Biol.*, no. 22, "Diversity and Stability in Ecological Systems," pp. 38–47, 1969.

Wright, M. J., and K. L. Davison: Nitrate Accumulation in Crops and Nitrate Poisoning in Animals, *Advan. Agron.*, **16:**201–256 (1964).

Wurster, C. F.: Chlorinated Hydrocarbon Insecticides and the World Ecosystem, *Biol. Conserv.*, **1969:**123–129.

Zar, D. H.: Calculation and Miscalculation of the Allometric Equation as a Model in Biological Data, *Bioscience*, **18:**1118–1120 (1968).

4 mechanisms of self-regulation in ecological systems

Within any population of living organisms, individuals vary enormously with respect to their morphological, physiological, and behavioral characteristics. All such characteristics constitute the *phenotype* of an organism. Organisms vary also with respect to their potentiality or genetic characteristics, the *genotype*. Those individuals that are best fitted to the requirements of the habitat where they live, survive and reproduce. Since the children of the best-fitted individuals resemble their parents, it is they who survive from one generation to the next. Therefore we would expect a constant improvement in the matching of phenotype to environment, for the phenotype is an expression of the genotype. Thus, after a long series of generations in which there had been constant selection for phenotype and genotype that best matched the environmental requirements, we would expect to find the precision of the matching to be quite remarkable. We do. Further, we would expect to find that evolution selects for the ability to adjust quickly to changes in the environment, if these changes are of a type that occur with any frequency or regularity. These adaptations are also found, and again we are impressed with the precision with which they respond to demands of the environment.

All adaptations can be thought of as homeostatic, or self-regulatory responses of organisms to their environment. Over geological time, a particular environment is responded to by evolution, which gradually matches the species more and more precisely to its environment. There is constant feedback control: the environment acting in one generation produces both genotype and phenotype frequency distributions for the following generation. The precision of response is measured by the proportion of each genotype that survives. The precision is best of all measured by the proportion of mutations that are harmful. A change in the environment produces a compensatory change in the frequency distribution of different genotypes and phenotypes in the population. Adaptations are usually recognized as attributes of an individual. However, there has always been controversy, since the idea of evolution took hold, as to whether natural selection operates on levels of organization higher than the individual. One can often sidestep this argument by noting that the best-adapted species populations will typically be those which have characteristics that lead to community integration. A species that destroys everything around it sows the seeds of its own destruction. Thus, those adaptations that are most useful from the standpoint of the population tend to be those that are most useful from the standpoint of the community. Adaptations can occur in response to the physical environment or the biological environment.

In this chapter, we will deal with several different kinds of evidence of self-regulation in biological systems: morphological, physiological, and behavioral adjustments of individual animals to their environment, and regulatory mechanisms operating at the following levels: the individual, the population, the two-species interacting system, and the community of species.

MORPHOLOGICAL ADAPTATIONS TO THE ENVIRONMENT

Many different attributes of animals provide obvious evidence of their adjustment to a particular mode of life: the jumping legs of grassland animals such as rabbits, kangaroos, grasshoppers, kangaroo mice, and frogs; the long snouts of all the ant and termite eaters; the expendable tails of lizards which yield only a limited morsel to their predators; and the serpentlike or owllike eyespots on the wings of many butterflies and moths. However, no examples of morphological adaptations are so awesomely precise as those involving camouflage coloration. Plate 3 demonstrates a particularly startling example.

PHYSIOLOGICAL ADAPTATIONS TO THE ENVIRONMENT

One of the most important types of adaptations found in animals is the set of mechanisms which allow adjustment to changes in temperature. There is a wide variety of such mechanisms. A major difference between different kinds of animals in this regard is that some maintain a constant temperature, like the birds and mammals, the *homoiotherms*, and others have a body temperature that drifts up and down in response to changes in the temperature of a simple laboratory environment, the *poikilotherms*. Figure 4-1 diagrams the difference. The temperature of lizards, fish, insects, and some other animals varies to a considerable extent with external temperature variations. The temperature of birds and mammals shows only a slight downward drift at very low external temperatures, and a slight upward drift at very high external temperatures. The temperature of homoiotherms is maintained by regulation of heat loss and heat production. Heat loss is regulated by fur and feathers, by piloerection to increase insulation, by vasodilatation or vasoconstriction of the capillaries near the skin to increase or decrease heat loss from the blood, by panting, and by behavioral adjustments such as hiding in the shade to avoid overheating or in a hole to avoid heat loss. Heat produc-

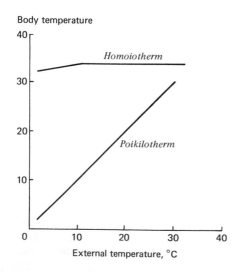

FIGURE 4-1 The responses of homoiotherms and poikilotherms to changes in the external temperature. (*Redrawn from Martin,* 1930.)

tion is varied by physical *thermogenesis*, the creation of heat through muscular contraction or shivering, and chemical thermogenesis, increase in chemical heat production in the liver and brain, particularly. In general, the metabolic response of organisms to changes in temperature is least in those organisms that live in an environment where temperature normally fluctuates a great deal (Fig. 4-2). In general, this graph implies that heat regulation in tropical animals is by means of chemical thermogenesis, primarily. That is, cooling the external temperature of a tropical homoiotherm is responded to by an increase in heat production within the body core of the animal. In arctic homoiotherms, on the other hand, decreasing the external temperature is responded to primarily by physical control of heat loss (piloerection of the fur, vasoconstriction of the surface capillaries). This makes sense, because an arctic animal could not afford the energy expenditure involved in trying to compensate for cold temperature by increasing the rate of biochemical heat production. A tropical homoiotherm rarely needs to call on this mechanism to maintain body heat, and so the high cost of having to use it when it is needed is tolerable. On the other hand, thick fur or feathers in a tropical animal would operate against heat dissipation, which is needed in the tropics.

Another mechanism for dealing with very cold arctic temperatures or very hot desert ones or any period during which enough energy cannot be found to make activity profitable, is *diapause*, or *torpor*. The former mechanism for poikilotherms and the latter for homoiotherms are comparable. In each case, the response to a long period of rigorous weather is the animal's going into a stage of arrested development, or dormancy, in which there is little or no muscular activity, and the animal exists in a virtually comatose state with very little biochemical activity apart from that required to maintain essential body functions, and consequently very little energy loss. In the case of insects, this adaptation to cold weather includes the feature that development will not be initiated again unless and until there has been an extended period of cold weather.

Cold-blooded animals respond to changes in temperature by changing the boundaries of their upper and lower lethal zones. If the

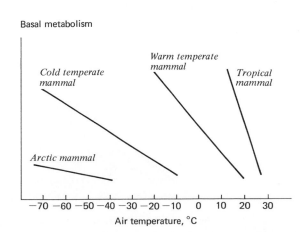

FIGURE 4-2 Heat regulation and temperature sensitivity in arctic, cold temperate, warm temperate, and tropical mammals. (*Redrawn from Scholander et al.,* 1950.)

environment in which the animal has been living warms up, the zone of tolerance moves upwards in the temperature scale. This may be expressed in a polygon relating the upper and lower lethal temperatures to the temperature to which the animal has become acclimated, as in Fig. 4-3.

Another group of important adaptations are those that maintain population stability through regulation of reproduction, growth, and survival. On the average, over time, a species survives because it maintains its biomass from one generation to another. However, the mechanisms used are quite different in different species, so that we may say, in effect, that different species are playing different games of survival against nature. The variables which measure the way this game is played are as follows.

1   The average number of young per female. The more uncertain the survival of each young, the greater this number will be (Table 4-1). Thus, the sheep liver fluke, which has a long and complex life history with an extraordinarily high probability of death at each of several different steps, has up to 500,000 eggs, each of which may produce 300 larvae.

2   The compressibility of the reproductive rate. That is, the extent to which the reproductive rate can be reduced by intraspecific competition. Presumably this reduction would be very small in blue whales, but might be 95 percent in blowfly larvae on carrion.

Lethal temperature, °C

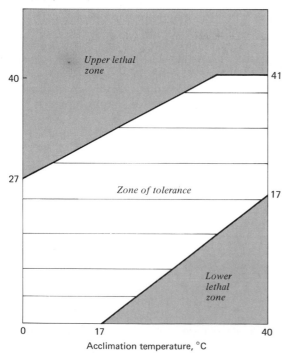

FIGURE 4-3 The relation between lethal temperature and acclimation temperature in goldfish. (*Redrawn from Fry, Brett, and Clausen,* 1942.)

3 The mean weight of each young at birth, relative to the weight of the mother. This is inversely related to 1.

4 The compressibility of this weight.

5 The compressibility of the territory required by each individual. In rookery birds, for example, territory sizes seem to be similar from bird to bird, and constant in size from one year to another, but in some other animals, such as the field mouse *Microtus arvalis*, the home range compresses greatly when population densities increase. Thus the species appears to regulate its biomass from generation to generation by reducing numbers reproduced in response to increased density, or by reducing the territory of each of the young, or both.

6 The growth rate from birth to puberty.

7 Compressibility of the growth rate.

8 The survival of the young as affected by the stress to which their mother was subjected.

9 The survival of the young as a function of the stress to which they are subjected.

10 Age at puberty.

Blue whales maintain their biomass from one generation to another by maximizing 3, 6, and 9. Herring maintain their stability by maximizing 1, 2, 6, and 7. The mouse *Microtus arvalis* maximizes 1, 3, and 5. The blowflies in the genus *Lucilia* maximize 1, 2, and 4. Most species have an Achilles' heel, or vulnerable point, in this list. For the snowshoe rabbit, it is 8. For blue whales, it is 1 and 10. For lake trout, sturgeon, and muskellunge, it is 10. That is why these large fish species are the first to show reduction in numbers when fishing pressure becomes intense enough to drive some species to extinction.

Further, species maintain parsimony. That is, each species can do no more than maximize a part of the list to maintain stability, since some of the above ten variables will show inverse correlations within sets of species being compared. For example, 1 and 3 are inversely correlated. Not only does natural selection select for homeostatic ability, the ability to prevent the species from becoming extinct, but also no more energy is expended by the species than necessary to maintain this homeostasis.

TABLE 4-1 Birth rates, life expectancies, and probabilities of survival in different species, as evidence of the rigor and uncertainty of the environment and adjustments of different species thereto

species	birth rate per female per year	birth rate per individual per year	implied mean life expectancy at birth, years	implied probability of surviving 1 year
Man	0.0282	0.0141	71	.9859
Blue whale	0.2	0.1	10	.9
Robin	4	2	0.5	.5
Spruce budworm	200	100	0.01	.01
Lake trout	10,000	5,000	0.0002	.0002
Lamprey	62,000	31,000	0.000032	.000032
Cod	6,000,000	3,000,000	0.0000003	.0000003
Sheep liver fluke	150,000,000	150,000,000	0.000000026	.0000000026

Thus, profligate apparent wastage of energy, as in the extraordinary infant mortality of sardines, herring, cod, and sheep liver flukes is necessitated by the life cycle and is compensated for by the very low number of calories expended by the female to make each egg.

The allocations of the calories expended can be explained in part by another principle, due to Margalef (1968).

PRINCIPLE 25 Large, unpredictable environmental variability selects for prolific species.

This type of compensatory mechanism also shows up in such activities as hunting. The respiratory output required by hunting is compensated for by reduced probability of local extinction, if the species spends energy to hunt for food, over an extensive area. High-energy expenditures for food and mate searching allow a species to maintain homeostasis at low population densities.

Another important group of adaptations to the environment are those observed in *xerophytes*, or plants adapted to extreme drought. These plants have smaller cells than most plants, thick cell walls, a dense network of veins, and various other adaptations. In general, desert plants can lose a much higher percentage of their water before they wilt than plants in humid environments can. Also, since biologically available energy in the desert is scarce, and a half-bitten leaf would cause a plant to lose the moisture it needs for survival, desert plants often have tough, sharp spines which discourage prospective leaf eaters.

An important part of the biological machinery for coping with stress in higher animals is the endocrine system. For example, if the animal is confronted with stress (the need to run or fight immediately, for example, or respond quickly under pressure) the adrenal medulla secretes adrenaline, which raises the blood pressure and the heart rate so that the body can work faster. However, if the body is stressed repeatedly, a set of physiological symptoms of the overstrain result: swollen and drained adrenal glands, shrunken lymphatic tissue, gastrointestinal ulcers, aggressive behavior, loss of parental behavior, and a number of other changes. All together, this syndrome constitutes the "general adaptation syndrome," which has been made famous by the work of Selye (1956). One of the most important consequences of overstressing a mother animal is a hormonal and endocrine change. Since fetuses have a common blood circulation with their mother, there may be changes in the bodies of the young as a consequence of stresses to the mother which show up in the following generation. The next chapter will present evidence showing that this is probably the case.

A different type of adaptation is the antigen-antibody system. If an antigen, or foreign protein, enters the body as part of a pathogen, such as a virus or bacteria, an antibody, or blood protein, combines with the antigen so that its danger to the body is removed. Becoming immune to an infectious disease involves forming these antibodies.

The kidneys represent still another type of homeostatic mechanism, which regulates the chemical composition of the blood even though there may be changes in the salt or other content of the food or liquid taken into the body.

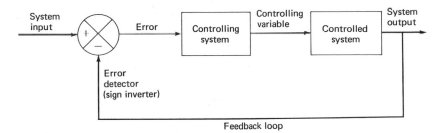

Error = set point–system output

FIGURE 4-4 The logical components of a feedback control system.

BEHAVIORAL ADAPTATIONS TO THE ENVIRONMENT

Of the various behavioral adaptations to environmental stimuli, a single one will be described in detail in order to give some understanding of the elements common to all such phenomena. In all cases, the key to understanding homeostatic mechanisms in animal behavior or plant responses (or in any other homeostatic phenomenon) is the concept of the feedback control system (Fig. 4-4). Such systems explain the behavior of a very wide range of phenomena. However, in all cases, the essential elements are those depicted in Fig. 4-4. The central feature of this system is that some type of message is fed back to an error detector which detects departure of system behavior from a set point. This detector modifies the sign of the stimulus which operates on the system, and the system performance is corrected thereby. The term *negative feedback* comes from the fact that the sign is changed by the error detector. Figure 4-5 shows how this type of conceptual model might be applied to the searching and pursuit of invertebrate prey by an invertebrate predator. There is no need for a predator to waste energy stalking and striking at prey when it is not hungry; but if a predator does not stalk and strike at prey when it is hungry, it will starve. Clearly, in this, as in most other phenomena, there is need for some kind of controlling mechanism which adjusts the predator's behavior to its need for food.

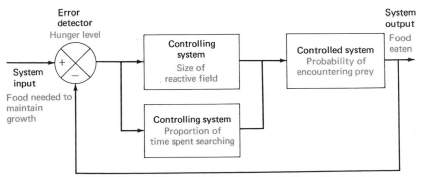

Feedback loop: effect of amount of food eaten on hunger level

FIGURE 4-5 The behavioral responses of an invertebrate predator to hunger level and prey density as a negative feedback system.

First, we will discuss the evidence on which the conceptual model is based. Holling (1966) has studied the behavior of the praying mantid in reacting to its prey, the housefly. The maximum distance a mantid would react to a fly was measured by dangling a fly on a thread and gradually moving it closer to the mantid. When any one of the four walking legs moved or the mantid struck at the fly, this was taken as a measure of the distance of reaction of the mantid to the fly. The hypothesis here is that all predators have a sensory field which surrounds them, contracts when they are satiated, and expands when they are hungry. When a prospective prey just touches the boundary of this field, on the average, the predator will attack it. Figure 4-6 shows the result of this experiment. Unless the mantid had been starved for about 9 hours before the start of the experiment, it was so satiated that its reactive field was zero. That is, it would not strike at a fly, no matter how close the fly was. (This may explain the observation of lions and prospective prey drinking nearby at a water hole: the prey may be able to detect that the lion is satiated.) However, beyond 9 hours, the longer the mantid had been starved, the hungrier it was and the larger its reactive field became. If it had been starved for 48 hours, it was quite hungry and would react to the fly if it was about 6.3 centimeters away. Since the reaction of the mantid to the fly did not begin at all until the hunger level of the mantid reached a certain minimum threshold level, it is reasonable to assume that the mantid would spend no time searching unless its hunger level passed some threshold.

Figure 4-5 includes two subcomponents of the predator behavior system: the size of the reactive field and the proportion of the time spent searching. Suppose the hunger level of the mantid is low. The size of the reactive field is small, and the mantid is unlikely to be searching at all. As a consequence, the probability of encountering prey is very small, and no food is eaten. Because of this the feedback control system adjusts the sign of the error detector so that instead of reading "not hungry," it now

Maximum distance of reaction, centimeters

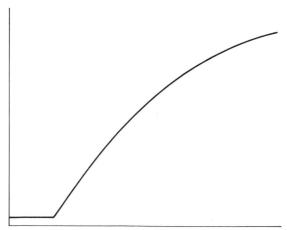

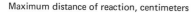

Time deprived of food, hours

FIGURE 4-6 Effect of time of food deprivation on the maximum distance that a praying mantid stalked or struck at flies (average of 12 replicates). (*Graph from Holling*, 1966.)

reads "hungry." Because of this, the size of the reactive field increases, the predator now spends an increasing amount of its time searching for prey, the probability of encountering a prey rises sharply, and soon a prey is caught and eaten. If the prey is large enough to satiate the predator, the error detector sign is changed again. Now it reads "not hungry," the size of the reactive field contracts, and the proportion of the time spent searching decreases, perhaps to zero. It will be clear to the reader that mechanisms of this general type, in which the key feature is negative feedback control, must govern the behavior of a wide variety of homeostatic (self-regulatory) systems in biology.

SELF-REGULATION IN SINGLE-SPECIES POPULATIONS

Species are regulated in numbers by density-dependent control. This control operates on the reproductive rate, the survival rate, the growth rate, and the dispersal rate. There has been immense controversy in the history of ecology about the validity of the preceding statement, but the controversy can be resolved as follows. It is entirely true, as in the case of the smallmouth bass in the preceding chapter, that a rigorous physical environment does not allow some populations to become dense enough for density-dependent controls to operate. However, it is still true, that if ever there were enough favorable years consecutively so that the population could build up, then density-dependent factors would operate. Density-dependent factors must be the ultimate controlling agent for any population. A more potent argument is quantitative: over the course of time, any population exhibits a mean net reproductive rate. If this rate were strictly density-independent, then each population would either grow *infinitely* or become extinct. But one observes neither. Only density-dependent regulation can account for that.

Figures 4-7 to 4-10 give examples of the evidence demonstrating the role of density as a population controlling agent. Figure 4-7 shows the effect of population density on mean oviposition rate per female in an experimental population of insects. Several different types of curves are found relating density to reproductive rate, but they all have in common a declining reproductive rate when density passes over some optimal level. Figure 4-8 shows the effect of population density on survival in another set of experimental insect populations. Figure 4-9 shows the effect of population density on the maximum length attained by natural

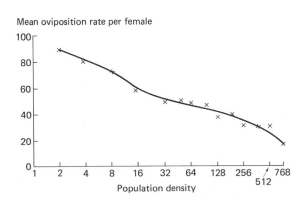

FIGURE 4-7 The effect of population density on mean oviposition rate per female at 30.4°C and 76 percent relative humidity in *Callosobruchus chinensis* (L.). (*Data from Utida,* 1941.)

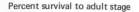

Percent survival to adult stage

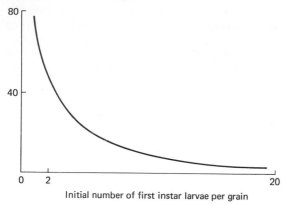

Initial number of first instar larvae per grain

FIGURE 4-8 The effect of population density on survival rate in experimental populations of *Rhizopertha* and *Sitotroga*. (*Data from Crombie, 1944.*)

populations of haddock during their first year of life. Figure 4-10 shows the effect of population density on the dispersal of caddisfly larvae in bowls.

Several points may be made about all these curves. First, none of them is an isolated instance. In all cases, similar patterns can be found throughout the natural world. Second, in all four graphs, it will be noted that the effect is not slight, or insignificant; on the contrary, in all cases it is quite marked. Also, in three of the four cases, the effect is nonlinear; that is, increasing the density fourfold does not necessarily have twice the effect of doubling the density.

Taking all these data together, we see that the density of any species population has a powerful effect in determining subsequent net reproductive success of that species at a particular place. If density is very high, it leads to reduced survival, reproduction, and growth, and may also cause emigration.

However, it is also clear that for many species density could get too low. For nongregarious species with two sexes, the density could get so low that prospective mates would have difficulty finding each other.

Maximum length of one-year-old haddock, centimeters

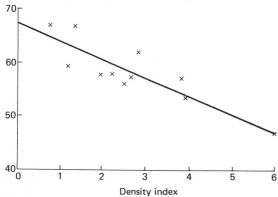

Density index

FIGURE 4-9 The effect of density on the maximum length attained in the first age group of haddock. (*Data from Raitt, 1939.*)

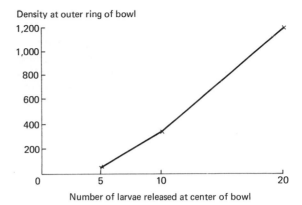

FIGURE 4-10 The effect of density on the emigration of caddisfly larvae from the center of a bowl to the outer ring. Dispersion with no central refuge of pebbles. (*Data from Glass and Bovbjerg*, 1969.)

Thus, some intermediate population density is probably optimal for most species.

We can combine all the material just presented into a single conceptual model, as in Fig. 4-11. In this figure, the numbers in the offspring population are plotted against the numbers in the parental population. Included in the diagram is a straight line rising at a 45° angle from the origin. Points falling on this line indicate that the offspring population is exactly the same size as the parental population, and consequently the population is in equilibrium. The actual relation for a given species is typically described by the curve. We can see that the curve crosses the line only at the two points A and B. From the standpoint of increasing N_O to exceed N_P, which is what is wanted if we wish to maximize fish production, for example, N_P optimal (on the figure) is the best parental population density to leave behind in the water after fishing each year. However, any population in nature will rock around the stable equilibrium point B. The reason for this is that if the population exceeds B, the population the subsequent year will be smaller, because of a decrease in survival and reproductive rate (or emigration). If N_P is smaller than B, the resultant N_O will be larger than N_P. Thus, population densities of N_P at either side of B tend to produce N_O densities closer to B. The farther away from B a population is, the more likely it is to return to B. This is not the case with A, however. It will be seen from Fig. 4-11 that population densities of N_P at

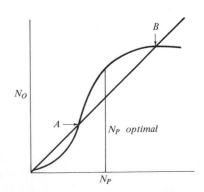

FIGURE 4-11 Relation between numbers in the offspring population and numbers in the parental population.

either side of A tend to produce N_O densities even farther away. Thus, although A is an equilibrium in the sense that the curved line describing the relation between N_O and N_P crosses the 45° line at A, it is an unstable equilibrium. B, however, is a stable equilibrium point and results in a steady state. In fact, Fig. 4-11 is describing a negative feedback system of the type diagramed for predator searching in Fig. 4-5. The population, like the individual, is a feedback control system. A is a positive feedback, B is a negative feedback.

One major complication concerning the species population has not yet been dealt with in the discussion. The greater the number of age classes in a population, the greater the resulting population stability will be. The reason for this is that all attributes of organisms are functions of their age. The death rate, the growth rate, the emigration rate, and the reproductive rate, together with the susceptibility to different kinds of disease, are all functions of age in animals and plants. What this means, then, is that the more age classes there are coexisting in a species, the less likelihood there is that any catastrophe will affect them all. For example, the spruce budworm tends to wipe out only forests of old trees. Thus, if the forest is a mixture of trees of many ages, all the undermature trees will survive. Similarly, cold temperature wipes out only bass in their first year of life; if there are many different year classes of bass present in cold water during a year when it is unusually cold, the older individuals will survive to carry on the species. The same phenomenon operates in man. It is highly unlikely that the human population will change in any way very rapidly, because it is unlikely that any social, biological, or physical factor will affect all ages equally. Thus, variety of ages is a stabilizing influence, just as diversity of species is a stabilizing influence. (We assume, optimistically, that such catastrophes as a nuclear war will not violate the assumption here.)

REGULATORY MECHANISMS IN INTERACTIONS BETWEEN TWO SPECIES POPULATIONS

A gradient of kinds of interactions between species is possible. The general term for all such interactions is *symbiosis*, meaning "living together." The gradient runs from relations in which the benefit appears to go all one way, at least at first inspection, to those in which both species benefit. At one extreme we have competition, where two species compete for a limited resource, and it can be demonstrated that the effects of competition are harmful to both species. Intermediate is exploitation (e.g., parasitism and predation), in which one species is hurt. Closer to the other extreme is commensalism, in which one species benefits, but the other is unhurt or unaffected; for example, epiphytes, such as Spanish moss, hang on other plants, and hermit crabs utilize empty mollusk shells. The extreme is beneficial mutualism, a relation between two species in which both benefit. There are many examples of this. For example, some fish clean the parasites out of other fishes' gills, then eat the parasites. One species gets cleaned; the other species gets lunch.

Most people think of attacking species as being large and vicious, like a tiger or lion, or offensive, like a tapeworm. In fact, parasitism and predation are common forms of activity in the animal kingdom, and they turn up in a remarkable variety of guises. Some striking examples of symbiotic relations are illustrated in Plate 4. Predators can be the slow-moving but very effective snails, starfish, and conch shells, or the slow-moving sloths, which hang upside down in trees and prey on birds. The insect orders Hymenoptera (wasps, ants, bees, and their relatives) and Diptera (flies) include a great many different kinds of parasites and predators, e.g., parasites of other insects' eggs (the minute wasps *Trichogramma*), predators of other insects' eggs (the bee fly larvae Bombyliidae), and even such exotica as parasites of other parasites. There are, for example, small wasps which lay their eggs in grasshoppers; the eggs hatch into larvae which eat the larvae of flies already parasitizing the grasshopper. There is a deep moral in the number and variety of these relations. *Nature is immensely complex, often in ways about which our ignorance is awesome. Consequently, when we make major changes in the world, deliberately or, as is more frequently the case, accidentally, we are modifying enormously complex homeostatic mechanisms in such a way that we may not be able to foresee the consequences. This is why such surprising things sometimes happen.*

For example, it is widely believed that insecticides control insect pests. Little advertising is devoted to well-documented cases in which an insect pest disappeared when spraying stopped and reappeared when spraying began again. For example, Conway and Wood (1964) describe a situation in which major reduction in pest damage was achieved after spraying was discontinued, not at the time of spraying. To make sure that this was not a coincidence, which seemed to implicate pesticides as the cause of pest outbreak, an experiment was conducted. In an oil palm stand where bagworms occurred in low numbers, an area of 2 acres was given a number of very light mist sprays with an insecticide often recommended to control the pest. A bagworm increase then occurred over the sprayed and surrounding areas, and counts across the sprayed plot and for up to half a mile away in six different directions clearly showed that the sprayed central area was the point of heaviest buildup, with a gradual decline away from the center in all directions.

The explanation is that the pesticides increase the death rates of carnivores (parasites and predators) as well as of the pests. However, as indicated previously in discussion of the pyramid of numbers in a community, the number of carnivores is always less than the number of herbivores they eat. Hence, the same percentage reduction in carnivores and herbivores leaves the former in a more precarious situation in terms of absolute numbers. Also, in the tropics, where this incident occurred, there are several generations of insect pests a year, and consequently a speedier evolution of strains resistant to pesticides.

INTERSPECIFIC COMPETITION   *Interspecific competition* is the phenomenon that occurs when two or more species require the same resource. If the combined population densities are great enough, and if the resource is scarce, then sooner or later, the interspecific competition will have a deleterious effect on one or both species, resulting in reduced survival, growth, and reproduction and, possibly, in emigration.

This is a subject about which a great deal has been written, but two recent reviews of the literature will convince the interested student that much that has been written is confusing and contradictory (De Bach, 1966; Milne, 1961). The reason for the confusion, as Milne pointed out, is that some writers did not read Darwin carefully, neglected the etymology of the word *competition*, and confused competition with its results. Indeed, in the natural world, competition often occurs without a result, because it occurs only fleetingly: the two species involved are kept apart most of the time by preferences for different times of the day or year, and somewhat different places, and they differ in their preferences for the resources they use.

Why has so much been written about this subject? It is immensely important, from a practical as well as theoretical standpoint. Here are examples of practical questions about the theory of interspecific competition. Suppose I have rangeland on which I can grow x pounds of meat a year using species A or y pounds a year using species B. Can I grow $x + y$ if I grow both? Is there any additional information which would tell me what the sum of both would be? Supposing I have a farm fishpond, or a lake, or a reservoir, what stocking policy will yield me the most pounds of fish per year—planting one species, two, ten, or thirty? Supposing I have an insect pest I want to control, do I release one species of parasite or predator to control it, or several? Should I grow date palms and citrus together or separately?

From a theoretical standpoint interspecific competition is the phenomenon at the heart of research in zoogeography and botanical geography, in evolution and speciation (the process of species formation), and in understanding such questions as why there are so many different species, if competition between species can be harmful. Competition between species is related to two principles.

PRINCIPLE 26   Different species having identical ecological niches cannot exist for long in the same habitat. (This is called the competitive exclusion principle, or Gause's principle, and has been stated by many authors, although the exact wording used here is due to De Bach and Sundby, 1963.)

PRINCIPLE 27   Different species which coexist indefinitely in the same habitat must have different ecological niches. (De Bach, 1966.)

Rather than a superficial treatment of several different sets of evidence in support of each principle, only one set in support of each will be presented. The data selected are about the best available to make each case.

A great number of studies have been conducted over a period of almost four decades by Park and his associates on competition in laboratory populations of flour beetles (see Leslie, Park, and Mertz, 1968, and Park, 1954, to key into this literature). These studies have left absolutely no doubt about the validity of Principle 26. However, they have also shown that interspecific competition is a complex and not entirely predictable subject for research. To illustrate, we will consider the results of a very large-scale experiment reported on by Park (1954, 1955), which has immediate implications for the study we will consider in support of the next principle.

The fundamental design of this experiment is indicated by Table 4-2: experiments were conducted at three temperatures for each of two relative humidities. Within each of these six combinations, three sets of replicates were censused every 30 days for 1,860 days. The first set consisted of 20 replicates of *Tribolium confusum* populations alone, the second consisted of 20 replicates of *Tribolium castaneum* alone, and the third of 30 replicates of both species together. Each replicate consisted of a shell vial of 8 grams of medium: 95 percent sifted whole wheat flour and 5 percent brewers' yeast. The initial population in each experiment was eight young adult beetles in equal sex ratio. The medium was replaced at the time all larvae, pupae, and adults were counted, every 30 days.

The most important conclusions from this experiment are as follows:

1 Whereas either species persists indefinitely when alone (with one exception), invariably when both species are husbanded together, one species eventually wins out and the other is eliminated.

2 The population density of beetles in the single-species cultures is different, on the average, for each combination of temperature and humidity. Also, the pattern of the effect is different for the two species. Thus, for example, although the carrying capacity for both species is lowered by lowering the relative humidity, it is lowered more for *T. castaneum* than for *T. confusum* (Table 4-2).

3 The species that wins out most frequently in the two-species cultures is not always the species that one would expect to win out from the single-species cultures. For example, at 70 percent relative humidity and 24°C, *T. castaneum* has a higher density on the average than *T. confusum*, and the difference is highly significant statistically. Yet under these conditions, when both species are husbanded together, *T. con-*

TABLE 4-2 The consequences of interspecific competition between *Tribolium confusum* and *Tribolium castaneum*[*]

relative humidity, percent	temperature, °C		
	24	29	34
30	26.04 ± 0.93	29.65 ± 1.28	23.73 ± 1.23
	2.63 ± 0.35	18.79 ± 1.32	9.61 ± 1.07
	T. confusum wins in all replicates	*T. confusum* wins in 87 percent of replicates	*T. confusum* wins in 90 percent of replicates
70	28.18 ± 1.70	32.94 ± 1.99	41.15 ± 3.04
	45.15 ± 2.77	50.11 ± 3.40	38.25 ± 1.53
	T. confusum wins in 71 percent of replicates	*T. castaneum* wins in 86 percent of replicates	*T. castaneum* wins in all replicates

SOURCE: Park (1954, 1955).
[*] The top figure in each box is the mean population density for *T. confusum* populations when reared by themselves over a 720-day interval. The second figure is the mean for *T. castaneum* when reared by themselves over a 720-day interval (except for the 2.63 in the first box: this is the average density over the period of persistence for each culture). The third piece of information in each box is the percentage of the two-species competing population systems in which the winning population replaced the loser. The figures after each mean are standard errors.

fusum wins in 71 percent of the cases. Even more baffling is the 70 per-cent–34°C combination. Here *T. confusum* does slightly better than *T. castaneum* in single-species cultures, and yet it invariably loses out to the latter in two-species competition. One must conclude that in inter-specific competition, there are phenomena that are quite unpredictable from analysis of single-species population dynamics.

What can one conclude from these findings? Certainly, if two species with identical requirements are forced to compete with each other for a limited resource, one will invariably eliminate the other, some-how: by predation, interference, or being more successful with respect to some other factor or combination of factors, such as predation, parasit-ism, or physical factors. However, the fact that these two species have different optima (hot and damp for *T. confusum*, warm and damp for *T. castaneum*) suggests what happens in interspecific competition in na-ture. There are three possibilities. First, the species might be found in entirely different geographic ranges. Second, they might be found in overlapping ranges, such that the physical conditions are optimal for each species at some part of its range where competition from the other species is minimal or lacking. Third, the two species coexist in areas where there are constantly fluctuating conditions which sometimes favor one species and sometimes favor the other, and one or both exhibit diapause during their unfavorable period.

This leads us to the next study, which establishes the link between Park's laboratory work and the natural world. The field work was con-ducted by a large team of *Drosophila* population geneticists and ecol-ogists and yeast biologists; as in Park's research, a massive volume of data was collected (Cooper et al., 1956; Phaff et al., 1956; Carson et al., 1956; Dobzhansky et al., 1956). This research was conducted in the Yosemite region of California, on fourteen different species of the fruit fly genus *Drosophila* known to inhabit the area. Specimens of adults were collected by exposing bait of bananas fermented with yeast or, in a smaller number of cases, at slime fluxes on oak trees or on fungi. About 95,000 specimens were collected and identified from nine different localities, representing different weather conditions (Cooper and Dobz-hansky, 1956). Yosemite Creek, at 7,200 feet above sea level, is a site colder than the altitude would suggest, because of cold air drainage over the site. In the other cases, the altitude is a good index of the climate at the site. For those nine species of which 95 or more specimens were collected, we can derive a height-frequency curve of some reliability; the collection data are summarized in Table 4-3.

A central question in this type of research is: How can there be so many closely related species in nature? Table 4-3 provides one of the several keys to this question. We notice that there are several different kinds of frequency distributions with respect to altitude. There are dif-ferences both with respect to the mean altitude favored by a species and with respect to the variance about that mean. For example, both *D. obs-cura* and *D. azteca* are most abundant at the 4,800-foot level, but of the latter, a much smaller proportion of the total collected are found at other levels. Ecological theoreticians, such as Levins (1968), have given considerable attention to this question of the variance of selected microhabitats by a species. In general, the wider the variety of niches a species will fill, the greater is its *niche breadth*. Theoreticians like Levins speculate that wider niche breadth tends to be associated with greater abundance in a species, and is an evolved protection against environmental uncertainty. There is not a particularly distinct relation between niche breadth and abundance in these data, but it is noteworthy that of the five species not abundant enough to be included, all seemed to have very narrow niches, and all three of the most abundant species have broad niches.

However, these species differ in many more respects than just their selected altitudes. At the 4,800-foot altitude the species which appear to complete most directly in fact do not, for other reasons. *D. azteca* is most abundant in September, *D. obscura* is most abundant in July and August, and *D. pinicola* at that altitude is most abundant in early June. Also, there are differences in the activity of species at different times of day. What I have called *D. pseudoobscura* in Table 4-3 is actually two almost identical species, *pseudoobscura* and *persimilis*. The latter is the more abundant of the pair in the morning, throughout the year, and the former is the more abundant of the pair in the evening, throughout the summer (Dobzhansky et al., 1956). Also, when nine different yeast species were exposed, differences could be detected between the fly species with respect to their preferences for the yeasts.

In short, direct, head-on competition between two species is difficult to observe in nature. Where it has been encountered, it has not lasted, because one of the species eventually excludes the other.

TABLE 4-3 The proportions of different species of *Drosophila* collected at different stations

station and altitude, feet	obscura	pseudo-obscura	azteca	miranda	occiden-talis	pinicola	melano-gaster	californica	montana
Timberline 10,000	0.002	0.004	0.000	0.004	0.001	0.001	0.000	0.000	0.000
Porcupine Flat 8,200	0.002	0.003	0.000	0.007	0.007	0.026	0.000	0.005	0.054
Yosemite Creek 7,200	0.001	0.001	0.000	0.006	0.014	0.005	0.000	0.000	0.798
White Wolf 8,000	0.002	0.003	0.000	0.018	0.004	0.013	0.000	0.010	0.054
Smoky Jack 7,500	0.012	0.014	0.001	0.113	0.311	0.170	0.051	0.013	0.010
Aspen Stream 6,400	0.037	0.027	0.002	0.223	0.539	0.280	0.000	0.044	0.074
Aspen Station 6,000	0.089	0.087	0.061	0.340	0.054	0.346	0.001	0.613	0.000
Mather 4,800	0.847	0.855	0.926	0.289	0.069	0.158	0.948	0.312	0.010
Lost Claim 3,000	0.008	0.006	0.010	0.000	0.001	0.001	0.000	0.003	0.000
Total	1.000	1.000	1.000	1.000	1.000	1.000	1.000	1.000	1.000
Total specimens collected	16,664	10,367	10,891	1,634	35,843	16,240	3,089	382	95

SOURCE: Cooper and Dobzhansky (1956).

These studies lead us to answers to the questions posed at the beginning of this section. In using biological control agents, for example, no gain is made, typically, by adding a new species to a situation in which the prospective niche is already being filled by another species, unless it is known that the new species is superior in some sense to the old one. However, it is worthwhile to add a species to an existing niche provided the new species differs somewhat in requirements from the species already in the niche, and consequently can be expected to coexist with it. However, whether or not unexpected and deleterious coactions between the species might occur is a matter that should be resolved by trials in limited and enclosed areas before the main release of the new species is conducted.

EXPLOITATION   Parasitism and predation are two particular examples of a general class of phenomena that might be subsumed under the general heading of exploitation. Whenever we have one group vulnerable to attack and another group attempting to attack—whether the systems involve mis-

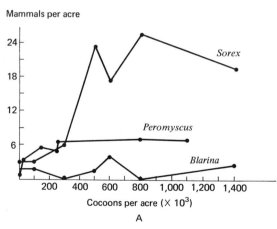

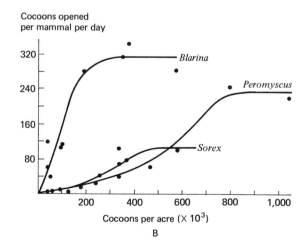

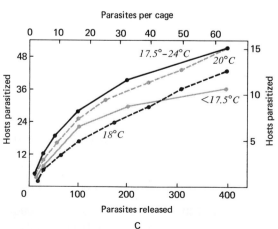

FIGURE 4-12 Three basic components of all attack phenomena. **A** The numerical response. **B** Saturation of attack capacity and functional response. **C** Intraspecific competition among the attackers. *Sources*: A *Holling*, 1959, B *Holling*, 1959, C *Burnett*, 1956.

siles; parasitic wasps or flies attacking caterpillars, sawfly larvae, or grasshoppers; lynxes attacking hares; or banditry—three features are typically present (Fig. 4-12 and Table 4-4):

1   If the numbers of the group vulnerable to attack increase, the growth rate of the attackers' density will increase (Fig. 4-12A).

2   The proportion of the numbers that are vulnerable to attack and succumb decreases as the numbers vulnerable to attack increase. That is, the attack capacity of the attacking population is saturated when the numbers vulnerable to attack are in excess of PK, where K is the average maximum number of attacks that can be generated per attacker and P is the number of attackers. K is determined by the number of eggs a parasite can lay or the number of attacks a predator can generate, given the limitations on its energy for pursuit and capture (Fig. 4-12B).

3   The effectiveness of attackers per capita decreases progressively with increasing size of the attacker population. This is caused by interference between attackers, fighting, or simply inefficiency because the same territory is searched over more than once (Fig. 4-12C).

The first and second points above lead to a new principle, which makes an important distinction (first stressed by Solomon, 1949).

PRINCIPLE 28   Organisms respond to an increase in the availability of any food species in two ways: by increasing its representation in their diet (the functional response) and by increasing in numbers with the result of more completely utilizing the increased food supply (the numerical response).

TABLE 4-4 Regulatory properties endogenous to the host-parasite and prey-predator systems operating against extinction of either or both species

density of attacker species	density of attacked species	
	low	high
Mechanisms preventing extinction of the attacked species		
Low	Low probability of attacker species encountering attacked species; decreased intraspecific competition in attacked species	Saturation of attack capacity in attacking species
High	Loss of search image (forgetting to look for rare host or prey); functional response to alternative host or prey species	Intraspecific competition among attackers
Mechanisms preventing extinction of the attacker species		
Low	Lowered interattacker competition	Increased attack rate per attacker
High	Switch to other foods; lowered attacker efficiency prevents annihilation of this food	Intraspecific competition and interspecific competition among attackers; switch to other hosts or prey

However, these are only a few of the mechanisms which operate to promote homeostasis in host-parasite and predator-prey systems. Holling (1968) has noticed several others, including the following:

PRINCIPLE 29 Predators lower their metabolic rate when their prey are scarce. (Thus, a higher proportion of the energy eaten goes to reproduction, and there is higher energy efficiency.)

PRINCIPLE 30 Predator-prey systems operate so as to maximize their energy efficiency. When prey is dense, more of each prey is wasted by the predator.

PRINCIPLE 31 Hyperparasites (parasites of parasites) or secondary predators (predators of predators) limit the buildup of parasite or predator populations and consequently dampen oscillations in the three-species system.

PRINCIPLE 32 Parasites distribute their eggs contagiously (in clumps) and thus dampen oscillations in the host parasite system.

This means that instead of covering a large area evenly in the next generation, the host will cover it patchily. Distribution and dispersal losses of the host dampen host rate of buildup consequently.

REGULATORY MECHANISMS AT THE COMMUNITY LEVEL

As communities develop through succession, their increase in diversity itself promotes still more increase in diversity. The greater the diversity of plants, the greater the diversity of niches created for animals to feed upon them. Also, taller plants create more layers of microhabitats, and the greater the number of such layers, the greater the number of species that can live in the community. MacArthur (1958) has shown with field data that increases in the vertical zonation of plants support greater bird species diversity. But this self-augmenting diversity becomes self-augmenting stability, for reasons pointed out in Chap. 3. The greater the number of competitors for an array of food items, the greater will be the population stability within each of the species of competitors.

All types of structural complexity in the environment constitute a regulatory mechanism that dampens out oscillations in species populations in communities, and prevents wide-amplitude oscillations in inter-species systems. Mountain passes, valleys, dense vegetation, altitudinal gradients, and rivers all add to the structural diversity of the habitat. Any child knows that when you play hide-and-go-seek, the game lasts longer and is more interesting in a large house with lots of passageways and rooms, and several different staircases connecting each pair of floors. Similarly, where there is a rich community of herbivores, parasites, predators, and other species, the chance that any species is going to be annihilated by its enemies or competitors is diminished if the environment is more structurally complicated, and hence there is greater difficulty in finding the organisms being sought.

In short, the more one studies nature, the more one is impressed not only with its orderliness, but also with the amazingly numerous mechanisms which operate to maintain balance. However, as we shall show in Chaps. 5, 13, and 14, there are two sets of forces operating on the planet which can deal severe blows to all this order: a very old set of forces—geophysical forces and the weather—and a very new set due to the increasing influence of man.

SUGGESTIONS FOR INDIVIDUAL AND GROUP PROJECTS

Much of the material in this chapter can be studied in the laboratory, using small animals or plants that grow quickly and can be cultured at high densities (flour beetles, fruit flies, grasses, grains, or carrots, for example). One type of experiment is designed to test the effect of density on survival or growth rates. Set up 100 experimental universes (bottle, vial, pot, or flat of plants), ten at each of ten densities (1, 2, 4, 8, 16, 24, 32, 48, 64, 96). At the end of an appropriate growing period, measure the survival or growth (or both) in each universe. Plot mean survival or growth against density. Is this system strongly density-dependent? How does the variation between the ten experiments at each density compare with the variation between the means of ten?

Such experiments can be used to explore the effect of density on reproduction or dispersal. Also, interspecific relations such as parasitism, predation, or competition can be studied experimentally.

REFERENCES

Burnett, T.: Effects of Natural Temperatures on Oviposition of Various Numbers of an Insect Parasite (Hymenoptera, Chalcididae, Tenthredinidae), *Ann. Entomol. Soc. Am.*, **48**:55–59 (1956).

Carson, H. L., E. P. Knapp, and H. J. Phaff: Studies on the Ecology of *Drosophila* in the Yosemite Region of California. III. The Yeast Flora of the Natural Breeding Sites of Some Species of *Drosophila*, *Ecology*, **37**:538–544 (1956).

Conway, G. R., and B. J. Wood: Pesticide Chemicals—Help or Hindrance in Malaysian Agriculture? *Malayan Nature J.* **18**:111–119 (1964).

Cooper, D. M., and Th. Dobzhansky: Studies on the Ecology of *Drosophila* in the Yosemite Region of California. I. The Occurrence of Species of *Drosophila* in Different Life Zones and at Different Seasons, *Ecology*, **37**:526–533 (1956).

Crombie, A. C.: On Intraspecific and Interspecific Competition in Larvae of Graminivorous Insects, *J. Exptl. Biol.*, **20**:135–151 (1944).

De Bach, P.: The Competitive Displacement and Coexistence Principles, *Ann. Rev. Entomol.*, **11**:183–212 (1966).

—— and R. A. Sundby: Competitive Displacement between Ecological Homologues, *Hilgardia*, **34**:105–166 (1963).

Dobzhansky, Th., D. M. Cooper, H. J. Phaff, E. P. Knapp, and H. L. Carson: Studies on the Ecology of *Drosophila* in the Yosemite Region of California. IV. Differential Attraction of Species of *Drosophila* to Different Species of Yeasts, *Ecology*, **37**:544–550 (1956).

Fry, F. E. J., J. R. Brett, and G. H. Clausen: Lethal Limits of Temperature for Young Goldfish, *Rev. Can. Biol.*, **1**:50–56 (1942).

Glass, L. W., and R. V. Bovbjerg: Density and Dispersion in Laboratory Populations of Caddisfly Larvae (Cheumatopsyche, Hydropsychidae), *Ecology*, **50**:1082–1084 (1969).

Holling, C. S.: The Components of Predation as Revealed by a Study of Small Mammal Predation of the European Pine Sawfly, *Can. Entomologist*, **91**:293–320 (1959).

————: The Functional Response of Invertebrate Predators to Prey Density, *Mem. Entomol. Soc. Can.*, no. 48, 1966.

————: The Tactics of a Predator, in T. R. E. Southwood (ed.), "Insect Abundance," *Symp. Roy. Entomol. Soc. London*, no. 4, pp. 47–48, 1968.

Leslie, P. H., T. Park, and D. B. Mertz: The Effect of Varying the Initial Numbers on the Outcome of Competition between *Tribolium* Species, *J. Animal Ecol.*, **37**:9–23 (1968).

Levins, R.: "Evolution in Changing Environments," Monographs in Population Biology, no. 2, Princeton University Press, Princeton, N.J., 1968.

MacArthur, R.: Population Ecology of Some Warblers of Northern Coniferous Forests, *Ecology*, **39**:599–619 (1958).

Margalef, R.: "Perspectives in Ecological Theory," University of Chicago Press, Chicago, 1968.

Martin, C. J.: Thermal Adjustment of Man and Animals to External Conditions, *Lancet*, **108**:561–567 (1930).

Milne, A.: Mechanisms in Biological Competition: Definition of Competition among Animals, *Symp. Soc. Exptl. Biol.*, **15**:40–61 (1961).

Park, T.: Experimental Studies of Interspecies Competition. II. Temperature, Humidity, and Competition in Two Species of *Tribolium*, *Physiol. Zool.*, **27**:177–238 (1954).

————: Experimental Competition in Beetles, with Some General Implications, in "The Numbers of Man and Animals," pp. 69–82, Oliver & Boyd, London, 1955.

Phaff, H. J., M. W. Miller, J. A. Recca, M. Shifrine, and E. M. Mrak: Studies on the Ecology of *Drosophila* in the Yosemite Region of California. II. Yeasts Found in the Alimentary Canal of *Drosophila*, *Ecology*, **37**:533–538 (1956).

Raitt, D. S.: The Rate of Mortality of the Haddock of the North Sea Stock, 1919–1938, *Rappt. Proces-Verbaux Reunions, Conseil Perm. Intern. Exploration Mer*, **110**:65–79 (1939).

Scholander, P. F., R. Hock, V. Walters, F. Johnson, and L. Irving: Heat Regulation in Some Arctic and Tropical Mammals and Birds, *Biol. Bull.*, **99**:237–258 (1950).

Selye, H.: "The Stress of Life," McGraw-Hill, New York, 1956.

Solomon, M. E.: The Natural Control of Animal Populations, *J. Animal Ecol.*, **18**:1035 (1949).

Utida, S.: Studies on Experimental Population of the Azuki Bean Weevil, *Callosobruchus chinensis* (L.). III. The Effect of Population Density upon the Mortalities of Different Stages of Life Cycle, *Mem. Coll. Agr. Kyoto Univ.*, **49**:21–42 (1941).

5 perturbation of ecological systems by weather

TYPES OF CLIMATIC REGIMES
Several types of descriptors are required for the climatic regime at any point on the earth's surface. In the first place, we need to know the average effect of temperature, rain, hail, snow, wind, light, and evaporation during any time interval. We need to know also the magnitude and pattern of variation from time to time for each of these variables. For different sites, the pattern will be characteristic for variation within the day, within the year, or from year to year for a given time of the year.

Figure 3-6 illustrated the seasonal and latitudinal patterns of variation in incident solar radiation. At the equator, the amount of radiation received is very similar throughout the year. The closer a site is to the poles, the greater will be the difference in the solar radiation received at the hottest and coldest times of the year. As we might expect, this difference in solar radiation is reflected in the temperature patterns. In order to explain them we must define a commonly used meteorological statistic that will be used throughout this chapter. By convention, the *mean monthly temperature for the month* is obtained by adding the maximum and minimum temperatures for each day of the month, and dividing the sum by twice the number of days in the month.

Our description of various climatic regimes will be organized as follows. We will discuss first temperature, then moisture. Under each of these headings, in turn, we will discuss (1) day-to-night variations, (2) seasonal variations, (3) variations from year to year for a given season, (4) interactions of the preceding three categories of variation, and (5) maps of geographical distribution of weather patterns. This descriptive treatment sets the stage for the next section of this chapter, on the causes of perturbations in weather regimes.

In general, very small variations in temperature within the day-night cycle are characteristic of places close to the equator and close to large bodies of water. An important general principle about climatic regimes is that proximity to large bodies of water tends to prevent large temperature differences within the day, within the year, and from year to year for a given time of year. This is because water has a very high specific heat: it takes about five times as much heat to raise the temperature of a unit mass of water 1° as it does to raise the temperature of air 1° (for the same unit mass). The specific heat of water is two to thirty times higher than that of most common gases, liquids, and solids.

This means that it takes a great deal of heat to raise the temperature of water, or that water must give off a great deal of heat before its temperature drops. The significance of this is that water masses have a tremendously important role in minimizing temperature fluctuations in nature, and the larger the mass of water, the more important the effect. Thus, small oceanic islands or low-altitude sites adjacent to oceans tend to have very small variations in temperature throughout the day-night cycle, the annual cycle, and from year to year for a given time of year. Very small variations in temperature from the hottest time of the day to the coldest time of the night are found in places like Singapore, Trinidad, and Georgetown, British Guiana. All are close to the equator and are close to or surrounded by the ocean.

Large masses of vegetation also are important in keeping down the amplitude of the temperature difference from the hottest time of day to the coldest time at night. Thus, deserts can be very hot at the hottest time of the day, but because there is no vegetation to prevent reradiation at night, the heat is lost to the sky and the coldest part of the night can be quite cold compared with the hottest part of the day. The very high temperature in the middle of the day is made possible because there are no trees to block incoming radiation, and also to lower the temperature, which they do through the heat loss to latent heat of vaporization absorbed at the surface of the leaves when water is converted to water vapor through evaporation.

Seasonal variations in temperature increase with increasing latitude, altitude, and distance from the nearest large body of water. The effect of proximity to water on both seasonal temperature variation and variation from year to year of temperature for a given season is brought out by the data for Canada, because most important Canadian weather stations lie within a rather narrow latitudinal band. Table 5-1 gives for each of eleven Canadian weather stations the latitude and longitude, to the nearest degree, the means, over 20 years, of the mean monthly temperatures, and the standard errors of mean monthly temperatures over years. This last statistic measures the variability in temperature for a given month from one year to the next, for a long series of years.

TABLE 5-1 The effect of proximity to large water masses on seasonal temperature variation, and variability of mean monthly temperature from year to year, illustrated by data on eleven Canadian weather stations in a 9° latitudinal band. All temperature statistics are in °F.

means of mean monthly temperatures

city	latitude, °N	longitude, °W	Jan	Feb	Mar	Apr	May	Jun	Jul	Aug	Sep	Oct	Nov	Dec
Vancouver	49	123	36.1	39.5	42.3	48.1	54.8	59.5	63.7	62.7	57.7	49.8	42.4	39.2
Summerland	50	120	24.7	30.8	38.1	48.1	56.9	63.1	69.4	67.6	59.4	47.6	35.9	30.3
Edmonton	53	113	6.1	10.8	22.1	38.3	51.8	58.3	63.1	60.3	51.3	40.9	25.0	13.7
Lethbridge	50	113	15.8	20.2	27.4	41.1	51.5	58.9	65.7	63.6	54.4	44.7	31.1	23.4
Regina	51	105	1.2	5.2	17.0	37.2	51.2	59.7	65.8	64.5	52.7	40.9	22.4	8.7
Winnipeg	50	97	−0.9	4.5	18.4	37.3	51.4	61.9	67.6	66.3	54.4	43.7	24.1	7.6
North Bay	46	78	8.9	12.9	23.3	37.9	50.1	60.3	64.9	63.4	54.4	44.8	30.8	15.5
Toronto	44	79	21.3	22.6	30.9	43.9	54.2	65.1	69.9	68.9	61.2	50.6	38.2	26.6
Montreal	45	74	14.6	17.0	28.5	43.1	55.2	65.2	70.3	68.6	60.1	49.2	36.4	21.5
Fredericton	46	67	16.1	17.9	28.4	40.2	51.4	60.4	66.7	64.9	57.2	46.5	35.7	22.3
Kentville	45	64	23.3	23.0	30.3	40.5	50.5	59.6	66.1	64.8	57.9	48.3	39.5	28.2

SOURCE: Data based on computer analysis of monthly weather reports from meteorological Branch, Canada Department of Transport.

Examination of this table brings out the following points.

First, during January, the coldest month, the four mildest cities are those adjacent to large bodies of water: Vancouver, Summerland, Toronto, and Kentville. This effect is not due to latitude, for although Vancouver, Summerland, Lethbridge, Regina, and Winnipeg are all in the 49 to 51° N latitudinal band, Regina and Winnipeg are the two coldest cities in January, of this group of eleven weather stations.

Second, during July, the hottest month for all these cities, Vancouver is the second coolest city, even though it was the warmest city during the winter. This is the meaning of maritime climate: the immense specific heat of the ocean warms the city in winter and keeps it cool in summer. Also, even though Winnipeg was the coldest city in January, it was the fourth warmest in July. This is the typical continental climate expected in a region far removed from the ameliorating effects of large water masses on temperature fluctuations.

Third, the cities with weather regimes influenced by lakes and large oceanic water masses have not only the mildest weather regimes, but the least variation in mean monthly temperature from year to year. The smaller the standard error of mean monthly temperature, the less the year-to-year variation in mean monthly temperature.

Fourth, however, it is important to note that there is not an exact correspondence between the average severity of the temperature and the stability of the temperature (or magnitude of the fluctuations about the mean). This point is brought out by examining the means and the variation of the mean monthly temperatures for January for Winnipeg and Edmonton. The mean monthly temperature for January is colder for Winnipeg than for Edmonton; yet the variability is higher at Edmonton because Edmonton is more arid. This illustrates a point made by Slobodkin and Sanders (1969): an environment can confront living organisms with either or both of two kinds of problems. It may be *severe*, or it may be *unstable*, or it may be both. A better intuitive understanding of the difference between severity and instability can be gained by comparing A and B of Fig. 5-1. A is a plot of mean monthly tempera-

standard errors of mean monthly temperatures

Jan	Feb	Mar	Apr	May	Jun	Jul	Aug	Sep	Oct	Nov	Dec	city
1.3	0.7	0.5	0.4	0.5	0.5	0.5	0.3	0.3	0.4	0.7	0.6	**Vancouver**
1.8	1.1	0.7	0.5	0.7	0.7	0.6	0.7	0.5	0.5	0.9	1.0	**Summerland**
2.4	1.6	1.4	1.3	0.6	0.7	0.4	0.6	0.5	0.9	2.1	2.0	**Edmonton**
2.8	1.7	1.5	1.1	0.7	0.7	0.6	0.6	0.6	1.1	1.8	1.8	**Lethbridge**
1.9	1.6	1.7	1.4	0.7	0.8	0.5	0.7	0.6	0.8	1.7	1.7	**Regina**
1.4	1.5	1.5	1.3	0.8	0.7	0.5	0.6	0.6	0.9	1.2	1.4	**Winnipeg**
1.0	1.2	1.3	0.8	0.8	0.5	0.5	0.6	0.7	0.8	0.9	0.9	**North Bay**
1.1	0.9	1.2	0.7	0.7	0.6	0.6	0.6	0.7	0.7	0.7	0.9	**Toronto**
1.2	1.0	1.1	0.6	0.8	0.5	0.5	0.6	0.6	0.5	0.8	1.1	**Montreal**
1.2	1.1	0.9	0.6	0.6	0.5	0.7	0.4	0.5	0.5	0.6	1.0	**Fredericton**
0.9	1.1	0.7	0.6	0.5	0.4	0.6	0.5	0.5	0.4	0.6	1.2	**Kentville**

Mean monthly January temperatures at Edmonton, °F

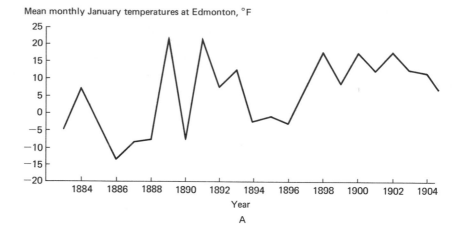

Year

A

Mean monthly January temperatures at Winnipeg, °F

Year

B

FIGURE 5-1 The difference between severity and instability. **A** Plot of mean monthly January temperatures at Edmonton. **B** Corresponding plot for Winnipeg. (*Data from the Meteorological Branch, Canada Department of Transport.*)

tures for January at Edmonton, and B is the corresponding data for Winnipeg. It will be noticed that the mean for Winnipeg is lower, but the amplitude of fluctuation is greater at Edmonton. Thus, although Winnipeg has the more severe winter environment, in terms of temperature, Edmonton is the more unstable. A main theme of this chapter will be that for living organisms the problems of instability are at least as rigorous as those of severity, and in many cases instability is more difficult to adapt to.

A fifth point made by Table 5-1 is that there can be interaction effects between severity and instability. The city that is most unstable in January is not necessarily the city that is most unstable in July. Lethbridge is most unstable in January, but Frederiction is in July.

Information like that in Table 5-1 can be plotted on maps for a very large number of weather stations. Maps can be made of either mean monthly temperatures or measures of variation in mean monthly temperature. Then all points with similar values can be connected, to give maps of isolines of either temperature means or measures to year-to-year variability. Figure 5-2 is an isoline map of standard deviations of mean

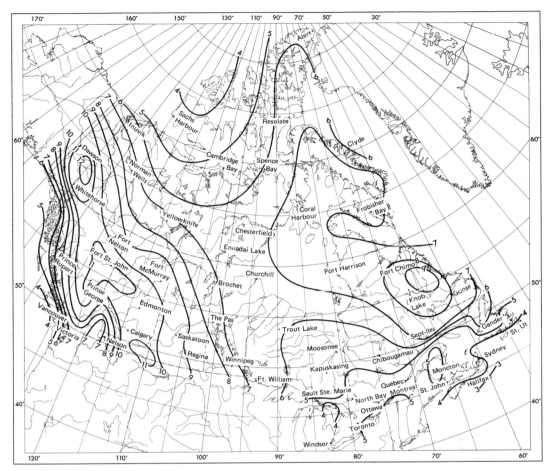

FIGURE 5-2 Isolines of standard deviations of mean monthly temperatures in °F for January in Canada. (*Kendall and Anderson*, 1966.)

monthly temperatures for January for Canada, and Fig. 5-3 is the corresponding map for April, to show how much the situation can change in just 3 months. The significance of such maps, as will be brought out later in this chapter, is that they may help us to explain why certain areas are the typical epicenters for particular biological events repeatedly, cycle after cycle. That is, a particular region may trigger a particular kind of biological event repeatedly, because that region has the largest-amplitude fluctuations in a particular weather variable to which the biological process in question is unusually sensitive at the time of year when the meteorological variable is fluctuating the most.

The same types of principles that apply to temperature variation apply to variation in precipitation. First, there is great place-to-place variation in the average amount of precipitation experienced at any time of year, and second, there is great place-to-place variation in the year-to-year variability in the amount of precipitation found at any time of year.

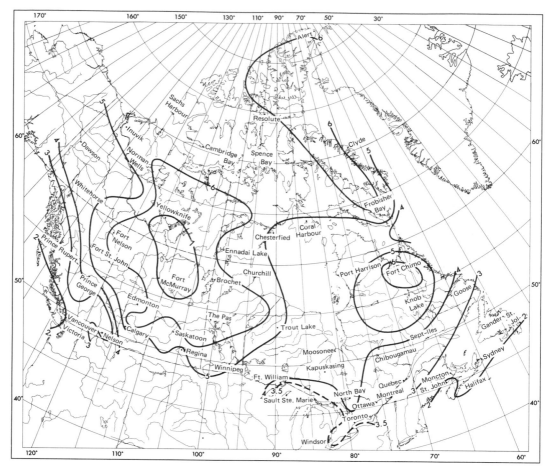

FIGURE 5-3 Isolines of standard deviations of mean monthly temperatures in °F for April in Canada. (*Kendall and Anderson,* 1966.)

Again, this information can be summarized in the form of maps of isolines. Figure 5-4 shows isolines for the median depth of Canadian snowfall. Again, the point made by such a map is that since there are great place-to-place differences not only with respect to the severity, but also with respect to the predictability of precipitation, the intensity of the problems biological systems must cope with varies from place to place. Also, since the magnitude of the year-to-year variation is different in different places, the possibility is suggested that certain places could be epicenters for variation in biological systems.

Since sites of greatest variability for weather variables shift geographically from month to month, we choose which place is an epicenter for a particular biological process on the basis of knowledge of the month that is most critical for the process.

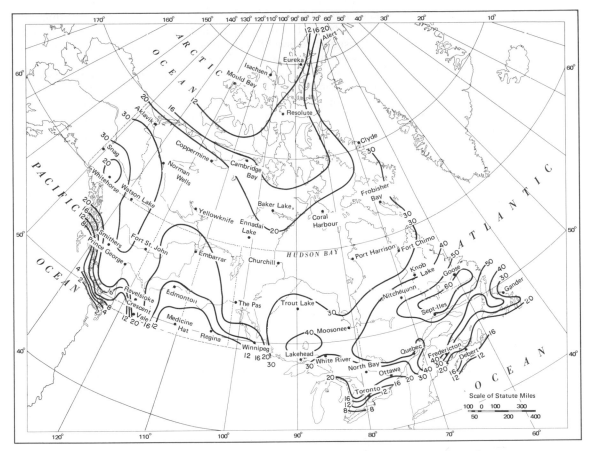

FIGURE 5-4 Isolines of median depth in inches of maximum snow covers over 20 winters in Canada. (*Potter*, 1965).

CAUSES OF YEAR-TO-YEAR FLUCTUATIONS IN METEOROLOGICAL VARIABLES

Figure 5-1 illustrates the rather large fluctuations in meteorological variables that can occur from one year to another. The variation can be large for different variables: at some sites the largest year-to-year variation will occur in January mean monthly temperature, whereas at other sites the most variable factor from one year to another will be summer rainfall. Since the beginning of civilization, men have speculated about the causes of such fluctuations. For a long time, the most common opinion was that the principal class of phenomena affecting year-to-year fluctuations in the earth's weather was extraterrestrial. Ultimately, men came to believe that one of the most important factors operating on such fluctuations was variation in the intensity of radiation coming from the sun. This led to a great deal of research on the sunspot cycle, which was discovered to have a mean wavelength of about 11 years. (Sunspots, measured by astronomers, are related to the intensity of radiation being emitted by the sun.) Many investigators then attempted to account for variation in all kinds of variables by performing statistical analyses to determine the correlation between the variable in question and the sunspot cycle, but often this research showed that there was no correlation. The reason was that another phenomenon can override the sunspot cycle.

One of the first men to realize that the earth's weather was subject to influence by variations in the turbidity of the upper atmosphere was Benjamin Franklin. He noticed that in the fall of 1783, a dry haze covered North America, and this haze so obscured the sun that he could not create a fire by focusing the rays of the sun with a magnifying glass. He correctly reasoned that this haze was the product of a gigantic volcanic eruption, probably off the coast of Iceland. He was half correct: in 1783 two of the three largest volcanoes of the last three centuries erupted. One was Skaptar Jökull in Iceland, and the other was Asamayama in Japan. The third of the three largest volcanoes of the last three centuries was Tambora, in the Dutch East Indies, which erupted in 1815.

However, the possible geophysical significance of volcanic eruptions did not attract the attention of modern science until much later, when Krakatoa blew up in 1883. Ironically, this was probably only about one-quarter of the magnitude of Tambora, in terms of the total mass of ash ejected into the atmosphere, Lamb (1972). After the eruption of Katmai in Alaska in 1912, scientists were thoroughly alerted to the fact that volcanoes could eject enough fine ash to affect the amount of energy from the sun that could penetrate the earth's atmosphere, and serious work on the possible effects of volcanoes on global weather fluctuations developed very rapidly.

One of the pioneers in such research was Humphreys (1913), who quickly discovered that large volcanoes had such a great effect on the ability of sunlight to reach the earth's surface that they could override the effect of the 11-year sunspot cycle. Humphreys demonstrated this by plotting global mean temperature deviations and sunspot numbers against each year for over two centuries of observations. His graph showed that global temperature departures did not always follow the sunspot cycle, and deviations could be explained on the basis of unusually large volcanic eruptions. The most important part of his graph is reproduced here as Fig. 5-5; it shows the effect of the 1783 and 1815 eruptions on global weather. These perturbations in global weather due to volcanic eruptions are so large that they can be detected statistically in biological time series for biological variables that have been measured over very long periods of time, as we shall see.

THE EFFECTS OF PERTURBATIONS OF WEATHER ON BIOLOGICAL SYSTEMS

Before we consider how meteorological factors operate on biological systems, it is necessary to examine the types of patterns that are observed in biological systems which fluctuate through time. All types of patterns which have ever been observed are like one of the three types depicted in Fig. 5-6, or some variant intermediate between two of them. A depicts fluctuations in the number of lynx fur returns through the Mackenzie River District of the Hudson's Bay Company over 78 years. This is typical of the highly regular patterns of fluctuation found in high latitudes or altitudes in populations of furbearers and certain insect pests. Most probably, the system is largely under endogenous control,

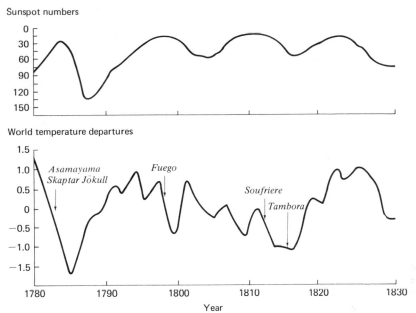

FIGURE 5-5 The effect of variations in solar radiation intensity fluctuations and volcanic eruptions as determinants of mean annual global temperature (*redrawn from Humphreys,* 1913). The global mean temperature departures are smoothed annual departures as computed from Köppen's actual annual departures, using all world weather station data. During periods with little volcanic activity, global temperature increases with decreasing sunspot activity, and decreases with increasing sunspot activity. The fairly close matching of sunspot numbers and world weather trend found in periods free of volcanic activity is exemplified by the period 1820 to 1930. For this entire period the peaks in sunspot numbers fall at 1788, 1804, 1816, and 1829, giving an average sunspot cycle length of (16 + 12 + 13)/3 = 14 years, which is somewhat longer than the average we would obtain for a period of several centuries. Note that the major eruptions of Asamayama and Skaptar Jökull in 1783 produced a global temperature nadir in 1785, one year after the time when a peak in global temperature would have been expected on the basis of the sunspot cycle. Also, the Tambora eruption of 1815 coincided with the time a nadir would be expected from the sunspot cycle; the combined effect of the eruption and the sunspot cycle produced the exceptionally deep and broad depression in global mean temperature of 1815–1816. The two deep troughs in global mean temperature depicted in this figure were two of the three deepest over the entire period 1750 to 1913. The third, in 1767, has been attributed to the eruption of Mayon (Lamb, 1972). We conclude that although the sunspot cycle typically is an important determinant, it can be overriden by volcanic eruptions, which eject vast masses of fine ash into the upper atmosphere and lower the earth's temperature by reflecting back into space a significant proportion of incoming solar radiation. It should be pointed out that the influence of sunspots on global weather would be more apparent during a period more free of major volcanic eruptions.

and the role of weather perturbations will only be to produce departures from this strongly density-dependent pattern. This is the type of pattern described by Principle 14.

Figure 5-6B, describing trends in the incidence of influenza deaths in England and Wales, shows an endemic state which is upset by some changed circumstance. The higher, epidemic state shows fluctuations about a gradually dropping trend line until once again the endemic state is reached. The physical environment could operate on such a system in either of two ways: by producing the transition from endemic to epidemic state and by determining the timing of fluctuations in the epidemic state.

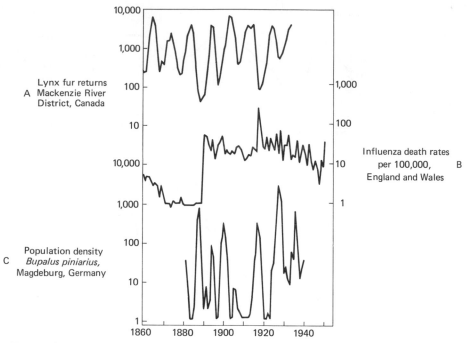

FIGURE 5-6 The pattern of fluctuation in examples of three different types of fluctuating biological systems. **A** Fluctuations in lynx fur returns
through the Mackenzie River District of the
Hudson's Bay Company, illustrating the cyclical
nature of fluctuations in certain furbearer and insect populations living in high-latitude or high-altitude forests (*data from Elton and Nicholson,* 1942).
B Fluctuations in influenza mortality rates per
100,000 population in England and Wales, illustrating the breakout from endemic to epidemic
phase in infectious diseases (*data for years up to*
1919 *from Andrewes,* 1953; *data after* 1919 *from
Deutschman,* 1953). **C** Fluctuations in population
density of the pine looper *Bupalus piniarius* L. in a
thicket near Magdeburg, Germany, illustrating the
irregular pattern of fluctuations found in insect
populations living in an unstable climatic zone (*data
from Schwerdtfeger,* 1941). (*Figure from Watt,* 1969.)

Figure 5-6C shows very little evidence of any endogenous control.
This pattern suggests the behavior of a biological system that is remarkably responsive to short-term fluctuations in one or more environmental
factors. The difference betwee A and C is that the former suggests that
the primary controlling influence is density-dependent factors, whereas
in C the primary influence is density-independent factors.

Any attempt to explore possible effects of weather perturbations
on biological systems should begin by determining which of these types
of fluctuating patterns we are dealing with.

It will facilitate the analysis of these various patterns of fluctuation
if we explain them in terms of two descriptors: stability and predictability. The first is expressed by the standard error of logarithms of
counts about their mean value, higher standard errors implying lower
stability. The second is expressed by the degree of pattern regularity.
The four possible combinations of high and low stability with high and
low predictability are illustrated by Fig. 5-7.

How can we describe quantitatively the mechanisms giving rise to
these different patterns? Stability is determined by the degree of
density-dependent control, and predictability by the importance of lag
effects with two or more steps (Principle 14). In the bottom panels of Fig.

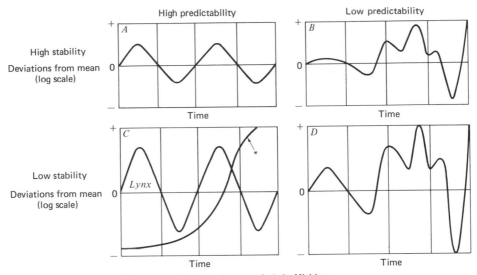

*Human populations or lampreys in Lake Michigan

FIGURE 5-7 Trends in population numbers expected for the four possible combinations of high and low stability with high and low predictability in the mechanism giving rise to the trends.

5-7, density-dependent control is weak: population densities are able to deviate far above or below the long-term mean value before reaching the probability of returning to the mean. In the left-hand panels of Fig. 5-7, the numbers at each time interval are strongly influenced by the numbers up to five or more time steps back. This type of situation has been documented by Moran (1953) in connection with the Canadian lynx cycle.

Thus, to understand the causal mechanism producing the lynx fluctuations (Fig. 5-6A), we must separate the mechanism into its two components: the density-dependent mechanism with delayed (lagged) effect that produces the high predictability and also the relatively high stability (relative to *Bupalus*, for example), and the small modifying effect of weather fluctuations which introduce minor unpredictability into the lynx trends (variations in nadirs, for example).

This separation has been effected by Moran (1953). He found that if he converted the number of lynx trapped in any year t in the Mackenzie River District to logarithms to the base 10, so that x_t represents the logarithm of the numbers trapped in year t, x_t could be well described by the formula

$$x_t = 2.9036 + 1.4101(x_{t-1} - 2.9036) - 0.7734(x_{t-2} - 2.9036)$$

That is, statistical regression analysis shows that lynx fur catches constitute a time series, in which the numbers caught in any year can be predicted by the numbers caught in the two previous years. Thus the lynx catches constitute a strongly density-dependent system, in which high catches in any year tend to follow high catches the previous year, but low catches two years back. The fundamental mechanism operating here is a somewhat more complicated version of that depicted in Fig. 4-11, because in this case at least two species and probably more are involved.

Lynxes eat hares, which eat plants that use soil nutrients that fluctuate in availability. Changes in lynx density are in part a response to changes in the density of hares and the supply of plant tissue and soil nutrients.

The weather perturbations caused by volcanic eruptions introduce the unpredictability. To demonstrate this, we must compute the residuals, that is, the value of x_t as computed from the above equation, subtracted from the observed values of x_t. Moran made an interesting discovery about the set of 112 residuals for the values of x_t from 1823 to 1934, inclusive. Exactly 56 residuals corresponded to values of x_t greater than the mean (2.9036), and 56 corresponded to values less than the mean. However, and this is the interesting point, the variation in the magnitude of the former set was only 45 percent of the variation among the latter set. That means that within the set of values of x_t below the mean value for the 112-year series, there seems to be an unusual number of very large negative departures from the mean. Examining the 112 residuals, we then notice that these large values seem to occur in clusters, and the clusters of years in which lynx catches were much lower than predicted by the time series were in all cases between 2 and 8 years after major volcanic eruptions, such as Krakatoa in 1883, Cosegüina in 1835, and Katmai in 1912.

Two kinds of argument can be used to illustrate this statistical relation between volcanoes and the lynx cycle. First, Fig. 5-5 demonstrates the relation between global mean weather and volcanic eruptions. If a statistical relation can be discovered between global mean weather and the lynx cycle, then this suggests a relation between the volcanoes and the lynx cycle. In fact, if we examine the relation between the residuals from Moran's equation from the lynx series, and Humphreys's global temperature departures 3 years previously, we find a correlation that would have occurred by chance alone only one time in fifty. Even more interesting, if we make a graph of residuals from the lynx series plotted against world temperature departures 3 years previously, we discover that the six lowest temperatures were all produced by volcanic eruptions, and all corresponded to cases where Moran's equation overpredicted lynx catches (Fig. 5-8).

One guess we can make about how volcanoes affect the lynx cycle goes as follows:

1   Volcanoes lower global temperature and consequently increase snowfall.
2   Increased snowfall has an effect on adult female hares, the food of lynxes, by sharply reducing the probability of survival of their offspring. This is reasonable, because Cox (1936) and Butler (1953) both noticed that the muskrat, a marsh animal, and the hare, a dryland animal, fluctuated out of phase, implicating precipitation as a factor in the fluctuations.

The reduction of food for lynx results in a declining lynx population, relative to the basic delayed density-dependent trend we would have expected.

The second assumption can be tested. Green and Evans (1940) kept detailed records on wild hare populations in a study plot at Lake Alexander, Minnesota, for 8 years, from a period of high hare population densities through a population decline to subsequent population

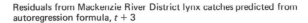

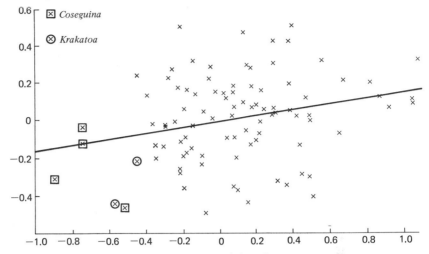

FIGURE 5-8 Relation between residuals from Moran's equation to account for lynx catch cycle and Humphreys's index of global mean temperature departures. The six lowest temperature indices are associated with volcanoes and correspond to unusually low lynx catches according to Moran's equation.

buildup. They found that the reproductive rate of adult female hares does not vary over a wide enough range of values to account for the great change in hare population densities from the high- to the low-density phases of the cycle.

What they did discover, however, was that the percentage survival of the young hares from the time of their birth to the following February did vary over a very wide range, and this could account for the changes in hare population densities encountered throughout the cycle. For example, form 1937 to 1938 the proportion of young hares surviving changed from 8 to 91 percent. Thus, if we are to account for the great variations in population densities of hares (and the resultant population density changes in lynxes), we must account for this great change in year-to-year survival of young-of-the-year hares.

Unfortunately, Green and Evans did not publish Lake Alexander weather records, but snowfall records were collected at Brainerd, 23 miles from Lake Alexander, and Duluth, 125 miles away. Figure 5-9 is a plot of Lake Alexander hare survivals from Green and Evans against Brainerd and Duluth snowfall data for December, January, and February of the previous winter. The expected relation between increased snowfall and decreased survival in the young of the females subjected to the snowfall is found.

A principal motive for introducing this example is to encourage the student to adopt a systems point of view. The student is encouraged to draw for himself a diagram of the volcano-temperature-snow-hare-lynx-trapper system, in the format used for Fig. 4-5.

It is of interest to note in passing that major changes in weather may have effects on economic, social, and political systems as well as

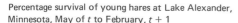

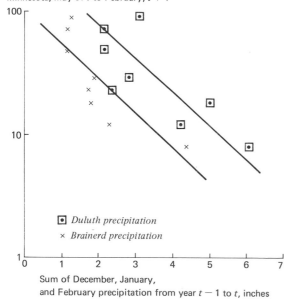

Percentage survival of young hares at Lake Alexander,
Minnesota, May of t to February, $t + 1$

⊡ Duluth precipitation

× Brainerd precipitation

Sum of December, January,
and February precipitation from year $t − 1$ to t, inches

FIGURE 5-9 The relation between survival of
young hares at Lake Alexander, Minnesota (*Green
and Evans*, 1940), and sums of winter precipitation
at Duluth and Brainerd. Locations of sites:

Lake Alexander   46°15′N,94°30′W
Brainerd   46°22′N,94°11′W
Duluth   46°47′N,92°6′W

The precipitation sums are for December of
year $t − 1$ and January and February of year t. The
percentage survivals are for the period May of
year t to February of year $t + 1$. (*Weather data from
U.S. Department of Commerce.*)

biological, including epidemiological, systems. Figure 5-10 shows the
relation between the highest asking price for a sack of flour on the
London commodities market and the weather as affected by the Tambora eruption. Newspapers of that period make it clear that the volcano
had a large enough effect on world climate to seriously disrupt farm
operations and crop growth in England, in the opposite hemisphere from
that in which the volcano erupted. Another motive for including this example is that volcanoes give us a foretaste of the impact that air pollution will have on crop growth if air pollution continues to build up on a
global scale, as it is at the moment. It is entirely possible that air pollution concentrations will have increased sufficiently over the next few
decades to produce crop shortages and resultant increases in crop
prices of the magnitude indicated in this figure.

Weather perturbations not only affect the hare-lynx system but also
are known to have effects on a wide variety of biological systems. A very
large literature exists indicating that weather variables have an overriding effect on the survival of commercial and sports fishing stocks, the
incidence of different kinds of infectious diseases, and the rate of
increase in density of insect pests. However, the particular nature of the
primary weather variable is different for different systems. Thus in the
case of plague, weather acts primarily through the effect of the drying
capacity of the air on larval rat fleas and other fleas which serve as
vectors for the plague bacterium. In the case of influenza, the key mechanism by which weather affects incidence seems to be depressing the
resistance of the host, as an effect of very cold weather.

In the case of fish populations, the most critical weather variable
seems to be water temperature at the time the eggs are about to hatch,

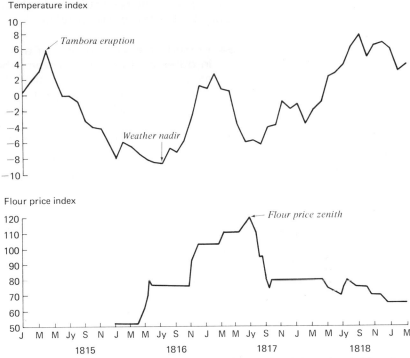

FIGURE 5-10 Possible effect of the Tambora eruption on the highest asking price for a sack of flour in the London, England, commodities market. The temperature indices are 5-month running sums of departures from the long-term mean for the corresponding month, using data from historical reconstruction of the central English climate by Manley (1959). Flour prices are from *Evans* *and Ruffy's Farmers' Journal and Agricultural Advertiser.* This is not a fortuitous relation, in all likelihood, because newspapers of the time describe the effect of inclement weather conditions on harvesting operations and crop growth following the eruption. Also, we expect a lag in the effect of weather on crop prices, because most countries have about a 1-year carry-over of grain in storage.

or later in the first year of life. In the case of insect populations, the situation is more complicated, and it is difficult to generalize because it depends on such conditions as the way the life cycle of the species is phased with respect to the time of year (whether the insect is an egg, a larva, a pupa, or an adult during the coldest time of year, for example).

The effects of perturbations in weather on ecological systems are specific examples of the operation of four important general principles relating to the effects of any type of perturbations.

PRINCIPLE 33   Perturbations applied to systems may have effects which do not show up until long after they have been applied. (This is the lag effect, due to time-delay mechanisms.)

This principle has a very important implication. Just because some environmental variable does not have an immediate deleterious effect does not mean that it will have no deleterious effect. In the case of many variables which are influenced by the state of the environment, such as emphysema, lung cancer, and the other cancers, the important factors operating on them may not fully reveal the magnitude of their effect for years after the cause was operating.

PRINCIPLE 34   Perturbations in one part of a system lead to perturbations in other parts of the system, the latter possibly being of much larger amplitude relative to long-term mean and standard deviation of those parts than the causative perturbation relative to its mean and standard deviation. (The more complex the system, the larger the number of state variables that can be affected. This is the "trigger with amplification" effect.)

A rather startling example of Principles 32 and 33 is illustrated in Fig. 5-11. This is a plot, against time in years, of the estimated number of 4-year-old smallmouth bass in South Bay, Manitoulin Island, and the temperature sums for the months June to October at North Bay during the year in which each year class was spawned. The remarkable feature of this graph is the difference in scales. A change of 5 in the temperature sum means that the average day over the 5-month period was 1°F warmer or colder. Thus, a temperature sum of 280 in 1951 resulted in only 1,000 estimated surviving 4-year-old bass in 1955, whereas a temperature sum of 298 in 1947 resulted in 44,900 4-year-old bass in 1951. This great dif-

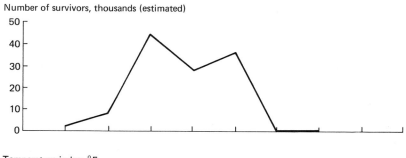

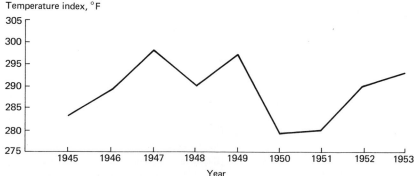

FIGURE 5-11 Large-amplitude changes in numbers of smallmouth bass surviving the first year of life produced by small changes in water temperature during the first year of life. The top line is estimated numbers of fish of each year class surviving to the fourth year of life in South Bay, Manitoulin Island (*Watt*, 1959). The bottom line is temperature sums (in °F) for the 5 months June to October during the season each year class was being spawned, from the North Bay weather station. Temperature sums are sums of the five mean monthly temperatures.

ference in biological productivity between the two spawning years was produced by a temperature sum difference of only $298 - 280 = 18$, representing a mean daily temperature difference throughout the growing season in the year of spawning of 3.6°F per day.

This type of large consequence from a small cause is a characteristic symptom indicating that a threshold effect is important for the variable affected. In this case, the species is clearly operating at the northern limit of the geographic range which has warm enough water for the young to survive in the year of spawning. Evidently the older fish, after the first year of life, are less sensitive to cold water, because the number of 4-year-old fish shows a correlation with the water temperature in the year of spawning, but none with the water temperature in subsequent years.

PRINCIPLE 35 The effects of a single perturbation can be much larger than expected if it is applied to a system repeatedly, either because of cumulative effects or because some threshold is finally exceeded.

The preceding example also illustrates this principle. It has already been mentioned that cold weather has a tremendous effect on the survival of fish in their first year of life, but a much smaller effect on older fish. This means that during a very cold year, even if almost all fish in their first year of life are wiped out, there will be enough survival of older fish to reproduce a large group of young in subsequent years. Also, even if enough fish die in the first year of life to virtually annihilate an entire year class, and if this happens for three consecutive years, this smallmouth bass population still will not be wiped out. The reason is that ten or more year classes of these fish are typically extant. Consequently, if three consecutive year classes were missing, by the time they would have been 5, 6, and 7 years of age, there would still be seven other year classes of adult fish in the water to maintain the spawning stock size. (This is an interesting variant of Principle 13. Diversity of year classes within a species population also promotes stability.) However, if ten successive year classes were eliminated, then the reproductive capacity of the population in this area would be destroyed, and the population would build up again only by recolonization by immigrants from more southern waters.

PRINCIPLE 36 Two or more separate perturbations applied to a biological system can produce much larger or much smaller perturbations than would be expected from the sum of the separate effects.

This is the synergism, or interaction, effect. This principle is operating when one environmental agent weakens organisms so that they become unusually vulnerable to the operation of another agent. For example, if cold weather weakens small organisms so that they are barely able to cling to the plant which normally supports them, and if this is followed by a high wind or torrential downpour which would tend to sweep them off the plant in any case, then losses may be very large.

Another example occurs when one toxic chemical in the air affects the body by predisposing it to succumb to a second chemical which otherwise would have had little or no effect.

PRINCIPLE 37 The speed with which a species population reacts to environmental changes favorable to the species is correlated with the turnover rate of individuals in the population.

This principle has been noted by many ecologists, but Margalef (1969) has stated it more formally by noting that there is a usually positive correlation between $r + m$ and m, where r is the intrinsic rate of natural increase and m is the instantaneous natural mortality rate. This principle can be illustrated by extending the argument associated with Table 4-1.

The largest rate of human population increase now occurring anywhere in the world is 7.6 percent per annum, in Kuwait, where the death rate is 6 persons per 1,000 per year and the birth rate is 52 per 1,000 per year (Ehrlich and Ehrlich, 1970). This just about sets a limit on the rate of human population increase, because Kuwait has a most extraordinary combination of a death rate lower than the United States, Canada, the United Kingdom, or Sweden, and a birth rate about three times that of the developed countries and equal to that of the developing or underdeveloped countries.

However, even this large rate of population increase is trivial compared with that of the organisms in the bottom half of Table 4-1. The spruce budworm, for example, has an average death rate of 0.99 per year, equivalent to 990 persons per 1,000 per year, but it can maintain a rate of population increase of 125 percent per year for many years in succession (Morris et al., 1963)!

Thus, the capacity of primitive organisms with high death rates to respond to favorable changes in weather is of an entirely different order of magnitude from that of man. Kuwait now has the most extraordinary set of conditions for human population growth ever known, and its population can only grow 1.076^5 over 5 years, or to 1.44 times the population at the beginning of the period. On the other hand, when the spruce budworm encountered weather conditions unusually favorable to it, it increased in density twentyfold during the period 1945 to 1950 (Morris et al., 1963). The lamprey, which has a much higher death rate, appears to have increased in Lake Michigan about 12,000-fold in a 13-year period (Applegate, 1950).

BIOLOGICAL IMPLICATIONS OF THE MAGNITUDE OF WEATHER PERTURBATIONS

To this point, we have been concerned primarily with the timing of perturbations in biological systems, as influenced by the timing of perturbations in weather systems. However, in the early part of this chapter, it was noted that there are large differences from place to place, not only in the severity of weather conditions, but also in the average magnitude of fluctuations (Table 5-1 and Fig. 5-1). The following principle relates to this distinction between the severity, on the average, over time at a site, and the magnitude of fluctuations at that site.

PRINCIPLE 38   The effect of climate and weather on populations can be separated into two components: either or both of severity or instability are deleterious.

This principle has been noticed by many ecologists, but it was stated formally in these terms by Slobodkin and Sanders (1969). Thus, for example, even though a deep benthic marine community may be subjected to cold temperatures, there may be a great deal of species diversity if the cold temperature is very stable and prevails over long periods of time.

Slobodkin and Sanders argue that the degree of stability of an environment will have effects on the characteristics of the species in that environment. Specifically, they argue that a stable environment will select for species with *stenotopy*, that is, complex behavior of rather specific and stereotyped kinds, a high degree of specialization, a high degree of accommodation to local competitive pressures, and a proneness to speciation. That is, a stable environment leads to a large number of species, each finely tuned to be able to respond with precision to interspecific competitive pressures within the environment. However, because of this fine tuning, the impact of a unit change in the physical environment would be more serious than we would expect in the species adapted to a highly unstable environment. On the other hand, they argue, unstable environments will select for *eurytopy*, or generalized and flexible behavior, a high intrinsic rate of natural increase, and a failure to specialize. Thus, we expect less species diversity in such unstable environments. Slobodkin (1968) has argued that in unstable environments, organisms would be expected to evolve mechanisms to lessen the stress produced by wide-amplitude fluctuations.

Fortunately, a body of data exists which is ideal for determining the quantitative impact of different degrees of instability in the environment. The Canada Department of Forestry for many years has operated a large-scale program to monitor the abundance of different species of insect pests in the forests of Canada. Data have now been published on indices of abundance of 989 forest Lepidoptera (butterflies and moths) over a period of many years. From 1945 on, the intensity of the monitoring effort was such that the indices have high enough statistical reliability to make them useful for various types of statistical analysis. Further, this program was conducted throughout Canada, and consequently we have available records on the changes in abundance of insects through time from a number of different regions that differ greatly with respect to both the severity and the stability of the weather regime (Table 5-1). Accordingly, a large-scale program of computer research was undertaken to determine the effect of these differences in different climatic regimes on the characteristics of the population dynamics of the species in different climatic regions (Watt, 1968).

This research was designed to answer various questions, of which one of the central ones was as follows. Given that there is great variation from region to region with respect to the amplitude of fluctuations in weather, what implications does this have for the amplitude of fluctuations in population density of the species living in the different regions? That is, do species living in regions where the weather fluctuates only a small amount respond more or less to a unit change in weather than species living in climatic regions where the weather fluctuates a great deal? Have species living in highly unpredictable environments evolved the lowered sensitivity to environmental stress expected by Slobodkin and other ecological theoreticians?

To answer these questions, data from the Forest Insect Survey (McGugan, 1958; Prentice, 1962, 1963, 1965) were coded onto punch cards, along with data on the monthly temperatures for eleven weather stations in Canada, selected as representative of the climatic regions in which the insects were collected. All population counts were transformed to logarithms, because all populations vary at a rate which is proportional to the numbers present at a given time. Also, conversion to logarithms means that the fluctuations in number of each species are being expressed in terms of the long-term mean for that species. Thus, if one species increases from an abundance index of 40 to one of 300, the computer in effect equated this to a change in another species from 400 to 3,000. Thus, this study is concerned with rates of change, not with the absolute amount of the change. This is an important distinction. What most people would be most likely to notice is a change in density from 4,000 to 20,000. However, from the point of view of the biological mechanisms for adaptation to changes in the physical environment, a change from 40 to 2,000 is ten times as significant.

One of the measures of each population was the sensitivity to changes in weather. This was computed as follows. The computer determined for each species the month for which fluctuations in mean monthly temperature were most important in accounting for fluctuations in the population indices for the species. At the same time, through regression analysis, a statistical technique, the computer determined the effect on the index of abundance for the species population owing to a unit change in mean monthly temperature for the month most important in determining population trends in the species. Many other measures of the behavior of each species were also computed, but it turned out that this measure of sensitivity was the most useful in accounting for the dynamics of different species.

The most important results are collated in Table 5-2. This shows that in relation to their means, species actually fluctuated less, rather than more, in those climatic regimes where the weather fluctuated more violently from year to year. The reason for this is that the sensitivity of species to a unit change in mean monthly temperature is less for species in a violently fluctuating environment than for species in a less violently fluctuating environment. In other words, Slobodkin and Sanders are right: species do evolve mechanisms for decreased sensitivity to stress from fluctuations in the physical environment in habitats where the physical environment fluctuates with wide amplitude.

Many of the other results in this study were predictable from those in Table 5-2. No matter whether we examine data within each climatic region separately, or for all regions pooled, the result is always the same. By far the most important factor in accounting for the amplitude of population fluctuations of species is the sensitivity of the species to weather changes.

There has been a classic controversy in population ecology for decades as to which is more important in regulating population fluctuations, density-dependent or density-independent factors. This controversy seems to be resolved by this computer analysis in a most curious way. It is the relative sensitivity of the species which determines how the population of that species is regulated. How much a population fluctuates depends on the interaction of two factors: the magnitude of fluctuations in weather and the sensitivity of the species to that degree of weather fluctuation. If weather fluctuates only a small amount and the species is insensitive to weather fluctuations, then this is the circumstance under which the relative importance of density dependence as a regulatory mechanism will be maximal. The greater the sensitivity of the species to fluctuations in the physical environment, and the greater the amplitude of those fluctuations, the more important density-independent mechanisms will be in relation to density-dependent mechanisms in modifying trends in the species population. Putting it in different language, the importance of endogenous factors relative to exogenous factors as regulators of biomass through time increases with increasing suitability of the environment.

Two specific instances among the species studied in this computer analysis contrast the two types of situations with respect to weather sensitivity.

Species 345 of the Canadian Forest Insect Survey is *Orgyia pseudotsugata*, a small moth whose larvae feed almost entirely on Douglas fir, and which is confined to the interior of southern British Columbia. This is obviously a highly specialized species with respect to both diet and evolved adjustment to climate. The computer revealed that in relation to its mean, this species had very violent population fluctuations: the standard error of logarithms of indices was 1.204; the average value of this

TABLE 5-2 The relation between amplitude of population fluctuations, amplitude of weather fluctuations, and sensitivity of a species population to a unit change in mean monthly temperature*

weather stations at center of climatic regions	number of species in analysis	mean, for all species, of standard errors of logarithms of population indices	mean, standard error of mean monthly temperature for the month most important in accounting for population fluctuations	mean, sensitivity to unit change in mean monthly temperature for most important month
Vancouver	14	0.48	0.59	0.55
Summerland	31	0.47	0.83	0.38
Lethbridge, Toronto, Montreal, Fredericton, and Kentville	84	0.38	0.80	0.31
Edmonton, Winnipeg, Regina, and North Bay	106	0.39	1.07	0.27

* The cities are not in groups which are geographically contiguous, but rather in groups which are similar with regard to severity and unpredictability of climatic regime.

index for all species in the analysis was only 0.402. The computer showed that the two most important months in terms of temperature accounting for population fluctuations in this species were July and August. In these months the insect is typically found as a larva, pupa, or adult (mostly as larvae in early July, pupae in the middle of the period, and adults by August). During July and August in the interior of British Columbia, the temperature varies very little from one year to another. During the period for which data were analyzed for this species, mean monthly temperatures for July and August respectively varied only from 66 to 71 and from 64 to 69 (all in °F). Thus this species showed a truly amazing sensitivity to the two mean monthly temperatures the computer selected as most important in determining its density.

At the other extreme is species 719, *Petrova albicapitana,* a small moth whose larvae feed on the bark and wood of lateral shoots and leaders of jack pine and five other species of pine. This species has been collected over almost all of Canada, but the bulk of the collections are from areas with very severe and unpredictable climates: northern Ontario and Saskatchewan, and Manitoba. The numbers of this species have been remarkably stable from one year to another. For collections around Winnipeg, the standard error of logarithms of population indices was 0.170, and for North Bay, 0.223.

At Winnipeg, the computer selected mean monthly temperatures in January and February as the most important factors accounting for changes in abundance in this species; however, the sensitivity of population density to those variables was very low. The mean monthly temperatures for those two months at Winnipeg during the relevant series of years varied from −16 to 10° and from −6 to 19°, respectively. Thus, the species showed remarkably little sensitivity to these mean monthly temperatures in the light of their great year-to-year variability.

This startling contrast illustrates why sensitivity to a unit change in the physical environmental factor most important for each species is so important in determining the amplitude of population fluctuations within a species.

One might well ask what would have happened to species 719, for example, if it had not evolved this low sensitivity to weather changes. The answer, of course, is that it would have run a great risk of going extinct. In general, then, this analysis supports the conclusion of Slobodkin (1964) that what natural selection was ultimately selecting for was homeostatic ability (self-regulatory ability).

The contrast between these two species illustrates another of the generalizations pointed out by Slobodkin and Sanders. They argued that unpredictable environments will select for eurytopy. This is another way of saying that the niche breadth will be great. It implies that we would expect to find that species collected in a large number of localities would be those least sensitive to weather. That is, if a species is able to withstand wide-amplitude fluctuations in weather conditions in any one place, then we would expect to find that it is able to live in a great many different places. This was also verified by the computer analysis of Canadian forest Lepidoptera. The following table makes the point:

number of different regions in which species is collected	average sensitivity to most important mean monthly temperature for species collected in this number of regions
1	0.445
2 or 3	0.352
4 or 5	0.347
6	0.265

Again illustrating the generalization in terms of the two contrasting species, species 345, which showed a striking response to weather variations, occurred only in one very small region of Canada, whereas species 719, which showed very little response, occurred throughout the country.

SUMMARY   The central arguments presented in this chapter are as follows:

1 There are many different types of climatic regimes, which vary not only in severity, but also in stability (the magnitude of fluctuations).

2 The timing of weather perturbations is related to a solar phenomenon, the 11-year cycle, and a terrestrial phenomenon, the degree of atmospheric turbidity, which can cut down the penetration of the atmosphere by incoming radiation sufficiently to override the effect of the sunspot cycle.

3 The timing of fluctuations of any biological systems can be related to the timing of weather fluctuations. In such cases as the hare-lynx system, weather fluctuations only modify a system basically under delayed density-dependent control.

4 Time lags from first cause to ultimate effect may be so long that the identity of the first cause is concealed.

5 The patterns in population fluctuations can be analyzed in two ways: as unpredictability and as stability.

6 The most severe and unstable environments contain species that are generalized, insensitive to environmental fluctuations, and unlikely to speciate. For this group, the species diversity in such environments is low. The least severe and most stable environments contain species that are specialized, sensitive to climatic fluctuations, and likely to speciate. Because of the possibility of narrower niche breadth of species in stable environments, such environments can contain a great many more species (the species diversity is greater). This is why there are so many species in coral reefs, tropical rain forests, forests in Mediterranean-type climatic regions, and the cold but stable ocean bottom in areas free from estuarine and other current activity.

SUGGESTIONS FOR INDIVIDUAL AND GROUP PROJECTS   Obtain data on some biological phenomenon (plant disease, insect pest density, numbers of mammals) that is characterized by year-to-year fluctuation in your area. Test the hypothesis that the fluctuations are accounted for by weather fluctuations. Make graphs to illustrate the relation. First make graphs like Fig. 5-11, then like Fig. 5-9. Which graphs reveal the relation most clearly?

What other factors play a large role in accounting for year-to-year fluctuations in the data you are examining?

Collect 60 or 70 years of weather data for the weather stations closest to your home that have such long series of data (the U. S. Department of Commerce publishes the data monthly, and the Smithsonian Institution publishes data in summarized form for each decade in *World Weather Records*). Do you observe any trends when you plot annual or monthly means? Do you notice any unusual short-term perturbations? To help understand trends, see the papers by Bryson and Wendland and by Mitchell in S. F. Singer (ed.), "Global Effects of Environmental Pollution," D. Reidel, Dordrecht, Holland, 1970. Also see *Annals of the New York Academy of Sciences*, **95**:1–740 (1961).

REFERENCES

Andrewes, C. H.: Epidemiology of Influenza, *Bull. World Health Organ.*, **8**:595–612 (1953).

Applegate, V. C.: Natural History of the Sea Lamprey, *Petromyzon marinus*, in Michigan, *U.S. Fish Wildlife Serv. Spec. Sci. Rept., Fisheries*, no. 55, 1–237, 1950.

Butler, L.: The Nature of Cycles in Populations of Canadian Mammals, *Can. J. Zool.*, **31**:242–262 (1953).

Christian, J. J.: The Adreno-pituitary System and Population Cycles in Mammals, *J. Mammalogy*, **31**:247–259 (1950).

Cox, W. T.: Snowshoe Rabbit Migration, Tick Infestation, and Weather Cycles, *J. Mammalogy*, **17**:216–221 (1936).

Deutschman, Z.: Trend of Influenza Mortality during the Period 1920–1951, *Bull. World Health Organ.*, **8**:633–645 (1953).

Ehrlich, P. R., and A. H. Ehrlich: "Population, Resources, Environment," Freeman, San Francisco, 1970.

Elton, C. S., and M. E. Nicholson: The Ten Year Cycle in Numbers of the Lynx in Canada, *J. Animal Ecol.*, **11**:215–244 (1942).

Green, R. G., and C. A. Evans: Studies on a Population Cycle of Snowshoe Hares on the Lake Alexander Area. III. Effect of Reproduction and Mortality of Young Hares on the Cycle, *J. Wildlife Management*, **4**:347–358 (1940).

Humphreys, W. J.: Volcanic Dust and Other Factors in the Production of Climatic Changes, and Their Possible Relation to Ice Ages, *J. Franklin Inst.*, **176**:131–172 (1913).

Keeley, K.: Prenatal Influence on Behavior of Offspring of Crowded Mice, *Science*, **134**:44–45 (1962).

Kendall, G. R., and S. R. Anderson: "Standard Deviations of Monthly and Annual Mean Temperatures," Climatological Studies, no. 4, Meteorological Branch, Canada Department of Transport, Toronto, 1966.

Lamb, H. H.: "Climate: Present, Past and Future," vol. 1, Methuen, London, 1972.

McGugan, B. M. (compiler): "Forest Lepidoptera of Canada," vol. 1, Forest Biology Division, Canada Department of Agriculture Publication 1034, 1958.

MacLulich, D. A.: The Place of Chance in Population Processes, *J. Wildlife Management*, **21**:293–299 (1957).

Manley, G: Temperature Trends in England, 1698–1957, *Arch. Meteorol., Geophys. Bioklimatol.*, **9**:413–433 (1959).

Margalef, R.: Diversity and Stability: A Practical Proposal and a Model of Interdependence, *Brookhaven Symp. Biol.*, no. 22, "Diversity and Stability in Ecological Systems," pp. 25–37, 1969.

Moran, P. A. P.: The Statistical Analysis of the Canadian Lynx Cycle. I. Structure and Prediction, *Australian J. Zool.*, **1**:163–173 (1953).

Morris, R. F. (ed.): The Dynamics of Epidemic Spruce Budworm Populations, *Mem. Entomol. Soc. Can.*, no. 31, pp. 1–332, 1963.

Potter, J. G.: "Snow Cover," Climatological Studies, no. 3, Meteorological Branch, Canada Department of Transport, Toronto, 1965.

Prentice, R. M. (compiler): "Forest Lepidoptera of Canada," vols. 2, 3, and 4, Canada Department of Forestry Bulletin 128, and Publications 1013 and 1142, 1962, 1963, 1965.

Schwerdtfeger F.: Über die Ursachen des Massenwechsels der Insekten, *Angew. Entomol.*, **28**:254–303 (1941).

Selye, H.: "The Stress of Life," McGraw-Hill, New York, 1956.

Slobodkin, L. B.: The Strategy of Evolution, *Am. Scientist*, **52**:342–357 (1964).

——: in R. Lewontin (ed.), "Population Biology and Evolution," Syracuse University Press, Syracuse, N. Y., 1968.

—— and H. L. Sanders: On the Contribution of Environmental Predictability to Species Diversity, *Brookhaven Symp. Biol.*, no. 22, "Diversity and Stability in Ecological Systems," pp. 82–95, 1969.

Watt, K. E. F.: Studies in Population Productivity. II. Factors Governing Productivity in a Population of Smallmouth Bass, *Ecol. Monographs*, **29**:367–392 (1959).

——: A Computer Approach to Analysis of Data on Weather, Population Fluctuations and Disease, in W. P. Lowry (ed.), "Biometeorology," pp. 145–159, Oregon State University Press, Corvallis, 1968.

——: A Comparative Study on the Meaning of Stability in Five Biological Systems: Insect and Furbearer Populations, Influenza, Thai Hemorrhagic Fever, and Plague, *Brookhaven Symp. Biol.*, no. 22, "Diversity and Stability in Ecological Systems," pp. 142–150, 1969.

6 man, fossil fuels, and the pollution, perturbation, and degradation of ecological systems

MAN'S EFFECT ON
THE NATURAL
WORLD IN TERMS
OF PRINCIPLES 3, 4,
10, 11, AND 14

The three previous chapters have indicated the principles which determine the mode of operation of the natural world. We now turn to a consideration of man's effect on the natural world in terms of the principles presented in Chap. 2.

Of the fourteen principles presented in that chapter, five are of great importance to modern technological civilization, largely because we act as if these principles were of no importance for the way the world works. Unless we come to see quickly how these five principles are relevant to how civilization operates, peril awaits us. We begin by giving an overview of the argument presented in this chapter, by indicating how each of the five principles mentioned is relevant to the affairs of man.

Principle 3 states that matter, energy, space, time, and diversity are all categories of resources. The discussion in this chapter will argue that many problems for mankind are being created, and will increasingly be created, because of a failure to recognize that space, time, and diversity are resources as well as matter and energy, and impose constraints on us which we cannot ignore. An important implication of this principle is that since matter cycles through ecosystems, there must be a means of allowing it to be broken down and converted to reusable material once it has passed through the ecosystem. Our pollution and garbage disposal problem is in part due to our lack of concern about the important function of recycling about thirty elements, which is attended to by the decomposer organisms in natural communities. Pollution is simply the technological counterpart of undegraded litter in the natural world.

Another very important implication of Principle 3 is that the availability of resources determines the carrying capacity of an environment for a population. Our increasing dependence on fossil fuels and nuclear fuels, which are stock resources, unlike the sun, a flow resource, means that we are building up the carrying capacity of the world for people. Insufficient attention has been given to the argument that this is a perilous activity if we should be unable to maintain such high carrying capacities for any reason. As will be demonstrated in Chap. 15, it is not clear that the general belief that we can maintain them is true. Thus, the carrying capacity of the planet could drop precipitously to that carrying capacity made possible by flow resources alone if we run out of stock resources.

Various problems are developing for mankind because we are diminishing the diversity in the nonurban part of the world. We have failed to see that diversity is a valuable resource which promotes stability.

Principle 4 states that for any resource in any process going on in one environment, there is an optimum level of resource availability. It is the principle which states that there are saturation and depletion effects limiting all processes because of the finite availability of any resource.

Thus, pollution can become especially dangerous if we supersaturate the ability of air or water to disperse a pollutant, or overtax the ability of decomposer organisms to remove some substance from the environment by altering it. Everything has an inherent limit on possible growth, whether it is due to limits on availability of an essential raw material, or truncation of growth in air travel due to saturation of the airspace at airports.

This principle also has an important implication for man in the form of the optimum yield problem. There is an optimum rate of exploitation for all the organisms we harvest. One of the resources is space: if we overharvest a commercially important species, the space per individual is too great, the density is too low for adequate reproduction rate in order to replace the harvested stock, and the species goes extinct. This is exactly the situation for several whale species at present, for example. In this case the probability that a female whale will find a mate is too low, and too high a proportion of the adult females fail to reproduce every second year.

Principle 10 concerns the increase in efficiency of energy use in communities after the pioneer stage. Man in fact operates in an opposite fashion: as technological society evolves, we actually become less, not more, efficient in our use of energy.

Principle 11 states that mature systems exploit immature systems. For this reason, new large cities cannot develop a complete range of services, industries, and cultural advantages close to an existing city which already has them. Thus, there are real problems with a "new towns" program, an attempt to start attractive new cities in new locations. Such an attempt ignores the extent to which the new cities would constantly be parasitized by existing cities. The well-known problems of economic development in the New England states and Appalachia are mute testimony to the way in which large metropolitan areas, in this case Boston, New York, and Baltimore-Washington, can parasitize the surrounding region so that it is constantly maintained in an immature state, with a low level of diversity.

Principle 14 states that lag effects operating on populations produce great momentum and hence regularity in patterns of population fluctuation. But man himself is the ultimate example of the population influenced by lag effects, and has been growing out of control by any force except those endogenous to human civilization. In this chapter, the nature of the forces creating this inertia will be analyzed.

We turn now to a detailed discussion of the way man through his activities ignores the five principles just mentioned.

principle 3   Man is running matter through the planetary ecosystem faster than the wastes produced can be decomposed. Consequently, there is a tremendous accumulation of material, or pollution, which cannot be readily assimilated for recycling through the system. To intensify the problem, much of the litter from civilization, like plastic, is nonbiodegradable, unlike the litter of natural ecosystems. The problem is going to get vastly more serious because no industry will grow much faster in the next few decades than the plastics industry.

Pollution is one of many deleterious side effects of the tremendous rate of energy use by modern civilization. Because this high rate is not spread evenly all over the planet, but is concentrated at certain sites (cities), the production of pollution also is enormously concentrated in space, and this makes the disposal problem more acute.

Two overall striking features characterize energy use trends by mankind in the last 120 years. The first is that there has been an incredible increase, not only in the total amount of energy used, but also in the energy use per person. The second is that there has been a tremendous shift in the source of supply, away from sources that were dependent on the energy that reached the earth each year from the sun (wood, animal, and human energy) toward energy sources that were stored long ago (coal, gas, petroleum, and nuclear fuels). A few figures will illustrate the magnitude of these two changes:

year	U.S. population, 10^6	total energy production from coal, oil, gas, and water power, 10^{15} Btu	energy production per person, 10^6 Btu	proportion of total energy use supplied by work animals
1850	23	0.3	13	0.50
1880	50	2	40	0.45
1910	92	14	152	0.25
1940	132	25	189	0.0045
1970	203	61	300	0.00008

The United States population was 8.82 times as great in 1970 as it was in 1850, but the total energy production was 203 times as great, and the energy production per person was 23 times as great. Also, our source of supply has shifted almost entirely from the sun, which produced the food that fed our work animals and us (humans supplied 23 percent of the energy in the United States in 1850, in addition to the 50 percent supplied by work animals), to fossil fuels. There is a terrible risk in this, however. Principle 3 implies that the rate of energy use determines the carrying capacity of the environment for people. The great increase in the planetary carrying capacity for people has been made possible by our use of fossil fuels. However, we are using up fossil fuels much more rapidly than most people realize, as will be demonstrated in Chap. 15. By using up some readily available sources of energy (crude oil and natural gas) to increase the planetary carrying capacity, man as a species runs the risk of a sharp drop in carrying capacity if coal or nuclear fuels are not or cannot be used as substitutes for crude oil and gas on a vast scale when supplies of the latter are depleted. This is the irreversibility principle. That is, it is not always possible to retrace the history of a system. In our case, readily accessible fuels were available once, but they will not be available again.

Another consequence of the tremendous increase in planetary energy flux is that because it is being unevenly partitioned among species, the diversity of the biotic world is decreasing. This means that the stability of the ecosystems on the planet is less, and there is a greater degree of vulnerability to wide-amplitude fluctuations due to such perturbations as plant diseases, insect pests, and weather changes. Specifically, we have increased the population densities of cattle, sheep, pigs,

A

B

FIGURE 6-1 Space as a resource for man. A Amsterdam is an example of a large city for which the use of space is carefully controlled by zoning. Consequently, as the city is approached from the east, the demarcation between the countryside and the city extending off to the horizon appears as a distinct line. There is no urban sprawl. B Between 1960 and 1970, San Jose, California, grew explosively, the population increasing by 114 percent in the decade. There was little attempt to control the use of space, and the result has been this intermixture of agricultural and urban land use which extends for miles, adding enormously to the overhead cost of services and transportation, and wasting some of the best agricultural land in the world.

corn, wheat, rice, and a few other species while sharply diminishing the population densities of very large numbers of other species of animals and plants.

A very high proportion of the total energy use has gone into transportation (24 percent in 1968). The significance of this is that human society is increasingly characterized by a greater interchange of individuals and materials between different locations than that found in the natural world. This interchange is paid for at a great price in transportation overhead, the most important energy cost of modern civilization. This transportation overhead is also the principal contributor to our pollution problem.

Space is a critical resource for man, but in many countries this has not yet been realized. There is a great deal of urban sprawl around cities, so that there is an extensive area at the urban fringe in which farmland and urban land are interspersed in a checkerboard fashion; this has several bad features. First, it leads to wasteful use of top-quality farmland, because once farmland is surrounded by subdivisions, it may be sold to land speculators who do not develop it immediately, but leave it undeveloped until the price goes up. Such land has no urban or farm use. This is particularly regrettable, because often the best farmland in a region surrounds cities—cities that evolved from farming villages once in the center of the best farmland in an entire region. Second, this checkerboard pattern creates a high cost of providing services per house, because it increases the mean distance between all the houses on the urban fringe, and consequently length of gas pipes, telephone and electric wires, etc. However, perhaps the worst feature of urban sprawl is the tremendous transportation overhead it imposes on all human activity, with resultant loss of time, energy, and space.

Some of the important facts about man's use of space as a resource are brought out by Figs. 6-1 and 6-2 and Plates 5 and 6. Figure 6-1 contrasts the appearance of the urban fringe in a city where there is very tight land use control (Amsterdam) and one where there has been

A

B

C

FIGURE 6-2 The effect of the automobile on space use in the urban core. **A** Cities in which the automobile is the dominant mode of transportation lose a high proportion of their ground area to freeways (Los Angeles). **B** A high proportion of the urban core can be lost to the freeway and roadway (Rotterdam). **C** The large amount of space required for automobile parking in the urban core militates against interesting architectural development (San Jose).

very loose land use control (San Jose). Plate 5 illustrates the difference between empty space and totally saturated space, and demonstrates the type of values that are lost because of the saturation. Plate 6 is typical of the character of urban areas which are designed for pedestrians, rather than motorists. Figure 6-2 indicates the effects on the character of a city that result from dependence on the automobile, which is very space-hungry, as the dominant mode of transportation, instead of on rapid mass transit, which uses a small fraction of the amount of space per passenger seat-mile used by the car.

The fact that space itself is a resource, independent of any other resource within the space, and that space can be limiting, is increasingly at the core of our problems concerning saturation of airspace near large airports, freeway congestion, and the difficulty in finding new sites for airports, garbage disposal, and electrical power generating plants. We are running out of certain varieties of space. It is not a question of running out of *all* space; rather, we are running out of useful space in heavily populated areas where space is particularly valuable to us. The fact that there is a lot of space available for airport or generating plant construction in the middle of Nevada, Kansas, or the Sahara Desert is of remarkably little importance to the residents of New York City, San Francisco, or Vancouver, all of which are in effect on islands or peninsulas, but are of especially great value because of their location at the mouths of great rivers or harbors.

Man repeatedly ignores the significance of time as a critically important resource. This means that we ignore the fact that the probability of an event occurring depends on the time available for it to occur. For example, it is widely believed that breeder reactors and the controlled fusion reaction will replace natural gas and crude oil before they are completely depleted; yet this belief ignores the very long lead times involved in developing new techniques and technologies, and building the tremendous new plants in the great numbers required to apply the new technology on a vast scale (the scale on which it will be necessary). To illustrate, it now takes 5 years from the time contracts are let to the time a nuclear power plant is completed and operational. Within a few years, the total time from conception of a giant power plant using the breeder or fusion reactions through planning, design, and construction will probably take an absolute minimum of 8 years. This means that unless society plans a long time ahead as we approach the date of depletion of natural gas and crude oil, there will be no nuclear substitute at the required time. (Coal may not be as useful or convenient as natural gas or crude oil in the future because of intensifying public pressure against the environmental degradation associated with strip mining.)

Our species has given inadequate attention to the significance of diversity as a valuable resource. We make ourselves vulnerable to a variety of economic as well as biological hazards by making the world simpler, less variable. Margalef (1963, 1968) has pointed out that the net effect of civilization is to accelerate the energy flow through biological systems and simplify their structure. This results in a destruction of homeostatic mechanisms of the type discussed in Chap. 4. There is a tremendous increase in the homogeneity within cities, of similarity between different cities, and the sameness of the countryside surrounding different cities. The entire landscape becomes more uniform, because of our replacement of forests by agricultural crops, and the process is taken to its ultimate limit when we try to get economies of scale in agriculture by planting large tracts of land out to the same species, or even strain, of crops. In general, man as a species is currently operating so as to increase the man-made order on the planet at the expense of natural order. That is, fossil fuels are being burned, plant and animal species are being eliminated, and community pattern and species diversity are being diminished by conversion of natural communities to large tracts planted to monoculture.

We will now illustrate the significance of this general proposition by four specific mechanisms: the effect of community simplification on disease and pests, the effect of agricultural trends to monoculture on economic stability, the effect of community simplification on marginal habitats, and the effect of economic lack of diversity on economic stagnation in cities.

Whenever a large tract of land is planted out to the same crop, the number of species of insects is reduced, but the average abundance per species increases, and the probability that any particular species will reach pest densities is consequently increased. This is because the remaining ones waste less time and energy in dispersal to appropriate areas. The whole world gets carpeted with the "right" kind of plant to eat. For example, Bey-Bienko (1961) shows that when a typical square meter of virgin grass steppe in the Soviet Union was converted to wheat, the total number of insect species decreased from 340 to 142, but the average abundance increased from 199 individuals to 351. Also, the average abundance of the most abundant species increased from 16.48 per square meter to 300.40 per square meter. The decision to convert the Russian steppes to wheat was made by Premier Nikita Khrushchev, and it may be significant that his removal from office in 1964 followed a collapse of his farm policies, with the resultant purchase by Russia of more than 10 million tons of wheat from the West in 1963. In 1970, a corn blight which affected great acreages of corn in the United States had far-reaching economic implications which ultimately affected the price of meat and the overall consumer price index.

Countries in which the economic stability depends largely on one crop are notoriously vulnerable to economic instability for this reason (Brazil and coffee, British Honduras and bananas, Argentina and wheat). The agricultural economy of Hawaii is largely related to world markets for sugarcane and pineapple. As Nicaragua and Taiwan were able to supply pineapple more cheaply because of lower labor costs, they could outcompete Hawaii for markets.

Community oversimplification in a marginal habitat, such as a desert or semiarid area, increases the vulnerability to the hazard of an insect or mammal herbivore that becomes a pest on the one or few remaining plant species. A multiplicity of species reduces the risk to any one species.

The same principle applies to the economic life of a city, as Jane Jacobs has pointed out (1969). A city which makes use of economies of scale and consequently has its economic fortunes inextricably interwoven with the fortunes of one large industry is unusually vulnerable to economic stagnation if anything happens to cut off or diminish the market for the one large industry. Detroit is unusually vulnerable to economic stagnation in the automobile industry, and the same is true of Seattle with respect to airplane manufacturing.

This same principle applies to diversity of age groups within a population. The greater the diversity of age groups, the mean age of a population, and the proportion of the population in older age groups, the greater will be the resilience and resistance of that population to environmental perturbations. The reason for this is that young individuals are typically the most likely to succumb to environmental perturbations. For example, it has been discovered that fluctuations in the numbers of smallmouth bass entering a fishery were almost entirely due to fluctuations in the survival of the bass in their first year of life, which in turn were due to temperature variations from year to year (Watt, 1959). Statistical analysis showed that 94 percent of the year-to-year variation in the number of fish old enough to enter the fishery was due to year-to-year variation in the summer water temperatures in the year of spawning.

6,700 feet

3,500 feet

1,000 feet

100 feet

PLATE 1 The effect of altitudinal gradients in mean annual rainfall and temperature on species composition of plant communities up the windward slope of Mauna Kea, Hawaii (13,796 feet high). Kodak Ektachrome Infrared Aero Film has been used with a medium yellow filter to cut haze and accentuate the vertical zonation of plant communities.

The pale red color is produced by sugar cane plantations, which extend up the sides of Mauna Kea and Mauna Loa to about 1,800 feet above sea level. Above that, the high rainfall makes possible a belt of trees, up to 120 feet in height, extending up to between 2,500 and 6,000 feet, depending on the position around the mountains of wind and rainfall patterns. This shows up in the infrared photograph as a purple band. The light-colored band above that is cool, dry country with little rainfall and a desolate appearance, with evergreens, ferns, and a few windblown trees with few leaves. All leaves at that altitude are adapted for extreme drying. The cold dry air is only one of the drying problems confronted by these high-altitude plants; the other is the extreme drainage owing to a substrate of unweathered ash and lava. The very dark blue belt at the top of the mountain is largely devoid of vegetation, and at

A 6,700 feet

B 3,500 feet

C 1,000 feet

D 100 feet

PLATE 1 continued   the extreme top there are snow and ice throughout the year. Thus, many of the climates on earth occur up the face of this mountain, and many of the remaining climatic zones (tropical deserts, for example) can be found on the other, leeward slope.

The topographic map below gives isolines of mean annual rainfall and indicates the position of the camera. Arrows and letters correspond to the ground views of plant communities on the facing page.

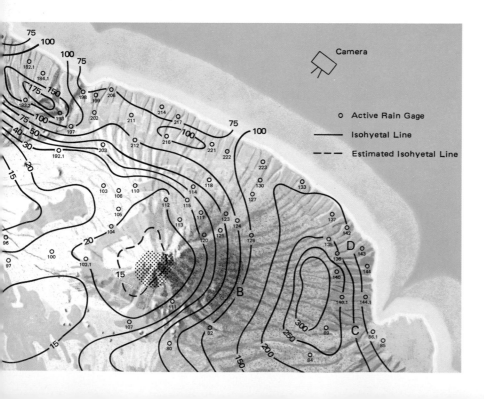

A   Vegetation adapted to cool dry climate at 6700 feet.

B   Tall trees in rain forest at 3500 feet.

C   Lush tropical vegetation in hot rainy habitat at 1000 feet.

D   Dense jungle growth at 100 feet where annual rainfall very high.

A

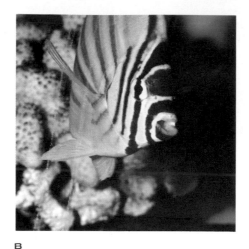

B

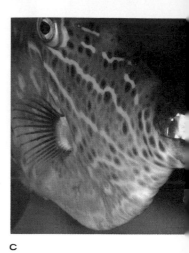

C

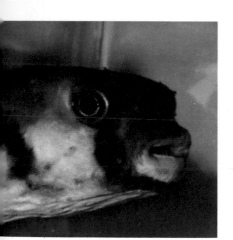

D

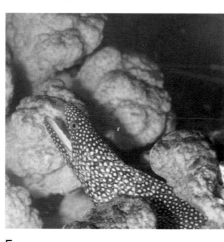

E

F

PLATE 2 Mouth structure in shallow tropical coral sea fish species as a clue to the feeding specialization which allows for great species diversity. **A** The very long protruding mouth of the longnose butterfly fish *Forcipiger longirostris* is used for picking minute food particles out of coral and crevices in rocks. **B** The orange-striped butterfly fish *Chaetodon ornatissimus* has flexible comblike teeth used to eat small invertebrates. **C** The file fish *Alutera scripta* uses its sharp pointed teeth to nibble fish, shrimp, and crabs. **D** The spiny puffer *Diodon hystrix* has powerful crushing teeth with which it can eat crabs. **E** The white-spotted moray eel *Gymnothorax meleagris* is a fish eater. Moray eels are known to attain sizes up to 10 feet in length. They have immensely powerful jaws, which in large species are full of fanglike teeth. Large species attack man if provoked. **F** The rectangular trigger fish *Rhinecanthus rectangulus* has a small number of flattened teeth, probably useful in crushing food.

These are only six diverse samples representing an enormous variety of mouth, jaw, and tooth designs found in the 584 or more fish species living in the ocean off the islands of Hawaii.

A

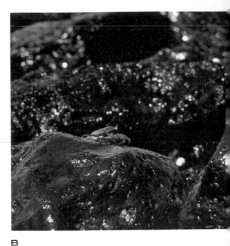

B

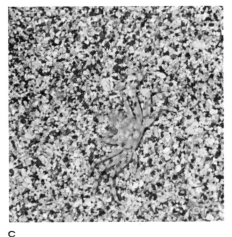

C

D

E

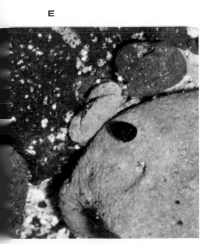

PLATE 3 The color matching of different species of crabs to their surroundings at Kona, Hawaii. **A** illustrates the nature of the habitat, and the coconut in the foreground indicates its size. All the species in the panels of this plate, except the last, were found in an area about 30 feet square. There are three distinct habitats in this area: the dark lava rocks, the fine, light-colored sand, and the pebbles and gravel at the interface between the water and sand. The crabs living in each of these habitats match the color of the background so precisely that they are almost invisible. **B** shows a rock crab. Its color and texture make it almost invisible, either dry or wet. **C** and **D** illustrate two different species of sand crabs: their color and texture are matched to the color and texture of the sand they inhabit. The crab in **D** runs so quickly and is so difficult to see against the sand background that it is called the ghost crab. This particular species is called the horn-eyed ghost crab, because of the horns on the end of the eyes. **E** to **K** illustrate different color adaptations in pebble crabs. The first, crab **E** under a green rock,

F

PLATE 3 continued

is green. The second **F**, flecked with black and white, appears like a small mound of gravel. The third, **G** living amid largely white gravel with a few black fragments, is white with black marks. The fourth, **H** adjacent to an orange rock, is orange, the fifth **I** is marbled green, like some of the pebbles in the environment, the sixth **J** is dull green, and the seventh **K** is very dark and dull. What is truly amazing is that all these small crabs live within a few feet of each other, and they all resemble some object in the environment. However, not all beach crabs look like the environment; **L** shows a vividly colored sand crab from Fiji.

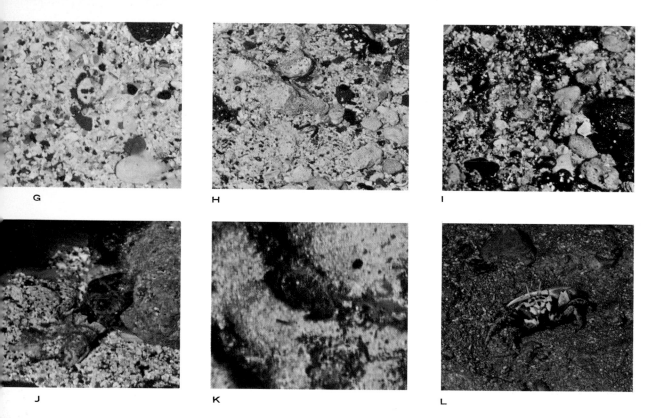

G

H

I

J

K

L

A

PLATE 4 Two examples of unusual predators. **A** View from the bottom of the crown-of-thorns starfish, which is eating up coral in the coral reefs of Australia, Fiji, Tahiti, Hawaii, and great numbers of oceanic islands. Sharp spines on the top make the starfish fairly invulnerable to attack by various other species, and the light-colored tube feet on the underside of the arms allow it to hang on to the coral, which it devours with its mouth (light-colored hole at center

PLATE 4 continued

of underside). **B** A brightly colored nudibranch inhabiting the ocean off the coast of Kauai. Various groups of nudibranchs are predators on such other types of animals as sponges, hydroids, jellyfish, corals, and barnacles. An interesting example of an unusual relation between animals occurs in the eating of hydroids by nudibranchs. The nudibranchs ingest the nematocysts of the hydroids, which are protective organs that produce a stinging sensation when a hydroid is contacted by another organism. These nematocysts then migrate to the surface of the nudibranch and become part of its protective apparatus.

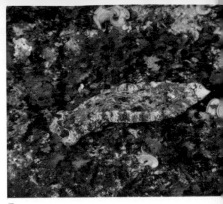

B

A

B

PLATE 5 Space as a resource for man. **A** This totally deserted beach clearly has very high value to man as a recreational resource. **B** This is an example of what is happening to the beaches of the world. Two points are made by this illustration. First, space itself is a resource which can become utterly saturated. Second, long before saturation is reached, certain values which a beach offers are no longer to be found there. If the human population exceeds the density at which some space on a beach is still available, then the surplus population can have no beach space at all.

PLATE 6 An example of a central street from which automobiles are excluded (Lausanne, Switzerland).

PLATE 7 Forest cover in a country largely untouched by man. The vegetation is dense enough to become an important factor in determination of the microclimate in which it grows (Fiji).

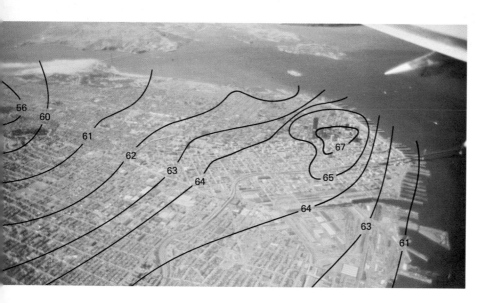

PLATE 8 The urban heat island effect as revealed by aerial photography with infrared aero film. The photograph was taken about noon. Isotherms superimposed on the photograph are based on readings taken about midnight at ground level by instrumented cruising cars, *reported by Duckworth and Sandberg 1954.*

PLATE 9 Treeless landscape in Malta.

Only 6 percent of the variation was due to all other factors operating in the first, second, third, and fourth years of life. Similarly, Green and Evans (1940) showed that the principal cause of great annual variations in the numbers of snowshoe hares was great year-to-year variation in the survival of young hares in the first year of life.

Similarly, optimal functioning of the human population is possible only with an approximately stable age distribution. In human society as in all others, the energy efficiency of the species is maximal with the stable age distribution. With a distorted age distribution, in which, for example, there are too many young individuals and too few adults, too much of the energy of the species goes into reproduction and caring for the young; and in the case of animals there is too much biomass loss in young ages, owing to competition, cannibalism, predation, parasitism, and starvation. In the case of humans, an excessive number of young creates too much of a tax and capital drain per taxpayer for educational costs.

Many examples illustrate the role of diversity as a resource which promotes stability by spreading risk. For example, the landscape in the United States has been simplified by the elimination of 11,428,000 workhorses between 1940 and 1960 (Landsberg et al., 1963). If it takes roughly three times the land to supply the food for a horse as to supply the food for a human in the United States, then land was freed to support another 34 million people simply by eliminating work animals. But this is not pure gain, because it could be argued that horses are like a giant condenser, storing a charge from the sun, which can be used as a source of stored energy in the case of an emergency. By killing off our draft animals, we eliminated one possible source of stored energy in an emergency. In other words, this is one more way in which our simplification of the planetary ecosystem allows for an increase in prospective instability.

principle 4   Pollution can become one factor imposing a limit on the human population size. That is, side effects of pollution on the weather, public health, and plant growth may be great enough to set limits on the maximum permissible amount of population. Figure 6-3 illustrates some of the effects of pollution. A particularly important effect is that of air pollutants on the proportion of the energy from the sun which is able to reach the surface of the planet. As demonstrated in Chap. 5, the amount of turbidity in the atmosphere, which varies with the magnitude and frequency of volcanic eruptions, has a sufficiently great effect on the proportion of the sun's energy that reaches the earth to override the effect of year-to-year fluctuations in the intensity of incoming solar radiation. This would lead us to suspect that particulate air pollution, another form of turbidity, might have a similar effect. In fact, recent analyses by Bryson and Wendland (1970) indicate that a global buildup in turbidity that is now occurring is too great to be explained by recent volcanic eruptions. Also, the global mean temperature has dropped slightly (about $\frac{1}{2}$°C) since 1950, although solar radiation intensity and increasing carbon dioxide content of the atmosphere would have led us to expect an increase in the global mean temperature. The only explanation would appear to be that the effect of

A

B

C

FIGURE 6-3 Air and water pollution. **A** Crude oil floating in to the beaches of Santa Barbara, California, as detected by color infrared film. **B** Air pollution over New York State between Albany and New York City, September 20, 1967, during an inversion. This sharp band between the polluted air below and cleaner air above is now observed over large sections of the planet. The most surprising characteristic of this line is the great altitudes at which it is observed: 5,000 to 5,500 feet in this case. Air pollution is often observed at seven times this altitude. I am indebted to Walter F. Dietz, Jr., Manager, Customer Relations, Mohawk Airlines, for information about the flight altitude. **C** The effect of jet contrails on cloud formation. Jets make a surprisingly high contribution to the total amount of cloud cover, but this fact is rarely noticed unless one watches an individual contrail for some time. In this photograph, three different contrails are shown in different stages of spread and diffusion. The thinnest has just been formed by a jet aircraft, and the widest is now difficult to identify as having been formed by a contrail.

an increase in turbidity, which causes planetary chilling, is currently overriding the heating effect of the increased solar radiation intensity and increasing concentration of carbon dioxide. Further, since volcanic eruptions do not account for all the buildup in turbidity, the discrepancy must be due to the manufacturing and transportation activity of man, with its resultant air pollution. If this interpretation is correct, it suggests the gravity of the greatly increased pollution which will occur in the future if present trends in energy use continue. A 4°C drop in global mean temperature, for example, would have extremely serious consequence for plant growth in most countries.

Figure 6-4 and Plate 7 illustrate another mechanism which may set a limit on how far man may be able to allow his own population to grow. Elimination of the forest cover from countries, in order to gain more land

A

B

C

FIGURE 6-4 Man's activities and plant growth. A Man has eliminated much of the original tree growth from the Mediterranean basin. This means that agriculture is conducted in a microclimate lacking in the ameliorating effect of trees on wind velocity, evaporation, and fluctuations in temperature and moisture content of the air and soil. B The dense tropical rain forest of Trinidad. C The soil under the forest shown in B. Although this forest appears almost indestructible, and immensely productive, the environment can in fact be destroyed quite quickly if the vegetation is removed before planting out a crop. The explanation lies in this soil, laterite, which turns to a concretelike material when exposed to the sun after deforestation. Therefore, remarks about the enormous agricultural potential of the great tropical rain forest areas should be interpreted with some caution.

for agriculture, may be self-defeating in the long run. Figure 6-4A and Plate 7 contrast two countries (Malta and Fiji): in the former there is almost no tree growth, and in the latter there is a dense forest cover over much of the country. The significance of this difference will be developed in Chap. 10; suffice it to say here that trees have a number of important effects on the microclimate within which crop plants grow, and therefore it may be anything but a pure gain to remove them. Figure 6-4B and C illustrates another problem related to the removal of forest cover. Many people have seen or heard about the fantastically lush forest growth in many tropical countries, as illustrated in Fig. 6-4B (Trinidad). What is not widely recognized is that the lateritic soils under such tropical forest quickly turn to something like concrete when allowed to bake under the tropical sun after the removal of the forest cover. Figure 6-4C is a photograph of the soil under the forest depicted in Fig. 6-4B in a spot where the vegetation had been removed. Clearly, when much of the plant cover has gone, the remaining soil is of very little use. This may well mean that much of the tropical rain forest areas may be difficult to use for any agricultural crops. Thus, the apparent immense agricultural prospects for some tropical areas may be illusory.

Several fundamental features of the environmental pollution and degradation produced by man are particularly noteworthy.

One property which pollutants may have is synergistic action; that is, the presence of one pollutant intensifies the effects of another. The net effect of several pollutants becomes more than the sum of their parts. The possibility is illustrated in Fig. 6-5. The implication is that the presence of a particular pollutant may have an unexpected significance because of the way it interacts with some other pollutant to produce a chemical product which is more lethal than either of its components. It is known, for example, that different components of air pollution react with each other to form new substances under the influence of light energy.

A second important characteristic of our degradation of the planet is that it often comes about not as a direct consequence of something we

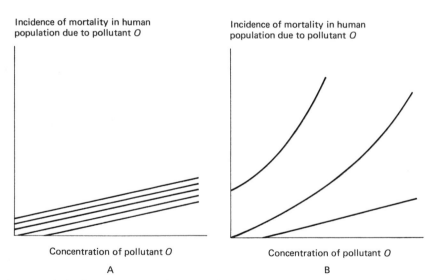

Incidence of mortality in human population due to pollutant O

Concentration of pollutant O

A

Incidence of mortality in human population due to pollutant O

Concentration of pollutant O

B

FIGURE 6-5 The danger of synergistic (interaction) effects of several pollutants occurring simultaneously. A represents the absence of any synergistic (interaction) effect due to pollutants O and P. B represents a very large synergistic effect due to pollutants O and P. In both panels, the bottom line represents the effect of pollutant O on mortality in the absence of pollutant P, the second line up represents a concentration of pollutant P equal to X, and the third line up represents a concentration equal to $2X$; in A, the fourth line up represents a concentration of pollutant P equal to $3X$, and the fifth line up represents a concentration of $4X$. The fact that the bottom line crosses the X axis some distance to the right of the origin means that no mortality at all is detectable as due to pollutant O until the concentration of O surpasses some threshold. B brings out the synergistic effects of pollutants O and P occurring together: for a given concentration of O, doubling the concentration of P more than doubles the mortality due to P, and for a given concentration of P, increasing concentrations of O increases the impact on mortality nonlinearly. By contrast, in A mortality increases a constant amount with increase in O, or with increase in P when both are present together; neither heightens the effect of the other.

do, but as an unexpected side effect of some activity unrelated to the resource which was destroyed. For example, in the primitive world, sheep, goats, rabbits, and other animals which crop grass close to the ground were presumably kept at population densities which the grass could support by large populations of predators (lions, mountain lions, lynxes, bobcats, and wolves, for example). However, man has hunted these predators down to very low population densities, with the result that herbivore populations could rise to population densities capable of destroying grasslands. Much of the world, particularly the semiarid world, has been destroyed or brought close to destruction by intensive close cropping from goats, sheep, and other herbivores. Another example is the poisoning of fish and shellfish waters (estuaries and shallow offshore shelves) off the coast of North America by the runoff of pesticides and fertilizers from heavily treated farmland. Still another example is the possible effect of chemical alteration of the environment on large-scale geochemical cycles on the planet. For example, it has been found that very small concentrations of the insecticide DDT are toxic to certain small marine organisms. Since these organisms are important in converting carbon dioxide to oxygen and utilizing the phosphate and potassium in the ocean to make it available to animals higher up in the food chain, this poisoning could have very serious consequences. We conclude that it has become important when evaluating the benefit/cost ratio for a proposed activity to consider the possibility of unexpected and deleterious side effects.

Pollution and degradation of the planet are inexorable consequences of very high and constantly increasing rates of energy flow through the human ecosystem. If projected rates of energy flux come about, then the problem of pollution will inevitably get much worse. The only way to deal with the problem is to recycle material or, preferably, slow down the rate of energy flux.

A very basic feature of human pollution and degradation of the planet is that temporarily it is completely out of control by density-dependent factors. The reason for this is that most populations of animals depend on a fixed resource base; consequently, when they overexploit it, it becomes depleted, the animals starve, and lowered birth and survival rates bring the population back into equilibrium with its resource base. But in the case of the present human population, it is able to keep increasing because it is no longer dependent on a fixed resource base. Instead, because of our conversion from solar energy to fossil and nuclear energy, we can keep the carrying capacity of the planet rising as long as we can keep increasing the use rate of these energy sources.

Also, man is able to outcompete and thus to reduce constantly the number of other species on the planet in a way that no other species has ever been able to before. Many other species of animals depend for their food on one other species, as the monarch butterfly on the milkweed plant. If a food species goes extinct because the herbivore overeats it, then the herbivore itself goes extinct. However, man is the ultimate omnivore: we eat such a variety of species in such a wide variety of habitats that we will have eliminated a great many other species of plants and animals from the planet before we ourselves are controlled by shortages.

It was pointed out in Chap. 2 that efficiency of energy utilization in ecosystems increases with increasing diversity. By our constant shift of agriculture toward more monoculture, we decrease energy efficiency in ecosystems and make the ecosystems more fertilizer-hungry in order to keep yields increasing. This means that we keep increasing our dependence on an energy subsidy to agriculture, (gasoline power, fertilizer industry, etc.) which in turn means that we keep increasing our dependence on fossil and, later, nuclear fuels. Consequently, our entire civilization is becoming increasingly vulnerable to catastrophe should there be any shortage of these fuels.

Finally, as in the case of the potato blight in Ireland in the mid-nineteenth century, simplification of ecosystems increases the improbability of instability. The reason for this is that one plant disease, insect, or other pest can quickly spread throughout an entire area when the entire area is the food of that disease or pest. Also monoculture places a premium on evolution of specialized strains of pests that are much more competent destroyers of crops. In fundamental terms, exploitation of ecosystems by humans places a high premium on high productivity per unit area. Since only immature ecosystems with a high productivity to biomass ratio give a satisfactorily high productivity, this is what we select for. However, the price is steep: these immature communities are also characterized by low stability. The low stability has several causes. The vulnerability to disease just mentioned is one. The smaller number of species, with consequent reduction in the number of community-level homeostatic mechanisms is another. Still another has to do with a drop in the mean age of plants and animals in the heavily harvested community. Intensive exploitation of populations drops the mean age of those populations, by decreasing the probability that an individual will live to an old age. Also, intensive exploitation of the landscape causes a shift to cultivation of more short-lived species, from trees to grains, and from buffalo to cattle. In general, younger individuals are more sensitive to environmental fluctuations, and so the lower the mean age of populations and communities, the greater will be the likelihood of the average animal to succumb to a wide-amplitude change in environmental circumstances. Thus, several mechanisms exist which produce a general tendency for community simplification to be related to vulnerability to instability.

In conclusion, it is clear that several mechanisms exist which can set limits on the carrying capacity of the planet for people; for us, as for all other species, there is an optimum level of abundance, as one would expect from Principle 4.

principle 10   It has been shown in Chap. 3 that natural communities tend to evolve toward greater energy use efficiency as they develop a higher degree of diversity. Since human society is accumulating biomass at a great rate, in the form of buildings and other artifacts, and diversity at a high rate, in the form of stored information, and is increasing diversity of occupations, equipment, and other products, it is reasonable to expect, from Principle 10, that our energy efficiency would also be increasing. Two measures of efficiency of particular interest are the energy use per person and the energy use per unit of gross national product. Trends in these two efficiency measures for the United States are collated in Table 6-1.

We conclude from this table that if we define energy efficiency in terms of the energy cost to produce a unit of gross national product, then our energy efficiency is increasing. However, if we define energy efficiency in terms of the energy cost to support one person, then the system is becoming less efficient. The explanation for this apparent contradiction is in the last column: our standards of living are rising very fast.

There are several basic explanations for the decreasing energy efficiency involved in supporting a person. To gain an insight into this problem, we must examine where our energy goes.

In 1968, 24 percent of the energy consumed in the United States went directly into transportation. However, an additional large proportion of the energy spent in society found its way into transportation in the form of the energy cost of making vehicles, airport runways, roads, and roadbeds, and the energy cost of extracting, manufacturing, and processing the raw materials that go into vehicles and equipment and buildings associated with the use of vehicles. Consequently, it appears that transportation is by far the largest single consumer of energy in society. Rather than attempting to analyze the fundamental aspects of each type of energy use, therefore, transportation only will be analyzed to bring out certain fundamental aspects of energy flow in human society.

It is extremely revealing to consider a comparison of how transportation energy flows operate in natural animal systems and how they operate in the human ecosystem. Hamilton et al. (1967, 1969) have counted the number of starlings flying outwards from a roost each day to forage for food. The number observed to pass concentric rings at different distances out from the roost falls off as a function of distance, and none was observed to make round trips over 120 miles per day. It is a reasonable supposition that the birds make a trade-off between a gain—the reduced competition pressure for food that results from flying farther out from the roost—and a loss—the increased transportation overhead and time lost to flying greater distances. Further, it is reasonable to suppose that at some maximum distance, the energy cost of flight is so great and the time remaining in which to feed is so small that no possible reduction of competition could compensate for any further distance

TABLE 6-1 Trends in the efficiency of energy use in the United States

year	population in millions	energy use in quadrillions of Btu	GNP in billions of dollars (1960 dollars)	energy use per person in millions of Btu	energy use per dollar of GNP in thousands of Btu	GNP per person in thousands of dollars
1940	132.1	25.8	234	195	110	1.77
1950	151.7	36.4	363	240	100	2.39
1960	179.9	45.3	503	252	90	2.80
1970	208	60.2	746	289	80	3.59
1980	245	79.2	1,060	323	75	4.33
1990	287	101.9	1,510	355	67	5.26
2000	331	135.2	2,200	408	61	6.64

SOURCE: Data on population, energy use, and GNP for this table were all obtained from Landsberg et al. (1963). Population data are from their table A1-1 (p. 516), and data on GNP and energy use are from their table 15-4 (p. 292).

flown. At the distance at which no further benefit is to be derived from flying farther out, in effect a ring is drawn around the foraging area of the starlings in the roost. It is interesting to discover whether a similar appearance of rationality can be discovered in the process by which human beings make choices about the mode of transportation.

To illustrate a general point, two pairs of comparisons will be made: between taking a cab and walking in a large metropolitan area, and between a high-speed train traveling from the downtown area of one city to the downtown area of another city and a jet aircraft traveling from airport to airport.

In the first comparison, we will assume that the first option involves walking from point to point at a constant velocity of 4 miles per hour, and the second option is taking a cab at rush hour, when it takes 15 minutes to hail or wait for a cab. In this instance, the true portal-to-portal velocity must include the time spent hailing or waiting for the cab; therefore, assuming a mean cab velocity in rush hour traffic of 10 miles per hour, we have an overall cab velocity for various distances as follows:

distance	portal-to-portal time	overall velocity in miles per hour
1 mile	$\frac{1}{4}$ hour $+ \frac{1}{10}$ hour	2.85
2 miles	$\frac{1}{4}$ hour $+ \frac{2}{10}$ hour	4.44

The results are plotted in Fig. 6-6, which shows that it is not worthwhile to take a cab unless the distance is at least 1.7 miles! (The same result is obtained also from a simple algebraic equation.)

Similar but somewhat more complicated calculations compare the portal-to-portal velocities of a 175-mile-per-hour train and a jet plane with a cruising velocity of 600 miles per hour. The complications arise here because the jet plane, particularly, will not reach cruising velocity immediately. Indeed, on very short distances it may never fly at anything like its rated cruising velocity. This leads to some rather amazing portal-

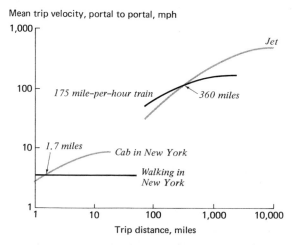

Mean trip velocity, portal to portal, mph

Trip distance, miles

Actual velocity of different modes of transport, by trip length

FIGURE 6-6 The effect of computing mean portal-to-portal velocities on choice of transportation mode.

to-portal mean trip velocities if one uses jet planes for short distances. Assume, for example, that it takes 1 hour from the time of leaving home to the time that a plane takes off, and another hour from the time the plane lands until one has reached his ultimate destination. Also, assume that for a 72-mile flight, 20 minutes elapse from the time the passengers are all seated in the plane and it leaves the airport, to the time the passengers begin disembarking. The total time for this 72-mile trip is then 2.33 hours, for a mean trip velocity of 31 miles per hour. A more typical jet flight is 2,600 miles and takes $2 + 5$ hours, for a mean portal-to-portal trip velocity of 371 miles per hour. Clearly, the cruising velocities of jets improve portal-to-portal velocity only when trips are many thousands of miles. For comparison, the figures plotted in Fig. 6-6 for a 175-mile-an-hour train show that if it travels from urban core to urban core, only $\frac{1}{2}$ hour is lost connecting with the train at each end. In this case, it is not worthwhile to take the faster mode of transport unless one is traveling at least 360 miles.

Figure 6-6 demonstrates that bad choices of mode of transportation are being made at both ends of the velocity scale. Our culture is so conditioned to the idea that faster is better, that most people fail to perceive that faster is often slower in the long run. Thus, time and energy are both being used wastefully.

It is interesting, before leaving this topic, to examine briefly fuel costs per passenger-mile for different types of transportation. The figures in Table 6-2 are all based on United States national averages for 1967, except those for the electric train, which were obtained either directly or by calculation from other data in Hay (1961).

The automobile with a typical load of 1.3 passengers is one of the worst ways to use fuel. Well-loaded ground public transportation is by all odds the most efficient way to use it. Jet aircraft use fuel less efficiently by far than ground public transportation, even when the latter is almost empty. This comparison raises some interesting questions about why the

TABLE 6-2*

vehicle	average number of miles per gallon	passengers	passenger-miles per gallon
Automobile	14.1	1.3	18.3
		4.0	56.4
Bus	5.36	5.0	26.8
		50.0	268
Jet aircraft	0.24	54	13
Electric train	1.88	600	1,130 (level grade)

* The figures in the table were derived as follows. The numbers 14.1 and 5.36, for the average numbers of miles per gallon traveled by cars and buses, respectively, in the United States in 1967, come from the *Statistical Abstract of the United States*, 90th ed., table 816. The two figures for numbers of passengers in cars and buses assume light loadings and heavy loadings. The figures in the third column are the product of the figures in the first and second columns. The figures for jet aircraft were calculated as follows from figures in table 856 of the same publication. The total fuel consumed by scheduled air carriers, domestic and international, in 1967 was 7.553×10^9 gallons, almost all in the form of jet fuel. The total number of domestic and international revenue miles flown was 1.813×10^9, so that miles per gallon were $\frac{1.813}{7.553}$, or 0.24, on the average. The revenue passenger-miles flown were 98.8×10^9 for domestic and international flights, and so the average number of passengers per trip was $\frac{98.8}{1.813} = 54$.

United States has drifted into a national transportation system in which regional airlines are using high-performance jet aircraft for short and intermediate hauls.

Another problem in the efficiency of energy use is clear from examination of aerial photographs of cities of different sizes. Such photographs show that in most large modern cities, the average distance between buildings increases with increased distance from the center of the city. Thus, as city size increases, the transportation overhead involved in daily living becomes worse for the average occupant, and the city as an integrated functional unit makes less efficient use of input energy. This suggests either that cities should increase in size by going up rather than out or, preferably, that the most efficient pattern of human social existence is to avoid having cities over a certain size. Thus, the most efficient distribution of mankind is in large numbers of small cities rather than a small number of large cities. Results of the United States 1970 census demonstrate that people actually prefer less dense living: the most dense urban areas in the country have experienced a net population outflow in the last 10 years.

A surprising contributor to low energy efficiency is the increasing interchange of materials between different and remote parts of the world. Thus raw material may travel several thousand miles to be processed, and then travel back to its ultimate destination to be sold. Food is being exchanged over vast distances. All this interchange adds to the transportation overhead of human existence.

The fundamental spatial organization of society has changed considerably in modern technological culture. Formerly, a high proportion of the population lived on farms, or in small villages that provided services for groups of farms. Under such circumstances, consumers are often very close to the production site for the products they consume. However, as society has evolved, people have moved great distances away from farms and small farm villages, and also the cities have enlarged enormously. This means that the average distances have become much larger between:

Fertilizer and farms that use it

Farms and the people that eat farm produce

People and the stores where they buy their produce

People and the places where they work

People and the places where they go for recreation

These are only a few examples. It takes little imagination to see that a rather different social spatial design could lead to a far greater efficiency in the use of energy, and a more pleasant, far cheaper existence because of less time and energy spent commuting.

The conversion from horses to tractors, already mentioned, and the increased need to maximize agricultural productivity by using greater amounts of fertilizer also decrease the energy efficiency of society.

Society has low energy efficiency because almost all passenger transportation is being performed by the most energy-inefficient modes of transportation, cars and planes, rather than the most energy-efficient modes, buses and trains.

There is insufficient staggering of working hours, and the consequence is that transit system rolling stock must lie idle most of the day, to receive maximum use only for a short time during the peak demand rush hours. Thus, the immense amount of matter and energy used in manufacturing transportation systems is inefficiently used.

In short, we live in a society which has never given adequate attention to efficiency in the utilization of energy, chiefly because it has not been generally perceived that energy could be scarce in the future or that wasteful use of energy was a primary cause of pollution.

principle 11   Margalef has for some time argued this principle, namely, that mature ecosystems exploit immature ecosystems, or, more generally, immature subsystems are exploited by their more mature, more highly organized counterparts.

He states that in general, if two subsystems differ in their degree of organization, the less organized gives energy to the more organized, and because of this, diversity in the less organized is destroyed, and diversity is gained by the already more organized (Margalef, 1968). He recognizes that this principle is remarkably ubiquitous in its operation, and in fact characterizes relations between human societies. It is quite dramatic, in fact, to what extent rapidly growing and highly organized cities suck energy out of other areas, thereby inhibiting their development.

One measure of this principle is the age composition of the population that immigrates to growing cities, compared with the age composition of the population from which they emigrate. If there is an unusually high proportion of people aged 20 to 35 in an immigrating group, then the city that receives this population is given a tremendous economic boost. This age group has already received its education at the expense of taxpayers in some other location, and it is too young to require high medical expenses or other social services per capita. These are the most vigorous, exploratory ages, important in the establishment and development of new industrial, business, professional, and educational activity.

Table 6-3 gives some idea of the extent to which this phenomenon operates in human society. The first column gives the age distribution of the United States population in 1960; the second gives the age distribution of the net migrants into Los Angeles and Orange Counties between 1955 and 1960. Net migrants means the immigrants less the emigrants. Los Angeles and Orange Counties were two of the fastest growing metropolitan areas in the United States during that period. The most significant measure of the difference between these two age distributions is the ratio of the number of people in nonworking age groups to people in working age groups. This is a measure of the tax drain per taxpayer for nonworking age people, and also is a measure of the availability of capital for capital investment in various new social developments. The difference between these ratios in the two columns is a measure of the extent to which energy is being drained out of a society as a whole to foster the development of rapidly growing urban areas, which, in Margalef's language, are already more organized.

Also astounding is the growth rates that can be supported by immigration. Between 1950 and 1960, the United States as a whole grew 18.4 percent. During that period Los Angeles grew 26 percent, and

Anaheim, in Orange County, grew an astounding 616 percent! It would appear that Margalef's principle is indeed applicable to human society. It also applies to relations between different countries. For example, the phrase "brain drain" signifies the extent to which developed countries help underdeveloped countries. In fact, they don't. Most of the help flows in the other direction.

principle 14   This principle states that populations regulated by causal systems with lag effects tend to have great regularity in their patterns of population fluctuation. In effect, such populations are characterized by a great deal of momentum: the population trend at any point in time is strongly determined by the history of the system. That is, a population will have highly regular patterns of change if there is any mechanism operating in the population or the population-environment system such that a growing population tends to keep growing, or a declining population tends to keep declining. Does the human population have any mechanism like this?

In fact, there is a mechanism operating in the human population which will tend to make a growing population keep growing. This mechanism is the result of the joint action of two phenomena: fertility rates in women are highest in young women, and the proportion of young women in a population increases when a population is growing. Thus, when a population is growing rapidly, young women are being added to the population faster than old women are being subtracted from it. Consequently, the proportion of young women in the population rises, but since young women have a higher fertility rate per woman, the population growth rate goes up. Thus, the more the human population increases, the more it will

TABLE 6-3 The age distributions of the United States in 1960, and the immigrants into Los Angeles and Orange Counties, 1955–1960

age group	proportion of U.S. population in this age group, 1960	proportion of net migrants into Los Angeles and Orange Counties in this age group, 1955–1960
Under 5	0.112	0.079
5–9	0.104	0.110
10–14	0.094	0.084
15–19	0.074	0.064
20–24	0.061	0.146
25–34	0.128	0.241
35–44	0.135	0.134
45–54	0.114	0.069
55–64	0.086	0.041
65 and over	0.092	0.032
Preworking years	0.384	0.337
Working years	0.524	0.631
Postworking years	0.092	0.032
Ratio of number in non-working years to number in working years	0.91	0.58

SOURCE: Statistics in column 1 are from Department of Commerce, Bureau of the Census; statistics in column 2 are from California Migration 1955–1960, Department of Finance, Budget Division, Financial and Population Research Section, Sacramento, 1964.

continue to increase, unless some other mechanism reverses the trend.

The United States population has gone through great changes in birth rate, median age, and resultant effect on future birth rate. In the early years of the country, median age was very low (16.0 years). Birth rate gradually decreased, until, by the Depression of 1929–1935, it was very low, and as a result, the median age of the population increased until by 1950 it was 30.2 years. Then the economic stimulation of the period 1945 to 1964 was associated with a great change in birth rate, which went from 79.9 per 1,000 women 15 to 44 years of age in 1940, to 118.0 in 1960. This increased birth rate in turn meant that the median age of the population dropped from 30.2 in 1950 to 27.8 in 1968. The great effect that this has on the overall fertility of the population is apparent by inspection of the 1967 age-specific live birth rates:

age of mother	live births per 1,000 women
15–19	67.9
20–24	174.0
25–29	142.6
30–34	79.3
35–39	38.5

Obviously, any population phenomenon which increases the proportion of the population in the 20 to 24 age bracket and decreases the proportion over 29 will have a large stimulating effect on overall population fertility.

The process which first limits population growth in a technologically advanced society is the state of the economy. For example, consider the live birth rate for women 20 to 24 years of age. In 1940, at the end of the Depression, this was only 135.6 per 1,000. In 1960 it was 258.1. By 1967 it had dropped to 174.0. It is reasonable to expect that before long the economy will not be the only major factor operating on the birth rates of women of a given age. Shortages of resources and a plethora of pollutants will also be reducing the age-specific birth rate.

This factor will operate jointly with the age distribution of reproducing women to determine the overall reproductive rate of the population.

SUMMARY   This chapter has considered human society in the light of five of the principles presented in Chap. 2. The overall conclusion is that these principles are relevant for human society, just as they are relevant for the natural world of plants and animals. The future of our society depends on our recognition of the significance of time, space, matter, energy, and diversity as essential resources, and the existence of optimum levels of availability for all of them, to us and to the organisms which we harvest. The efficiency with which society uses energy is at the core of a variety of problems in pollution, traffic congestion, and resource depletion. Mature systems parasitize immature systems in the world of human affairs as well as in planktonic systems or terrestrial communities of plants and animals. Finally, there is an important demographic mechanism which works against sharp decreases in the overall population birth rate, and hence tends to perpetuate growth in a growing population unless there is a sharp downturn in the economy to operate against it.

SUGGESTIONS FOR INDIVIDUAL AND GROUP PROJECTS

What evidence of pollution can you detect where you live? Is there an unusual amount of disease of any type that might be due to pollution? Compare county death rates with national rates. County rates will probably be available from your state department of public health, and national rates are available from the U.S. Department of Health, Education, and Welfare.

Are plants poisoned or dead in large numbers?

Does paint seem to be affected by pollution? What about metal surfaces?

Try picking up evidence of water pollution with panoramic views from high vantage points, using a wide-angle lens and either black and white or color infrared film, with a yellow, orange, or red filter for the lens.

Are your local transportation systems efficient in terms of energy, space, money, and material? Might an alternative system be more efficient?

Do you see any evidence that economic activity in some other city affects the city where you live, or is affected by it?

What is the age structure of the population where you live? Is this markedly different from the national average? Why?

REFERENCES

Bey-Bienko, G. Y.: On Some Regularities in the Changes of the Invertebrate Fauna during the Utilization of Virgin Steppe, *Rev. Entomol. U.R.S.S.*, **40**:763–775 (1961).

Bryson, R. A., and W. M. Wendland: Climatic Effects of Atmospheric Pollution, in S. F. Singer (ed.), "Global Effects of Environmental Pollution," pp. 130–138, D. Reidel, Dordrecht, Holland, 1970.

Green, R. G., and C. A. Evans: Studies on a Population Cycle of Snowshoe Hares on the Lake Alexander Area. III. Effect of Reproduction and Mortality of Young Hares on the Cycle, *J. Wildlife Management*, **4**:347–358 (1940).

Hamilton, W. J., III, W. M. Gilbert, F. H. Heppner, and R. J. Planck: Starling Roost Dispersal and a Hypothetical Mechanism Regulating Rhythmical Animal Movement to and from Dispersal Centers, *Ecology*, **48**:825–833 (1967).

——— and ———: Starling Dispersal from a Winter Roost, *Ecology*, **50**:886–898 (1969).

Hay, W. W.: Engineering Characteristics and National Policy, in Robert S. Nelson and E. M. Johnson (eds.), "Technological Change and the Future of the Railroads," pp. 15–38, Transportation Center, Northwestern University, Evanston, Ill., 1961.

Jacobs, J.: "The Economy of Cities," Random House, New York, 1969.

Landsberg, H. H., L. L. Fischman, and J. L. Fisher: "Resources in America's Future," Johns Hopkins, Baltimore, 1963.

Margalef, R.: On Certain Unifying Principles in Ecology, *Am. Naturalist*, **97**:357–374 (1963).

———: "Perspectives in Ecological Theory," University of Chicago Press, Chicago, 1968.

Watt, K. E. F.: Studies on Population Productivity. II. Factors Governing Productivity in a Population of Smallmouth Bass, *Ecol. Monographs*, **29**:367–392 (1959).

7 the lake: a typical exploitable ecosystem

There are many types of ecosystems: rivers, streams, the spruce forest, the desert, the tropical rain forest, prairies, coral reefs, the ocean bottom, and the hot spring are examples. Of all the kinds of ecosystems, the lake has been one of the most thoroughly studied, and also it has been thought of as a system for a very long time. In 1887, S. A. Forbes wrote an essay "The Lake as a Microcosm," which was one of the first works to recognize clearly that the plants and animals in a habitat are part of a dynamic interactive system, in which each part has various effects on other parts.

In this chapter we will use the lake to illustrate the operation of principles important in all ecosystems.

The first principle is that ecosystems are products of their history. By this we mean that those forces operating at any time in an ecosystem can have subtle, yet powerful effects which gradually change the character of the ecosystem. That is, the whole system undergoes succession (Principle 17). The lake is a particularly remarkable example of this. When we speak of succession in an ecosystem, we do not only mean that the individual species of plants and animals undergo constant genetic change, to allow them to adapt to a changed environment. Rather, we mean that because of the changes which occur, the ecosystem gradually becomes less suitable for some species and more suitable for other species. Because of this, the species composition of the lake changes. This process may involve several successive waves of species.

In this chapter we introduce the notion of the systems viewpoint in rational management of ecosystems. This viewpoint is illustrated by considering the problem of maximizing the productivity of food from a lake. It will be demonstrated that the entire approach to selection of the optimal strategy for this problem changes as one increases the number of aspects or components of the system being considered in developing an optimal harvesting strategy.

In this chapter we use the term *productivity* to describe the number of fish caught in a lake each year. *Standing crop* refers to the number of fish actually present in the lake. In the technical fisheries literature, the word *yield* is used to describe the catch, and *productivity* is used to describe the net production of living tissue in the fish population. In an underharvested fish population, yield would be less than productivity. Several successive years of overharvesting would wipe out the population, by reducing the breeding stock to a level at which it could no longer replace itself. This may already have happened to several whale species.

LAKE SUCCESSION   There are many ways in which a lake can originate. Very large lakes, such as Lake Superior, are produced by folding and faulting of the earth's surface, with subsequent modification of the basin by glaciation. Some lakes originate with volcanic craters. Glacial scouring, ox-bow river formations, or the sealing of sinks in limestone formations to make a depression that holds water are some of the processes that create lakes.

When a lake is first formed, it contains little organic matter, typically, and the water is often very clear. The density of plant and animal life in the lake is low, and because of the great water clarity light can penetrate to great depths, so that some photosynthesis can occur down to 60 meters or more. Because of the great depth, and because the heat-bearing (infrared) part of the energy spectrum from the sun does not penetrate far into water, the water is typically quite cold. The principal inhabitants are species that have a preference for cold water, rich in oxygen, such as lake trout. Such a lake is called *oligotrophic*, meaning that it has a small amount of food. There is little nutrient recycling because of the combination of great depth and low level of biological activity, so that the surface waters may be very deficient in phosphorus, nitrogen, and calcium. It will be noticed in Table 3-1 that these elements are all important constitutents of living organisms.

Unless a lake is in an extremely cold climate, or a geological formation very deficient in nutrients, the rate of biological activity in it gradually increases. Organic matter, such as algae (phytoplankton) and tiny animals (zooplankton) and debris, accumulates in the surface waters. Consequently, the clarity of the water decreases, at first very slowly, then more rapidly. This means that light cannot penetrate so far down into the lake as previously, and photosynthetic activity in the lake is concentrated more and more in the surface waters. As the total amount of biological activity in the lake per unit time per unit volume of water increases, the rate of production of debris from this activity increases. The debris drifts to the bottom of the lake, and combined with inflow from streams draining the surrounding watershed, it makes the lake begin to fill up and become shallower. Particularly in the shallow waters at the edge of the lake, the tempo and variety of biological activity become quite great. The oligotrophic lake has become a *mesotrophic* lake.

These sedimentation rates are very variable, and some mesotrophic lakes may change to the next state only very slowly, if at all. However, other mesotrophic lakes gradually fill up until they become quite shallow. Such a lake may now be only 10 to 30 feet deep in most places. The tempo of biological activity and the concentration of organisms, with resultant production of debris, become great enough for the water to become very turbid. There is very little light penetration below the top 10 feet of water. The lake has become much warmer, and there has been a great change in the species composition of all life in the lake. Warmwater fish such as the large- and smallmouth black bass, the sunfish, perch, and bullhead catfish are now obvious. As the lake becomes very shallow, only the perch and bullheads will be left. Very shallow bays of the lake will be obviously filling up with ooze and becoming swamplike. On balance, however, the lake is very rich in nutrients. Because it is shallow, making upwelling easy, and because there is a great deal of biological activity, there is an ample supply of nutrients in the surface waters. The lake is now called *eutrophic*.

The end of this process occurs when the lake becomes so shallow, and so choked with living matter, that it becomes *dystrophic*. A lake like this is the last aquatic step on the way to a marsh, and then to a terrestrial successional sequence. Semiaquatic and terrestrial plants invade the moist sediment which almost completely fills the lake basin by now. The decomposing matter on the lake bottom uses up oxygen faster than the plants can produce it, and the lake virtually chokes. The turbidity of the water is so great that light can penetrate only about 3 feet. The tempo of biological activity by aquatic organisms is sharply reduced, and finally the lake dies.

In fact, there is a physical measurement which succinctly expresses the difference between these different stages in lake succession. Goldman (1968a) has used a carbon 14 method to determine the rate at which carbon is fixed per unit time per unit volume of water by photosynthesis as a function of depth in lakes in different Trophic categories. Figure 7-1 contrasts his results for an oligotrophic lake, a mesotrophic lake, an extremely eutrophic lake, and a dystrophic lake. This graph shows how the increasing turbidity of the lake diminishes light penetration to great depths as the lake evolves through this sequence from oligotrophic to dystrophic. Also, it shows how the total amount of photosynthesis increases from oligotrophic through mesotrophic to eutrophic, and then declines again as the dystrophic stage is reached.

Man has an effect on the rate at which lakes go through this sequence by altering the rate at which nutrients, in the form of sewage or other products of civilization, enter lakes. Since lakes are formed at a very slow rate, determined by the low probability of occurrence of geological events, man's activities could remove lakes from the world stock of lakes faster than the number of lakes can be increased by geological activity and dam building. Consequently, the number of lakes could decrease. In view of the undeniable enjoyment that most people derive from lakes, it would not appear to be in our collective interest to allow

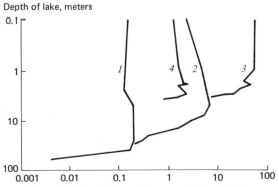

Depth of lake, meters

Photosynthetic activity, milligrams of carbon fixed per cubic meter per hour

FIGURE 7-1 Characteristic curves for photosynthetic activity as a function of depth in lakes at different stages in their successional history. (*Redrawn from Goldman*, 1968b.) 1. Lake Tahoe, a deep, oligotrophic lake with very great water clarity. 2. Castle Lake, a moderately deep, mesotrophic lake with moderate water clarity. 3. Clear Lake, a shallow, extremely eutrophic lake with turbid water. 4. Cedar Lake, a very shallow, dystrophic lake, in the process of becoming a bog.

The two main points to note in this graph are the sharp increase in lake productivity from 1 to 2 to 3, followed by a sharp reduction in productivity to 4, and the decreasing depth of light penetration from 1 to 2 to 3 resulting from increasing turbidity of the water.

this to happen. This means that we should take collective action to slow down the rate of eutrophication of lakes. Most people should see the advantage in keeping Lake Erie as a lake, not a vast swamp full of poisonous chemicals. To preserve lakes as lakes, vigorous and rigidly enforced government campaigns must prevent their being used as giant cesspools for the dumping of chemical wastes and sewage.

LAKES AS A SOURCE OF FOOD FOR MAN

Lakes have always been an important source of food for man, but the amount of food that can be obtained from a lake at any given stage in its succession varies enormously, depending on the sophistication of the strategy used in harvesting the lake. We will illustrate this argument by considering, first, a very simple approach to harvesting a lake, and then showing how an approved approach to harvesting and the amount that can be harvested depend on gradually increasing the sophistication of the harvesting strategy.

Suppose we have a lake in which the top predator is the bass, which eats perch and other smaller fish, which in turn eat a mixture of still smaller organisms such as insects or crustaceans, which in turn eat zooplankton, which in turn eat phytoplankton. The simplest way to harvest the lake is to decide that the only organism which man will use for food is the bass. Then harvesting strategy is designed so as to maximize the long-term annual sustained yield of bass from the lake. How should such a strategy be worked out? Suppose, to simplify the problem, we imagine that all the physical variables in the lake remain constant from one year to the next, and that there is always a surplus of food for the bass. Given these assumptions, we can assume that the only variables determining the rate at which biomass of bass is produced in the lake is the set of variables describing the harvesting strategy of man.

The line of reasoning used in developing an optimal harvesting strategy for the bass is based on consideration of the weight growth and survival of a typical year class of bass. From data in Watt (1959) based on analysis of a large volume of data for many year classes in a population of smallmouth bass, we can construct the history of a typical year class of smallmouth bass for a particular lake, as in Table 7-1. This table indicates the principal considerations involved in obtaining the maximum

TABLE 7-1 The weight and survival history of a typical year class of smallmouth black bass in South Bay, Manitoulin Island

age, years	mean weight, ounces	number still alive, thousands	biomass still alive, thousands of ounces
1	0.7	190	133
2	2.4	80	192
3	8.5	33	280
4	11.2	16	179
5	14.4	7.23	104
6	21.2	2.28	48
7	26.5	1.00	26.5
8	35	0.42	15
9	46	0.17	7.8
10	62	0.07	4

SOURCE: Most of these data are based on tables in Watt (1959), but some entries were obtained by extrapolation.

possible yield from a typical year class of fish. If, on the one hand, we harvest a year class when it is too young, the fish have not yet grown enough for the great number of fish still surviving to offset the very small average weight of the fish. Further, of course, if we harvest the year class at too early an age, the individual fish will be so small that the edible portion per fish will scarcely justify the effort in handling a fish. If, on the other hand, we harvest a year class when it is too old, the great weight of the average fish does not offset the sharp reduction, by constant erosion to natural mortality, of the numbers still alive in the year class. Thus we see that selection of an optimal harvesting strategy involves striking a balance between two extremes: not taking sufficient advantage of the growth potential of young fish, and fishing the year class so late in its life that there is too much wastage of biomass to natural mortality. Once again, this is an illustration of Principle 4: in this case fishing strategy should leave an optimal population density of fish in the water.

The optimal strategy therefore makes use of two variables: the age at which the typical year class is first harvested and the intensity of exploitation (the proportion of the stock harvested) at that age and all subsequent ages. The age at which the year class is first harvested is set high enough so that the growth potential of the young fish is adequately realized. The intensity of exploitation is set high enough to avoid excessive loss to natural mortality, without at the same time so depleting the stock that there are insufficient sexually mature adults to replace the fish lost to natural mortality and harvesting.

A highly sophisticated theory of harvesting fish populations has been developed, based on the idea of selecting the values of t'_p, the age at which a year class is first subjected to harvesting, and F, the fishing mortality coefficient, which maximize the yield from a year class. This theory is used to construct charts called yield isopleth diagrams, which show the yield from a typical year class, in weight, that will be obtained for different combinations of values of these two variables. Because fisheries management lends itself so readily to quantification (all the relevant variables can be obtained by weighing or counting fish, or determining their ages), the theory of fisheries management early became one of the most mathematical subdisciplines in all of modern biology. The best introduction to this subject for the interested student is the classic work in this area by Beverton and Holt (1957).

Is this the best possible way to deal with the problem of harvesting the food resources in a lake? Two lines of reasoning suggest that we may need a considerably more complicated approach if we are to make most effective use of the available food resources in a lake. The first is based on intensive research by Juday (1940) on Lake Mendota, Wisconsin. He measured the productivity at different levels in the trophic pyramid and obtained the following values for the lake:

	corrected productivity, gram-cal/sq cm/year
Solar radiation	119,000
Photosynthetic plants	480
Herbivores	42
First-stage carnivores	2.3
Top carnivores	0.3

Now with this tremendous reduction in productivity from plants to herbivores to first-stage carnivores to top carnivores, is harvesting directed at top carnivores an efficient way to use the lake? The answer, clearly, is no. If the pressure on the world's food resources is so light that we can afford to use lakes largely for entertainment, then it makes sense to harvest a small number of large fish, which are more fun for the angler to catch than a large number of small fish. However, if human population pressure on resources becomes considerably more intense than at present, then we will have to do without the luxury of using lakes for entertainment rather than food. From the perspective resulting from an overview of the entire ecosystem, the optimal strategy for harvesting a lake is to direct harvesting pressure at those trophic levels as close to the sun as possible. This would make use of Principle 2, the notion that there is inefficient conversion of energy between steps in a trophic pyramid.

However, this is still an oversimplified way of looking at the problem of harvesting a lake. Up to this point, we have not considered the implications of competition between species within each trophic level. How does this alter our plans for developing an optimal harvesting strategy? Is it better to have one species at each trophic level or several? If we have several species competing at each trophic level, should we direct the harvesting pressure at one, or a few species, or at all of them equally? This type of question has been examined by means of centuries of trial and error with different pond stocking strategies in the Orient. It has been demonstrated that by far the best means of maximizing the production of fish flesh from a body of water is to exploit a large number of competing species, rather than one or a small number of species. Carlander (1955) has collated a large body of data from Midwestern reservoirs bearing on this question and has arrived at the same conclusion (Fig.

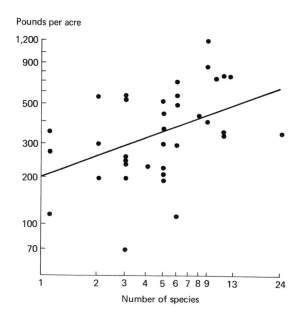

Pounds per acre

FIGURE 7-2 Relation between standing crops of fishes and numbers of species in Midwestern reservoirs. (*Carlander*, 1955.)

7-2). This illustrates Principle 9, the increase in the biomass/productivity ratio with increasing species diversity.

Extremely high yields of fish per acre are obtained in the Orient by using mixtures of up to nine species of carp, rather than by attempting to obtain high yields per acre with piscivorous fish, the usual policy in the United States.

The production of fish possible with intensive aquiculture is amazing. In open lakes, such as arms of Lake Huron, 4 kilograms per hectare per year is quite reasonable (Fig. 7-4). Lake Erie produces about 7 kilograms per hectare per year. By contrast, in Alabama unfertilized fish ponds produce 17 kilograms and fertilized fish ponds produce 169 kilograms per hectare. This is in amazing contrast to the following production figures from intensive aquaculture in the Orient[1]:

West Java carp	500 metric tons (500,000 kilograms) per hectare (raised in several meter cages; prorated on a per hectare basis)
Indonesian and Philippine milkfish	400 kilograms per hectare
Taiwanese milkfish	2,000 kilograms per hectare
Japanese yellowtails in intensive cultivation in a small portion of the ocean	280 metric tons per hectare
Japanese oyster culture (weight of meat only)	58 metric tons per hectare

The higher productivities of carp and other bottom feeders relative to those of piscivorous fish are reflected in the standing crop figures in Fig. 7-3.

THE RELATION BETWEEN PHYSICAL AND CHEMICAL CHARACTERISTICS OF LAKES AND THEIR BIOLOGICAL PRODUCTIVITY

In order to make optimum use of a fish stock, it is necessary to have realistic regulations governing the intensity of fishing by sports and commercial fisheries. On the one hand, if we have regulations that lead to underharvesting the resource, too much fish tissue is wasted to natural mortality and competition. On the other hand, if we have regulations that result in overharvesting, the fish resource will be depleted. Consequently, there is great interest in determining what level of fish production it is reasonable to expect from a lake, so that realistic fisheries regulations can be devised on this basis.

To this end, there has been considerable effort to account for differences in the fish production of different lakes in terms of physical and chemical characteristics of the lakes. Ryder (1965) reasoned that a useful predictor of fish production in a lake should be the total dissolved

[1] Data from J. E. Bardach, Aquiculture, *Science*, **161**:1098–1106, copyright 1968 by the American Association for the Advancement of Science.

Standing crop, pounds per acre

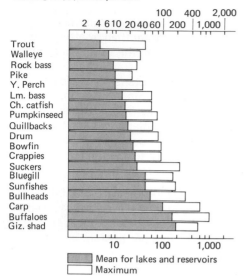

FIGURE 7-3 Standing crops of named fishes in North American lakes and reservoirs. Usually these fish were in combination with other species. The graph furnishes a rough approximation of the relative efficiency of the species listed. Note the high standing crops of the bottom feeders (suckers, carp, buffaloes) and the much lower standing crops of the piscivorous fish (pike, walleye). (*Carlander*, 1955.)

solids in parts per million as measured at the water surface T divided by the mean depth of the lake in feet D. Expecting that fish production would vary directly with T, but inversely with D, he used T/D as an index of expected fish production. When he plotted fish production in pounds per acre per year against T/D, using the data in Table 7-2, he obtained the relation in Fig. 7-4. This is another illustration of Principle 3, that the rate of energy flow through biological systems depends on resource availability, matter being the resource in the case of Ryder's T, and D being inversely proportional to energy availability per unit area of water.

Recently, there has been an elaborate effort to explain the differences between the productivities of different lakes in northwestern Ontario. Schindler (1971) has summarized the results and has at-

Fish production, pounds per acre per year

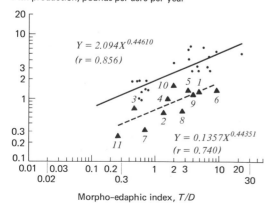

$$Y = 2.094X^{0.44610}$$
$$(r = 0.856)$$

$$Y = 0.1357X^{0.44351}$$
$$(r = 0.740)$$

Morpho-edaphic index, T/D

FIGURE 7-4 Plots of fish production against the index T/D (water surface divided by mean lake depth) for 34 lakes. The solid line is the best least squares fit for 23 moderately to intensively fished lakes. The broken line is for 11 lakes with restricted fisheries or incomplete records. (*Ryder*, 1965.)

tempted to explain them on the assumption that a good predictor of lake productivity would be the catchment area of the watershed for a lake, divided by the volume of the lake. The catchment area, he assumed, was $A_d + A_o$, where A_d was the terrestrial portion of the drainage basin and A_o was the surface area of the lake. Thus, the index of expected biological productivity was $(A_d + A_o)/V$. This is based on another variant of Principle 3: if there is a constant rate of nutrient acquisition per unit area of the total catchment area, then the critical resource is space: the greater the catchment area, the greater the nutrient availability for the lake. This turned out to be a good assumption, as can be seen in Fig. 7-5.

TABLE 7-2 Data on area, depth, dissolved solids, and fish production for 34 lakes.

lake	area, sq miles	mean depth, D (ft)	total dissolved solids, T (ppm)	morphoedaphic index, T/D	fish production, P (lb/acre/year)
Lakes moderately to intensely fished					
Superior	31,820	487	60	0.12	0.83
Huron	23,010	195	117	0.60	1.38
Michigan	22,400	276	118	0.43	1.59
Great Slave	10,500	204	150	0.72	1.44
Erie	9,930	58	196	3.38	7.30
Ontario	7,520	264	155	0.59	0.99
Athabaska	3,050	85	58	0.68	1.50
Reindeer	2,150	56	39	0.70	2.00
Nipigon	1,875	180	99	0.55	1.07
Lake of the Woods	1,378	26	83	3.19	4.60
Wollaston	796	68	35	0.51	2.00
St. Clair	460	10	208	20.80	5.72
La Ronge	455	42	149	3.55	3.00
Cree	446	49	27	0.55	2.00
Rainy	345	25	47	1.88	3.96
Big Peter Pond	213	45	138	3.07	6.60
Ile a la Crosse	172	27	185	6.85	3.00
Churchill	167	30	136	4.53	3.00
Little Peter Pond	73	17	144	8.47	6.00
Waskesiu	27	36	188	5.22	7.00
Mountain	20.8	38	229	6.03	3.50
Otter	20.6	34	146	4.29	3.50
Heming	1	9.5	61	6.42	5.00
Lakes with restricted fisheries or incomplete records					
Great Bear	11,800	400	99	0.25	0.27
Lac Seul	495	25	76	3.04	1.42
Big Trout	238	52	68	1.31	0.65
St. Joseph	199	24	61	2.54	0.72
North Caribou	131	31	56	1.81	1.74
Wunnummin	114	20	74	3.70	1.29
Attawapiskat	101	19	83	4.37	1.31
Deer	50	71	33	0.46	0.77
Onaman	45	9	77	8.55	1.53
Kakagi	38	57	87	1.53	1.08
Opeongo	23	49	33	0.67	0.34

SOURCE: Ryder (1965).

Midsummer phytoplankton biomass,
milligrams per cubic meter

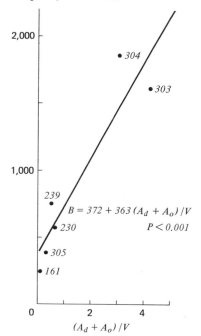

FIGURE 7-5 Plot of phytoplankton biomass against $(A_d + A_o)/V$. (*Schindler*, 1971.)

CONCLUSION

The nature of man's relations with lakes will depend on human population densities. If human population densities are low, then we can afford to have deep, oligotrophic lakes which provide us with great aesthetic pleasure. As the world population of humans builds up, however, it is inevitable that the chemical wastes and sewage effluent from civilization will increase in volume, speeding up the process of eutrophication, or conversion of oligotrophic lakes to eutrophic lakes, and finally to swamps.

Also, if human population densities are low, we can afford to use for food only those organisms living in lakes which we think of as having high-quality flesh. In the Orient, where human population densities have long been much higher than those now found in North America, it has been necessary for a long time to develop a taste for those organisms which could provide a very high productivity per unit area of water per year.

Thus, for lakes as for all other ecosystems which can be exploited by man, the quality of life is inevitably related to human population densities. If the density of people continues to rise, then finally we will hit a density at which the quality of life available to the average person must decline. This is an inexorable relation: there is no way to escape from it. Either we can drift by indecision, inaction, or irrational, unthinking behavior into a trap from which escape will be very difficult, or we can make the conscious decision not to get trapped and take the necessary actions to ensure this outcome.

SUGGESTIONS FOR INDIVIDUAL AND GROUP PROJECTS

Would you expect the fish production per acre in lakes in your area to fall near the low or high end of the production range in Table 7-2 and Fig. 7-4? Why? Explain in terms of the physical characteristics of the lakes and the diversity of the fish species in the lake.

REFERENCES

Bardach, J. E.: Aquiculture, *Science*, **161**:1098–1106 (1968).

Beverton, R. J. H., and S. J. Holt: "On the Dynamics of Exploited Fish Populations," Fishery Investigations Series II, vol. XIX, Ministry of Agriculture, Fisheries and Food, Her Majesty's Stationery Office, London, 1957.

Carlander, K. D.: The Standing Crop of Fish in Lakes, *J. Fisheries Res. Board Can.*, **12**:543–570 (1955).

Forbes, S. A.: The Lake as a Microcosm, *Bull. Sc. A. Peoria*, 77–87 (1887).

Goldman, C. R.: Absolute Activity of C^{14} for Eliminating Serious Errors in the Measurement of Primary Productivity, *J. Conseil Intern. Exploration Mer.*, **32**:172–179 (1968a).

———: Aquatic Primary Production, *Am. Zoologist*, **8**:31–42 (1968b)

Juday, C.: The Annual Energy Budget of an Inland Lake, *Ecology*, **21**:438–450 (1940).

Ryder, R. A.: A Method for Estimating the Potential Fish Production of North-temperate Lakes, *Am. Fisheries Soc.*, **94**:214–218 (1965).

Schindler, D. W.: A Hypothesis to Explain Differences and Similarities among Lakes in the Experimental Lakes Area, Northwestern Ontario, *J. Fisheries Res. Board Can.*, **28**:295–301 (1971).

Watt, K. E. F.: Studies in Population Productivity. II. Factors Governing Productivity in a Population of Smallmouth Bass, *Ecol. Monographs*, **29**:367–392 (1959).

8 the forest: a study in the determinants of ecosystem structure and dynamics

Man is by far the most important determinant of forest ecosystem structure and dynamics at present.

A comprehensive treatise on supply and demand trends for resources in the future of the United States brings out the reason for concern about the future of forests. In this country, as in all others, when total demand for land in all categories exceeds supply, forest land will be sacrificed so that other needs can be met. That is, as urban land needs grow, they are met by robbing from agricultural land, which then in turn robs forest land so that needs for food can be met. The treatise (Landsberg et al., 1963) shows, first, by comparing projected annual forest growth rates with projected annual demand for softwood to the year 2000, that under medium projections the growing stock in the eastern United States is wiped out by that year. Under medium projections, growing stock of softwood in the West is reduced to about half or less of its 1960 volume by 2000. Second, these authors show that by 2000 it would be tremendously difficult to remedy this situation, i.e., to increase the land under forests, by transferring land back to forests from other land use categories. The reason for this is that by 2000, the total demand for land in the United States in all land use categories will be 1.954 billion acres, 50 million more acres than there are in the United States. This assessment is based on medium projections of growth in demand, and it excludes Alaska and Hawaii. Thus, truly heroic efforts will be required to ensure that the United States keeps its forests, and—this is important—some other goals will have to be sacrificed if we are to keep them. It is absolutely not possible for us to have all the land we might want in all land use categories from now on, indefinitely. In countries which have had high human population densities for long periods of time the long-term fate of forests is evident. They have sharply diminished or vanished unless society has made an effort to save them at the expense of other goals.

Consequently, we must face up to the question whether forests are really worth anything to us. Do they have any aesthetic or ethical value? Are there any *hard reasons* for not cutting down large tracts of forest and using the cleared land for growing food?

There are two broad categories of questions that need to be examined in dealing with this issue. The first set has to do with how forests affect society directly, and the second has to do with the indirect effects, through the effect of forests on other resources.

Within recent years, economists have become much interested in the problem of the strategy of economic development. That is, if the economy of a developed country stagnates, or if the economy of an undeveloped country proves refractory to efforts to develop it, what is the best strategy for spending capital to get the economy moving? Chenery and Watanabe (1958) have developed an approach to the problem of determining which sector of the economy is most central and stimulating to other sectors. They utilize the theory of forward and backward linkages. That is, the degree of interdependence of various industries is assessed by determining for each industry the number of backward linkages it has (the number of other sectors of the economy from which it purchases its inputs) and its number of forward linkages (the number of other sectors of the economy to which it sells its outputs). This method of assessing the potential stimulating effect of a boom in one sector on all other sectors is only a very rough index, and needs much development (Hirschmann, 1958), but it does have some value. However, given this warning, it is interesting to note that for the average degree of interdependence between sectors in the economies of Italy, Japan, and the United States, paper and paper products have the third highest number of linkages. Forests appear in two other places in the table of numbers of linkages for all industries, and clearly they are quite central in economic development. Forestry sells to a great many other industries, the lumber and wood products industry buys from 61 other industries and sells to 38, and the paper and paper products industry buys from 57 other industries and sells to 78. Clark (1967) shows that each person in the world requires about 250 square meters continually under forest to supply his average annual wood and pulp requirements.

Now if it is true that forests have a stimulating effect on countries' economies, then it ought to be possible to demonstrate this, both by use of qualitative evidence from history and by statistical comparison of the behavior of similar areas with and without well-developed forest industries.

First, it appears that the power, influence, and cultural vitality of certain societies at various points in their evolution have been strongly influenced by the availability of forests. Athens and Sparta were strong city-states before the Christian era, but most Greek forests were cut by the fourth century B.C., which coincided with decline in influence of the Greek city-states. It appears that the great declines of many other cultures followed the demise of their forests. The dramatic decline in the influence of Spain as a world power in the seventeenth century, after its great successes of the three preceding centuries, was perhaps related to the depletion of its forests, which had provided the timber for shipbuilding as well as construction. The United States cut its forests fast, and the sites of maximum cut quickly moved south and west from New England, which had been the site of maximum cut up to the 1870s; New England supplied 54.5 percent of all United States lumber in 1850 (Ely and Wehrwein, 1940). What allowed the United States to continue moving ahead rapidly as a nation was the convenient discovery of coal, which surpassed wood in importance as a fuel source in 1886, and crude oil, which passed coal in turn by 1952 (Landsberg et al., 1963). Could the industrial might of the United States have grown as rapidly as it did without the forests? Probably not, and in any case the development should have come decades later than it did. However, other countries were often not so lucky when they ran out of trees, and it appears that the history of civilizations has been strongly affected by forest management practices, which in turn had an affect on soil stabilization and water runoff management (Lowdermilk, 1962).

A great deal of attention has been given to the possible effects of forests and forest industries on the economic fortunes of underdeveloped countries (Westoby, 1963; Leslie, 1968). Westoby points out that if an underdeveloped country does not develop its forest industries, its international balance of trade will be adversely affected. This is because the underdeveloped country will be exporting sawlogs, which are cheap, and then buying back saw wood, veneers, fiberboard, pulp, newsprint, and other paper and board, all of which are more expensive than the raw materials which were exported. Consequently, in the period 1957–1959 the less-developed countries had a net export balance of 112 million dollars from sale of sawlogs, but they bought back so much in wood products that they had a net trade deficit of 467 million dollars for all wood products. Newsprint and other paper and board alone left the less-developed countries with a trade deficit of 394 million dollars on the average for those 3 years.

Leslie (1968) also considered the problem of the influence of forests and forestry industries on economic development rates in different countries, but he concluded that such a large number of variables were operating simultaneously that it was difficult to be sure that any observed correlation between size of forest industries and rate of increase in gross national product per capita might have many causes. Consequently, he used a comparative experimental approach to measurement of the economic effect of forests. He compared in detail the histories of two similar regions in southern Australia.

In one region, of which Warrnambool was the principal city, there was almost no forest industry. In the other region, centered on Mount Gambier, there was a very large forest industry. What makes this comparison so fascinating is that both regions were completely lacking in native forest resources of commercial potential at the time of European settlement. The entire forest industry at Mount Gambier is based on a manmade, planted forest resource. Thus, we have a genuine, large-scale social and economic experiment in which, of two entirely comparable regions adjacent to each other, one planted forests and one did not.

Leslie followed the history of these two regions from 1954 to 1966. In Warrnambool, the percentage of the regional population employed in forest industries was always small: about 0.2 percent. In the Mount Gambier region, on the other hand, 4.8 percent of the regional population was employed in forest industries in 1954, and 6.1 percent in 1966. Two measures of the economic consequences of this difference are the population size and the degree of urbanization of the two regions. From 1934 to 1966, the Mount Gambier region showed a much faster rate of increase in both total population and degree of urbanization than the Warrnambool region.

This fascinating comparison is worthy of, and indeed requires, the most careful possible interpretation. First, the lesson to be drawn is not that planting forests is a good thing because it allows a high rate of population increase. Indeed, if the population of the afforested region continues to increase, then it will gradually outstrip the forest resource on which it depends, with the same results encountered historically all over the world when a country developed too large a human population for the carrying capacity set by the resource base: starvation and a very low gross national product per capita, as in India, Egypt, and Haiti. Rather, the point is that in an area with a poor resource base, planting forests and husbanding them carefully on a sustained-yield basis while also developing a forest products industry can make a large contribution to the economic health of a region or a nation. The experiment would go a step further if the population unwittingly killed the goose that laid the golden egg, by cutting the forests too fast, so that cut exceeded growth year after year. Then we would have a model system showing us the reason for the downfall of various Mediterranean basin, North African, Middle Eastern, and northern Indian cultures.

In short, the effects of forests on civilization illustrate Principle 3, for forests supply both matter and energy for man.

Forests have a variety of effects on other resources. These effects occur by means of three related groups of causal pathways: those in which forests affect climate, soil, and the average amount and variability of water available to other resources, such as agriculture.

An interesting question about the relation between forests and climate concerns ultimate causation. Why is there so little tree growth and rainfall in certain countries? Is the small amount of tree growth the reason for the small amount of rainfall, or is the small amount of rainfall the reason for the small amount of tree growth? A few authors have considered this question (Nicholson, 1930; Paterson, 1956), and it appears from comparison of cut and uncut areas that forests have an effect on rainfall. It rains less than expected where forests have been cut.

Chang (1968) has summarized information on the effect of forests and hedges on microclimate (Fig. 8-1). It is clear that trees cut down wind velocity and consequently decrease the evaporation from plants on the lee side of shelterbelts of trees. As we would expect, this means that plant growth of agricultural crops is higher adjacent to shelterbelts. A number of figures on the exact benefits of trees to agricultural production have been published, but these vary widely, depending on the crop, the trees, and other aspects of the situation. As a crude overall average, it appears that the net gain to agriculture in having shelterbelts of trees in areas of strong wind is about a 15 percent increase in yield, even after allowing for the fact that the trees themselves occupy acreage that would otherwise be used for more crops.

Also, trees have a moderating effect on climate. Elimination of forests produces greater-amplitude climatic variations from hot to cold and wet to dry, thereby making an area less habitable for future vegetation.

It has been known for a long time that trees have a great influence on soil structure and erosion, and hence on the amount and variability through time of the water available from a watershed. Lowdermilk (1930) pointed out that forest litter prevents raindrops from hitting soil with too much force. Without litter soil is compacted by rain and its absorbing capacity is reduced. If the forests are eliminated, dashing rains churn

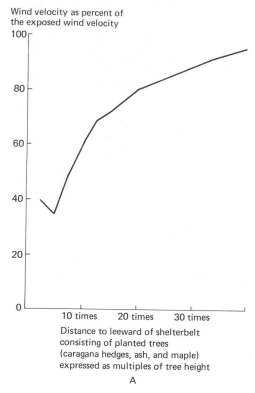

Distance to leeward of shelterbelt
consisting of planted trees
(caragana hedges, ash, and maple)
expressed as multiples of tree height

A

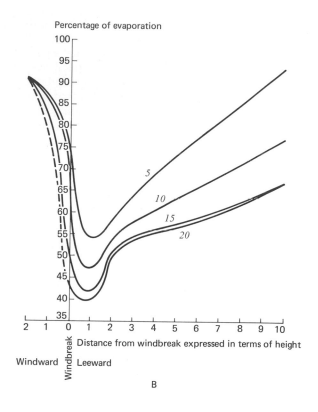

B

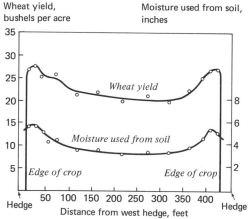

C

FIGURE 8-1 Examples of the evidence that vegetation has a strong effect on microclimate by moderating the influence of wind, evaporation, and
moisture. **A** The influence of planted tree shelterbelts on wind velocity in the Canadian prairie prov-
inces. (*Staple and Lehane*, 1955). **B** The effect of
cottonwood groves on evaporimeter readings leeward of the windbreak. (*Bates*, 1911.) **C** The effect of proximity to hedges on the moisture used
from the soil by wheat. This statistic obtained
from the difference between the soil moisture
content at spring and at harvest. (*Staple and Lehane*,
1955.)

bared soil into muddy suspensions. Percolation of water into the earth
filters out suspended soil particles and partially closes up the pore
spaces in the soil, and the soil becomes sealed so that it absorbs less
water than normally. This in turn means that when it rains, a higher than
normal proportion of the rain runs off the surface of the watershed

quickly, rather than being absorbed by the watershed. This is the type of situation that makes for flashflooding in desert areas: the soil has too little absorptive capacity, and almost all the water falling down runs off quickly, rather than being retained. We are describing a complex process, for the runoff leads to flooding of sufficient force to wash away the humus and topsoil, the very soil components that ensure high biological productivity.

In short, forests are important to agriculture in maintaining soil in a productive state, in evening out the flow of rainfall waters to agricultural areas, in ameliorating extreme wind, evaporation, temperature, and humidity conditions, in minimizing erosion and dust formation, and in promoting rainfall. It now appears that a principal cause of the drought-flood cycle of northern India is man-made: the historical elimination of vegetation by man (Bryson and Wendland, 1970). This elimination creates meteorological conditions which increase the dust concentration over the desert, and also promote aridity. Thus, a vicious circle develops: the more vegetation is removed, the more this process promotes meteorological conditions which make it difficult for most species of remaining vegetation to survive.

Clearly, vegetation and local climate constitute a circular feedback system: vegetation affects the climate, which in turn affects the vegetation. Once vegetation has been removed, regional climatic conditions are created in which it may be expensive to reestablish quickly the natural vegetation, or any other vegetation, unless there is the type of capital investment associated with intensive irrigation agriculture. This is an interesting variant of Principle 13. In this case, cutting down plant species diversity affects not only the stability of the community but also the stability of the physical environment.

THE EFFECT OF THE ENVIRONMENT OF FORESTS

It has been known for a long time that the climatic regime of a region is related to the type of vegetation in that region. Consequently an elaborate body of theory has been developed, designed to account for distribution patterns of different types of plants and forests on the basis of physical climatic factors. The result is a separate and very important branch of ecology, sometimes referred to as life zone ecology. This subject concerns the application of Principle 15, pertaining to zones of tolerance. This subject has had a long history, in which some of the important figures are Réaumur (1735), Merriam (1889), Thornthwaite (1948), Paterson (1956), and Holdridge (1967).

The significance of this topic is that it can serve as an important tool in national land use planning. If an objective basis can be developed for determining what kinds of plant associations can be grown most profitably on each site, it will ensure long-term most productive use of each site. Otherwise, people are trapped into trying to make a living out of agriculture practiced on soil and in a weather regime that is best-suited to a particular type of forest. The only possible result is personal tragedy.

Probably the most elaborate exposition of the relation of life zone ecology to national land use planning is by Holdridge (1967). He points out that two strategies are available to a country. The better strategy is to determine the human carrying capacity for each type of soil and climate and then adjust the number of humans accordingly, for a desired standard of living. This strategy leads to a high standard of living for the human population, because everyone has a pattern of living that makes best use of available resources, and no one is forced into a marginal existence by trying perennially to obtain a decent living through growing agricultural plants on soil not suitable for agriculture. The alternative strategy is to let the human population grow out of control and then use even the poor land for agriculture. This commits a segment of the population to a lifetime of hard labor in exchange for a scarcely marginal existence. Clearly, the former of the two strategies is rational and makes a human use of human beings, whereas the latter is not rational, but cruel. Misuse of land is based on a refusal to accept the validity of saturation-depletion effects as discussed in connection with Principle 4.

An immense amount of effort has been expended on development of the best possible technique for classifying each site so that the types of plants selected to be grown there make best use of the environmental resources. Each of the techniques is checked by making distribution maps of where a certain forest type of plant species ought to grow on the basis of some index, and then comparing this with the actual distribution of the plants. Discrepancies between predicted and actual distribution then become the basis for refining the technique of computing the index.

A number of recent tests by geographers suggest that of various methods devised to account for the distribution of different plant communities, the best available is by Holdridge (1959b, 1962, 1967). For example, his method of discriminating different types of life zones could be used to predict approximate rank orderings of different stream flows in the Colorado mountains (Thompson, 1966); it proved better than the Thornthwaite system in accounting for distribution of vegetation in the Mediterranean basin (Steila, 1966). Accordingly, Holdridge's method of defining life zones will be explained here. His system is based on the following assumptions:

1   The particular association of plant species found at any point on the earth's surface is fundamentally determined by only three factors: heat, precipitation, and moisture, the last being dependent on the other two. Groups of plant species associations can be defined on the basis of the three named climatic variables, and these groups are called life zones. These zones have a double meaning: they refer to a certain type of plants and also to the ranges of values of temperature and precipitation which produce them. Thus, one should be able to calculate, from weather records, maps of where different types of plant associations ought to occur, and find that they do occur there, and also determine the range of weather conditions operating in a certain site on the basis of field inspection of the plant associations there.

2   The equivalence among three critical climatic variables and certain types of plant associations is made possible by expressing heat, precipitation, and moisture in units that have the maximum possible biological relevance. Thus, for heat, the index used is the *mean annual biotemperature*. Holdridge defines biotemperatures as those falling in the range in

which vegetative growth occurs. It is generally believed that vegetation grows only in the range from 0 to 30°C. Holdridge obtains the mean annual biotemperature by adding up all the mean daily biotemperatures and dividing the sum by 365. Thus, days with a mean daily temperature less than 0° or greater than 30° are excluded from the sum. The precise biotemperature for the day is obtained by adding up all the hourly temperatures and dividing by 24, again excluding from the sum temperatures outside the range 0 to 30. Where only a daily maximum and a daily minimum are available, Holdridge recommends adding both figures and dividing by 2, again excluding from the sum negative temperatures or those above 30°. Clearly, the method is rather crude unless hourly temperatures are available.

The precipitation is expressed as annual amount of precipitation in millimeters. These two numbers, precipitation and biotemperature, are then sufficient to determine the position of a group of plant associations invented by Holdridge to define the relation between types of plant formations and their climatic determinants (see End paper).

3   The influence of temperature, precipitation, or potential evapotranspiration on plants is related to the logarithm of the measured value of any of those variables. It will be noted in End paper that a boundary of any hexagon corresponds to a value of one of the three climatic variables that is half or double the value of that variable for the opposite boundary.

4   Latitudinal regions have their equivalents in altitudinal belts. That is, if one climbs high enough up a mountainside in the tropics, he will finally encounter a band of vegetation that he would have expected to find under comparable precipitation conditions closer to the poles. This is the point of Plate 1, which shows clear altitudinal belts of plant formations up the side of Mauna Kea; all high tropical mountains show this same pattern.

5   The type of plants growing in any particular site is not uniquely determined by the life zone. In fact, Holdridge proposes a three-level hierarchical system for classifying plant environments. The highest, or grossest level is the life zones. Within these units is a finer set of subdivisions, which Holdridge calls *associations*. These associations are determined by modification of the effects of heat, precipitation, and moisture owing to additional azonal modifications in climate caused by strong winds, abundant mists, or pronounced variation in the seasonal precipitation pattern (Holdridge, 1966). The type of associations just mentioned he calls *atmospheric associations*. He also recognizes *edaphic associations*, which are modifications of those normally expected in a given climate, owing to topography, drainage, parent material of the soil, or soil age. Also, he recognizes *hydric associations*, in which the soil is covered with water for all or most of the year. Finally, he recognizes that within a life zone, and within a particular association within that life zone, there may be *third-order divisions*. These are further modifications produced by such factors as land use practices.

This system implies that one ought to be able to go anywhere in the world and, having sufficient information about the weather, the atmosphere, the soil, and topographic factors, predict what type of plant community will be growing naturally on a particular site and, especially, what type of plants man should be attempting to grow there.

Human history can be interpreted in a most interesting way using End paper. Holdridge (1959a) has pointed out that most civilizations began in the steppe-prairie life zone characterized by 250 to 500 millimeters of rainfall a year and a mean annual biotemperature of 6 to 12°C. Very early it was found possible to extend civilization into semiarid and arid regions by means of irrigation. This pattern of civilization, starting in grasslands, characterizes central Iraq, the Mayan culture in the savannas of Mexico and Guatemala, the Chibchas in the savannas of Bogotá, Colombia, and the Incas of Peru. Holdridge notes that the first of the American civilizations to succumb, even before the arrival of the Spaniards, was that of the Mayas, who extended from the grasslands into the forests. Earlier in this chapter the problems involved with use of forest soils for agriculture were noted.

In areas of the world where the climate is very similar over immense tracts, the applicability of the Holdridge system is not so apparent. However, anyone who has traveled in mountainous areas of the tropics and has been amazed at the tremendous differences in vegetation up and down mountainsides, or on the windward (wet) and lee (dry) sides of mountains, will readily perceive the correlations between climate and plant associations. Places where this correlation is particularly startling are the Mexico City area, Guatemala, the tropical Andes of South America, the Hawaiian Islands, Lebanon, and northern India. In each case, there are very steep altitudinal gradients in a tropical or subtropical latitude, with the result that a startling variety of plant associations are growing within a small distance of each other. In the vicinity of Mexico City, for example, the hot, wet climate around Fortín produces rain forest, the very high-altitude mountains around Mexico City are covered with pine forests, the moderate-altitude plateau is grassland, and the hot, low dry areas between Cuernavaca and Taxco are desert scrub or desert.

As Holdridge points out, constantly increasing human population densities increase the need to use all land so that its productivity is as high as possible. This is done by using each particular land area to grow those products that have the highest productivity. This might appear to be such transparently obvious common sense that it seems unthinkable that anyone should ever do anything else. Unfortunately, this is not the case. As Holdridge (1967) points out, in some areas of the world grazing is conducted where cultivated crops would be more productive, and there are other areas where both cultivation and grazing are practiced on land which would be more productive if used for forestry. In an ideally rational world, the resource management departments of national and regional governments would conduct their land use planning on the basis of maps employing the Holdridge system of life zones.

It has already been mentioned that the effect of the basic climatic factors can be modified to produce a variety of associations within each life zone. Figure 8-2 gives two interesting examples of this. A is a photograph taken at Kalalau Lookout on the island of Kauai, in the Hawaiian

A

B

FIGURE 8-2 Two illustrations of the effect of factors other than the three basic climatic variables in producing associations, or subdivisions of the life zones of Holdridge. **A** Profuse growth of epiphytes at Kalalau Lookout, Kauai, Hawaii, illustrating the effect of sea mists in increasing the moisture content of the air. **B** A cemetery near Manhattan, Kansas, illustrating how a natural area, the small tamarack forest in the cemetery, reveals the basic vegetation pattern of the life zone, which is, however, disturbed by man-managed grazing, or fire in the absence of man, so that the area is in grassland.

group. This place is at about 4,000 feet altitude, at a spot where mists often blow in from the ocean. A characteristic of such misty areas, often laden with moisture, is the great growth of epiphytes, or plants that grow high up on other plants on which they depend for their support. This illustration depicts the incredible tangle of vegetation which can be found in such circumstances. In general, epiphytes are characteristic of places where there is a very high humidity and little fluctuation in the humidity. Mists can produce an association where there is more evidence of moisture availability than one would expect simply from inspection of precipitation records. The presence of epiphytes where they would not be expected from precipitation records is one evidence for the frequent presence of such mists.

Figure 8-2B is a photograph of a cemetery near Manhattan, Kansas. The cemetery is the one place in the entire region which indicates what the natural vegetation would be in this life zone but for intensive management by man. The fact that the cemetery is a small tamarack forest, whereas the surrounding area is all grassland used for seasonal grazing by cattle, indicates that the climate of eastern Kansas is more conducive to tree growth than we would guess from the small number of trees in the area. Thus, the grassland is not natural this far east in Kansas. But for man, and fire, which would periodically operate in the absence of man, this area would be clearly in the life zone defined by considerable tree growth. In short, the activities of man and fire make this area appear to be in a drier life zone than that in which it actually occurs.

Conditions like this illustrate the role of factors other than the three basic climatic factors in producing associations within each life zone. Because of the large number of such factors and combinations of them, there can be many different associations within each life zone. However, it is possible to account for these and thus maintain the realism of the Holdridge system when mapping vegetation locations and distribution patterns.

FOREST SUCCESSION To this point we have considered for each life zone only those types of vegetation characteristic of that zone at the climax of succession. All such zones have a history, however, and a characteristic of this history is the waves of successive associations of plant species, each of which replaces another, as the successive waves modify the environment and make it more suitable for later waves of invading species.

One of the best known, most studied, and most widely distributed successional systems in North America is that depicted in Fig. 8-3. The dynamics of this system were first worked out by Cooper (1913) for Isle Royale in Lake Superior, but this same type of forest is found over much of Canada, the Lake states, and the northeastern United States. This figure is a simplified and modified diagrammatic summary of the system as studied by Cooper, but it illustrates the main features of that and other forest successional systems.

1 There is a general tendency in forest succession for small plants to be replaced by large plants, in a succession that goes from lichens, herbs, or aquatic plants to shrubs, then small trees, and finally large, tall trees.

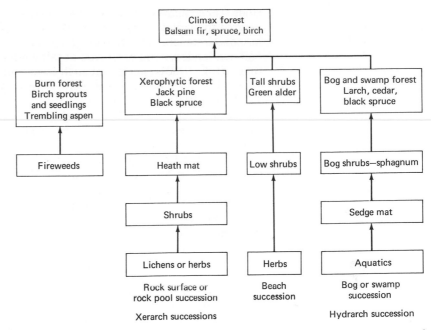

FIGURE 8-3 The various successional processes leading to the fir-spruce-birch climax forest type of Canada, the Lake states, and the Northeastern states. (*Modified and simplified diagrammatic summary of research by Cooper,* 1913.)

2 Several different pathways can end with the same climax forest. Thus, we may know what the end member of a series is, but we do not know for sure the process by which it has developed.

3 The same climax can be produced by progressions originating in quite different habitats (the xerarch, or dry successions, and the hydrarch, or aquatic successions). This is, incidentally, the most compelling argument in support of the Holdridge system. It indicates that it does not matter how a climax forest originates; for any origin, a particular type of climate, cold and wet in this case, produces a particular type of climax, or life zone.

4 At each stage of succession, for any association of species, there are recognizable dominant species that identify the association.

A number of points about the climax forest should be made. First, not all climax forests of the same type need always be in the same condition. Such a similar condition would be true only if the forest was a mixture of trees of different ages, so that new trees were maturing as old ones were dying and falling over. In many forests, such as the climax type depicted in Fig. 8-3, there is a strong tendency for all the trees to be the same age. Such uniform-age stands existed even before the advent of man, because after a forest had been intact for 35 or more years, it became overmature, and particularly because of flower production it became an optimal habitat for important insect defoliators, such as the spruce budworm, which can build up in numbers to tremendous densities. These budworms would then weaken the forest so that it was vulnerable to plant diseases and other hazards, such as fire. Because of a mix of circumstances, the great balsam fir–spruce forests of North America tended to be destroyed at fairly regular intervals. For example, Greenbank (1963) reports that such forests in New Brunswick had great spruce budworm outbreaks in periods beginning in 1770, 1806, 1878, 1912, and 1949. This means that the forest was being destroyed and replaced by an even-aged group of new trees after 36, 72, 34, and 37 years. If evidence of a break at about 1842 has been lost, then this appears remarkably like a 36-year cycle. Thus, the concept of "climax forest" does not refer to a static equilibrium system, but rather can be a regular cycle of birth, growth, death, and replacement of trees, most of which may be about the same age. Fire by itself can produce such even-aged stands, as in the case of ponderosa pine forest.

Also man, browsers, grazers, fire, and wind are very important determinants of the successional process. Large old trees can simply not be replaced if the environment is full of sheep, goats, rabbits, or deer that eat seedlings almost as soon as they appear. It has been discovered in many countries that the entire course of plant succession, or perhaps the extent to which succession can proceed, is tremendously affected by heavy feeding pressure from such animals.

Two recent studies on forest succession are particularly revealing as examples of some of the basic principles previously mentioned.

Loucks (1970) has studied the change in plant diversity through time in southern Wisconsin forest communities in which oak is important at 100 years of forest age, and the seedling layer is dominated by the shade-tolerant sugar maple at 200 years. He finds that at the seedling layer, diversity varies through time as in Fig. 8-4. This idealized chart is based on diversity measurements in various forests in which the age was determined from the age class of the oldest trees. Diversity in the seedling layer drops slightly at the beginning of each cycle (after fire) and then grows to a peak when the forest is about 100 years old. After this, however, diversity at the seedling layer drops because of the increased shading produced by the canopy of the dominant tree layer. (The diversity at the dominant layer increases for at least the first 100 years of the life of the stand.)

Evidently, fire is an integral part of the system which tends to keep

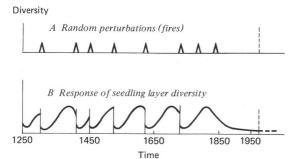

Diversity

A Random perturbations (fires)

B Response of seedling layer diversity

1250   1450   1650   1850 1950

Time

FIGURE 8-4 A model of the change in species diversity at the seedling layer through time in southern Wisconsin forests (*Loucks*, 1970). Note the role of forest fires in starting the successional sequence over and increasing diversity.

diversity up. Since man suppresses forest fires, he is interfering with this system.

This research illustrates Principle 7, in that the steady-state diversity tends to be higher some time (about 100 years) after plant succession has been operating in a forest, rather than a few years after an unpredictable event, a fire.

This pattern of a decline in diversity after 100 years may not be typical of all types of plant community succession. Reiners, Worley, and Lawrence (1970) have studied change in plant species diversity in sites 10 to 1,500 years after glaciers had melted off at Glacier Bay, Alaska. They showed that stand diversity increased in a roughly sigmoid fashion from 10 years to about 100 years of stand age, then remained approximately constant for the next 1,400 years. This successional sequence more nearly matches that expected from Principle 7, in that there is no simplification of the community owing to shading by dominant species after equilibrium has been reached.

STRATEGIES FOR MANAGING FORESTS

Two general questions arise in connection with forest management strategy. When, considering the age of the trees, should a forest be cut, and where should the trees be cut? The first problem is selecting what foresters call the *rotation age* for the forest. If the trees are too young when they are cut, the full growth potential of a year class of trees has not been realized, and the ground on which the trees were growing has not been used to best advantage. On the other hand, if the trees are too old when they are cut, they will have been left standing when they were overmature and when their maintenance metabolism was a much higher percentage of total production. Overmature stands are especially vulnerable to the ravages of insect pests and plant diseases. An even-aged forest stand grows through a natural cycle of birth, growth, and senescence, just like all other organisms, and the optimal time for harvest is before senescence. As for where trees should be cut, one is concerned about the spatial pattern of the cutting operation: should a very large contiguous stand be clear-cut, so that all the trees are cut at the same time? Or should a mosaic or checkerboard cutting policy be followed, in which only certain squares are cut each year? If the clear-cutting policy is followed, then the forest that replaces the cut forest will have an even-aged composition. Thus, a very large contiguous stand will all reach maximum vulnerability to pest ravages at the same time, and be an absolutely optimal environment for pests to build up to enormous densities. Clear-cutting is bad also from the point of view of soil and nutrient balance, and it makes expensive replanting necessary to leap over early stages of succession.

Indeed, just as the vulnerability to disease is a good index of ill health in many organisms, so the vulnerability to insect pests is a useful indicator of the advisability of different forest management policies. Figure 8-5 shows the population density per 10 square feet of branch surface of spruce budworm larvae in three different types of forest stands. The pest larvae reached the highest densities in mature plots that were part of a large contiguous stand. Intermediate densities were reached in mature but isolated stands. Lowest densities were reached in immature,

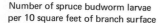

Number of spruce budworm larvae
per 10 square feet of branch surface

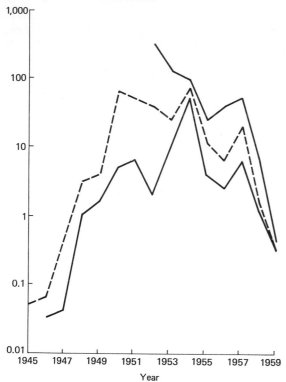

Year

FIGURE 8-5 Population density trends of the spruce budworm in three different types of balsam fir stands in New Brunswick. The top line represents trends in a mature plot that was part of a large contiguous stand. The middle line is for an isolated stand of mature trees. The bottom line is for an isolated plot of immature trees. (*This figure is a redrawing of fig.* 28.1 *in Morris et al., 1963.*)

isolated stands. These data have a high degree of precision and accuracy because they were based on an extremely sophisticated sampling scheme. Thus the differences as indicated in this chart are real, not the product of chance errors in sampling. Consequently, this chart has two important messages: forest management strategy should be designed to leave behind a checkerboard or mosaic pattern, in which different squares represent trees of different ages, and it is important to cut trees before they become too mature and vulnerable to pests. Thus, forest management illustrates Principle 24. We can diminish the vulnerability of forests to pests by creating an excessive transportation energy overhead on the pests when they disperse from stand to stand. In this way, we minimize the hazard from rapid spread of major outbreaks.

The exact details of the cutting strategy must be worked out by balancing off against each other the additional cost of forestry operations caused by not clear-cutting, which has the simplest logistics, and prospective tree losses and spraying costs associated with pest density buildups.

This has been an immensely simplified discussion of forestry from which many important issues, such as competition, the productivity of different tree species on the same site, and the effect of mixed-age stand composition on the efficiency of leaves in producing timber have been omitted. However, the student who is particularly interested in forestry can find an excellent synthesis and guide to the literature by Bakuzis (1969). We do not wish to leave the student with the impression that all forests should be managed, or that each tree should be managed in a forest. There are powerful motives for preserving a diversity of life forms in a forest, and a diversity of forest types (Principle 13). For example, many species depend upon rotten wood. Who will want to kill the last woodpecker and the other species dependent on rotten wood? Even within a managed forest, some trees should be left to die in order to support the organisms of decay and maintain the nutrient cycle and to support populations of pest control agents.

SUGGESTIONS FOR INDIVIDUAL AND GROUP PROJECTS

What life zone do you live in, on the basis of available weather records? Do you have the expected types of plants in your area? If not, why not? How can you explain the plant associations you find on the basis of man's activities, fog, smog, or intense grazing and browsing of seedlings and saplings?

Find out about the history of the forests, woodlots, or plant communities in your area. Has fire, plant disease, or pest damage played an important and recurring role in succession?

What can you discover about the successional sequences in your area by observing swamps, sand dunes, lava flows, or cutover areas?

Are the forests in your area in a stable state? Are the dominant members of the canopy being replaced by members of their own species?

What do you think of local forest management practices?

Explore the historical evidence on the relations between civilization, forest and land management practices, and climate, using the following sources. In particular, try to determine what effect civilization and climatic change have on the distribution of forests, and also try to determine the effect of forests on the vitality of particular civilizations.

Slatyer, R. O., and R. A. Perry (eds.): "Arid Lands of Australia," Australian National University Press, Canberra, A.C.T., 1969.

Wittfogel, K.: "Oriental Despotism: A Comparative Study of Total Power," Yale University Press, New Haven, Conn., 1957.

Jones, A. H. M.: "The Later Roman Empire, 284–602: A Social Economic and Administrative Survey," vol. II, Basil Blackwell, Oxford, 1964. See particularly the discussion of denudation on page 817. The Greek landscape has been made arid and rocky by excessive forest cutting, a failure to replant forests, and failure to prevent goats from eating young saplings before they had a chance to grow. On Mount Athos, goats have been excluded for a thousand years and the forests remain. Removal of forests in the arid Mediterranean region leads to erosion and the elimination of soil.

Rostovtzeff, M.: "The Social and Economic History of the Hellenistic World," vol. II, Clarendon Press, Oxford, 1941. See pages 1168–1170. This section documents the destruction of forests. Areas such as Cyprus are deforested now, not because of a climatic change, but because the forests were all cut down.

Rostovtzeff, M.: "The Social and Economic History of the Roman Empire," 2d ed., vol. II, Clarendon Press, Oxford, 1957. This book pays considerable attention to the relation between the strength of a society and the management of land and forests. On page 376 it is stated that the impoverishment of some districts in Italy was due to foolish deforestation and the neglect of drainage works. Page 476 indicates that a factor in the decline of the Roman Empire was the decreased general productivity owing to increased tracts of land running to waste as a consequence of neglecting irrigation and drainage systems. This neglect was not the driving force behind the decline of Rome; that was social decay. But once the social decay set in, the mismanagement of natural resources aggravated and accelerated an already marginal situation.

REFERENCES

Bakuzis, E. V.: Forestry Viewed in an Ecosystem Perspective, in G. M. Van Dyne (ed.), "The Ecosystem Concept in Natural Resource Management," pp. 189–258, Academic, New York, 1969.

Bates, Carlos G: Windbreaks: Their Influence and Value, *U.S. Dept. Agr. Forest Serv. Bull.*, no. 86, pp. 7–100, 1911.

Bryson, R. A., and W. M. Wendland: Climatic Effects of Atmospheric Pollution, in S. F. Singer (ed.), "Global Effects of Environmental Pollution," pp. 130–138, D. Reidel, Dordrecht, Holland, 1970.

Chang, Jen-Hu: "Climate and Agriculture," Aldine, Chicago, 1968.

Chenery, H. B., and T. Watanabe: International Comparisons of the Structure of Production, *Econometrica*, **26**:487–521 (1958).

Clark, C.: "Population Growth and Land Use," Macmillan, London, 1967.

Cooper, W. S.: The Climax Forest of Isle Royale, Lake Superior, and Its Development, *Botan. Gaz.*, **55**:1–44, 115–140, 189–235 (1913).

Ely, R. T., and G. S. Wehrwein: "Land Economics," University of Wisconsin Press, Madison, 1940.

Greenbank, D. O.: The Development of the Outbreak, in R. F. Morris (ed.), "The Dynamics of Epidemic Spruce Budworm Populations," *Mem. Entomol. Soc. Can.*, no. 31, pp. 19–23, 1963.

Hirschmann, A. O.: "The Strategy of Economic Development," Yale University Press, New Haven, Conn., 1958.

Holdridge, L. R.: Ecological Indications of the Need for a New Approach to Tropical Land Use, *Econ. Botany*, **13**:271–280 (1959a).

————: Simple Method for Determining Potential Evapotranspiration from Temperature Data, *Science*, **130**:572 (1959b).

————: The Determination of Atmospheric Water Movements, *Ecology*, **43**:1–9 (1962).

————: The Life Zone System, *Adansonia*, **6**:199–203 (1966).

————: "Life Zone Ecology," Tropical Science Center, San José, Costa Rica, 1967.

Landsberg, H. H., L. L. Fischman, and J. L. Fisher: "Resources in America's Future," Johns Hopkins, Baltimore, 1963.

Leslie, A. J.: Tropical Forestry and Economic Development, *Commonwealth Forestry Rev.*, **47**:40–51, 13 (1968).

Loucks, O. L.: Evolution of Diversity, Efficiency and Community Stability, *Am. Zoologist*, **10**:17–25 (1970).

Lowdermilk, W. C.: Influence of Forest Litter on Run-off, Percolation and Erosion, *J. Forestry*, **28**:474–491 (1930).

————: "History of Civilization without Soil and Water Management Planning," *Intern. Seminar Soil Water Util.*, South Dakota State College, Brookings, Washington, pp. 5–10, 1962.

Merriam, C. H.: Life Zones and Crop Zones, *Bull. Biol. Surv.*, **10**:9–79 (1899).

Morris, R. F. (ed.): The Dynamics of Epidemic Spruce Budworm Populations, *Mem. Entomol. Soc. Can.*, no. 31, 1963.

Nicholson, J. W.: Forest and Rainfall, *Empire Forestry J.* **9**:204–212 (1930).

Paterson, Sten S.: "The Forest Area of the World and Its Potential Productivity," Dept. of Geography, Royal University of Göteborg, Sweden, 1956.

Réaumur, R. A. F. de: Observations du thermomètre, *Mem. Acad. Royal Soc.*, Paris, pp. 545–576, 1735.

Reiners, W. A., I. A. Worley, and D. B. Lawrence: Plant diversity in a chronosequence at Glacier Bay, Alaska, *Ecology*, **52**:55–69 (1970).

Staple, W. J., and J. J. Lehane: The Influence of Field Shelterbelts on Wind Velocity, Evaporation, Soil Moisture and Crop Yield, *Canad. J. Agr. Sci.*, **35**:440–453 (1955).

Steila, D.: An Evaluation of the Thornthwaite and Holdridge Classifications as Applied to the Mediterranian Borderland, *Professional Geographer*, **18**:358–364 (1966).

Thompson, P. T.: A Test of the Holdridge Model in Midlatitude Mountains, *Professional Geographer*, **18**:286–292 (1966).

Thornthwaite, C. W.: Approach towards a Rational Classification of Climate, *Geograph. Rev.*, **38**:55–94 (1948).

Westoby, J. C.: The Role of Forest Industries in the Attack on Economic Underdevelopment, *Unasylva*, **16**:168–201 (1963).

9 the watershed: an example of multiple use and conflicting demands

A watershed is the terrestrial component of the precipitation catchment area from which water runs into a brook, stream, river, or lake. Since the watershed often drains into a lake and is often covered by forest, it is a focus of research interest combining the complications indicated in the two previous chapters. Students of watersheds are concerned with such matters as the effect of forest management practices on the proportion of rainfall that runs off the watershed immediately, rather than being absorbed into the hillside, and on the retention of nutrients in the watershed in proportion to their loss to the aquatic component of the catchment area.

Although watersheds illustrate many of the principles introduced earlier, they are particularly interesting examples of Principle 18, concerning the importance of mechanisms that allow for the availability of necessary resources. Thus, for example, maintenance of the soil nutrient level in a forest on a watershed can be deleteriously affected by clearcutting the forest (Bormann et al., 1968). This indicates that a pattern of cutting which leaves some of the trees standing in each small unit of area may be necessary to avoid depletion of soil nutrients.

Few subjects of interdisciplinary study in the environmental sciences can generate as much intense controversy as selection of optimal management strategies for watersheds. This is to be expected, because few types of land forms are of potential use for such a wide variety of purposes as the watershed. At least eleven distinct types of uses can be recognized:

1 Watersheds are clearly of immense economic importance as sites on which to grow trees to be used for pulpwood or timber.
2 Because watersheds are often in the form of a steeply sloping valley rising up from a river basin, they have in such cases immense potential value as sites of impoundments to store water used in hydroelectric power generation.
3 Watersheds may be of considerable agricultural importance, depending on steepness of slope and soil type. The soil in the valley bottom may be excellent for grazing, orchards, or market gardening.
4 Watersheds have great recreational value for hunting, angling, hiking, and camping.
5 An important real estate industry can develop for the sale of summer recreation homesites, second homes, or retirement homes in watersheds.
6 The rivers at the bottom of watersheds may be the spawning sites for fish of great commercial importance, such as salmon.
7 The same rivers may be the center of the traditional home of Indian populations for whom the fish stocks are an important source of food.

8 Watersheds may be the center for important mining operations.

9 A river at the bottom of a valley may provide an important site for water collection and storage in the interests of ameliorating flood damage. To this end, an impoundment evens out the flow of water within a year, and from year to year.

10 Also, an impoundment may be used to store water for subsequent shipment to a distant location which suffers from a chronic water shortage.

11 A watershed may have different values to naturalists and conservationists interested in the preservation of wildlife for its own sake. As we come to learn more about ecology, we may discover that wildlife preservation may be valuable for a number of other reasons which become apparent only with continuing accumulation of scientific knowledge. To illustrate this point, it may be true that unless some areas in a watershed provide suitable habitat for predators such as the coyote, mountain lion, or other cats, there is inadequate check on herbivore populations such as deer and rabbits, which consequently overeat the vegetation on their range and prevent the growth of young trees needed to regenerate cutover forest areas.

Obviously, several of the items in this list represent very big economic interests. Also, it is clear that several of these interests are in direct conflict with each other because a strategy beneficial to one group of users may be detrimental to another group of users of the same watershed.

Some of these conflicts are apparent; many others are not so obvious. For example, consider the question whether it is a good idea to cut trees at the bottom of the watershed, immediately adjacent to the stream or river. If we are primarily concerned with conserving water for shipment elsewhere, these trees should be cut, because they transpire the greatest amount of water per tree per unit time of all the trees in the watershed because water is more available to them. However, if fish in the watershed are an economically important resource, then it is a bad thing to cut the trees overhanging the water. These trees shade the stream or river and thus keep the water cold enough to support fish food and valuable commercial and sport fish species (Levno and Rothacher, 1968). Trout and salmon, for example, cannot live in warm water. Rough fish, on the other hand, can tolerate higher temperatures. Thus, the interests of the people who want the water in the watershed for shipment elsewhere are in direct conflict with the people who want to preserve salmon or trout populations in the water.

In fact, every one of the eleven uses mentioned is in direct conflict with one or more of the others. From the standpoint of forest cutting operations, logistics of cutting are simplest under a clear-cutting strategy, whereby all the trees in a large tract of land are cut simultaneously. However, this decreases the water absorptive capability of the watershed and increases the proportion of the rain in a rainstorm that runs off the surface immediately, rather than being held in the ground by living tree roots. Consequently, clear-cutting operates against the objective of flood control, which is to even out water flow from the watershed, rather than having big surges just after heavy rains and only a trickle later on in the season.

If flood control is a major objective, then we wish to maintain a high potential for water absorption in a watershed and minimize runoff. If, however, our principal concern is obtaining the maximum amount of rainfall as runoff for irrigation or water storage down the watershed, our strategy is different. In this use, as in many of the western states, the watershed will be managed to increase rather than decrease runoff. This is achieved by replacing trees and shrubs with grass, so as to prevent erosion but yield the maximum ratio of runoff to rainfall absorbed.

Some notion of the complexity of watershed ecosystems has been developed recently by Bormann et al. (1968), who have documented the mechanisms relating forest management strategy to the eutrophication of stream water. This research group measured partial nitrogen budgets for two small watersheds in New Hampshire. Watershed 6, the control, was 13.6 hectares and was not disturbed during 1965 and 1966. Watershed 2, the experimental, was 15.6 hectares and was clear-cut in the winter of 1965–1966.

The experimental procedure on watershed 2 involved leveling of the beech-maple-birch forest. All trees were cut and left in place, and great care was taken to minimize erosion of the soil surface. Regrowth of vegetation was inhibited by a herbicide.

Input of ions into the watersheds in the precipitation was monitored, and output in the stream water. From the difference between the input and the output, net gain or loss of nutrient to the watershed could be calculated. The design of the research allowed two separate measures of the effect of clear-cutting: after versus before, in the disturbed watershed, and disturbed versus undisturbed.

Two major observations were made in this experiment. First, the clear-cutting made the runoff increase 40 percent more than what had been expected. The authors attribute this to loss of transpiring leaf surfaces on the watershed. Second, there was a striking loss of nitrate nitrogen in the stream water of the clear-cut watershed. The loss that occurred was about 38 times the gain that should have occurred!

From these observations the authors concluded that clear-cutting tends to deplete the nutrients of a forest ecosystem, and that depletion is accomplished by the simultaneous combined effect of the following five processes:

1 Reducing transpiration, thus increasing the amount of water passing through the system
2 Reducing root surfaces able to remove nutrients from this water
3 Removing nutrients in forest products
4 Adding to the organic component of soil available for immediate mineralization
5 Sometimes, producing a microclimate more favorable to rapid mineralization, as by increasing the available substrate

The major consequence of these five processes is a very great increase in the nitrate concentrations in the stream flowing out of the watershed. These authors found that established pollution levels were exceeded for more than a year and algal blooms appeared during the summer.

Thus, forest cutting strategy in a watershed affects the physical and chemical characteristics of the water flowing out of the watershed. From Chap. 7, we would expect this in turn to have an effect on the composition of the fish stocks in the stream. Specifically, we would expect species such as salmon and trout, which require cool water rich in oxygen, to be replaced by game or rough fish able to tolerate warmer water with less oxygen, as well as all the other changes associated with eutrophication.

Forestry may be immensely beneficial to wildlife, which in turn is subsequently inimical to a regenerating forest. By cutting down an overmature forest, where a high proportion of the green tissue is too high up to be available to most wildlife, a tremendous regeneration of young, short plants is promoted, stimulating increase in the number of individuals in browsing and grazing species. However, this very numerical response creates a dense population of rabbits, deer, and other herbivores that make it difficult for young trees to grow to maturity.

Cutting down a forest will have an immediate detrimental effect on the recreational value of a watershed, but a beneficial subsequent effect. A partially mature forest is more biologically interesting than an overmature forest and certainly is a better game habitat for the naturalist or hunter.

Game is clearly inimical to agriculture in a watershed. Deer invade and damage orchards, and bighorn sheep and domestic sheep are competitors. Mountain lions are enemies of sheep. Weasels are enemies of chickens.

Important conflicts are developing because, although conservationists fought the use of valleys for hydroelectric impoundments, the valleys have subsequently been turned into vast tracts of identical suburban houses.

OBJECTIVE RESOLUTION OF CONFLICTS OF INTEREST

A technique is available for making an *objective* economic determination of the best use of a watershed, considering the views of all parties. This involves computer simulation of a number of alternate scenarios for a watershed. For each scenario all costs and benefits for each of a large number of years into the future are tallied, and their value is computed in terms of present dollars. Thus, if we assume that money can be invested and perpetually receive a 5.5 percent interest rate, the present value of a dollar earned t years in the future is

$$\frac{1.00}{1.055^t}$$

This type of comparative assessment of the benefit/cost ratio can be made very sophisticated, and it is a useful tool in demonstrating that certain strategies which appear highly profitable have a benefit/cost ratio more illusory than real when the time horizon is extended. However, the method has defects. Discounting the value of money, the highest benefit/cost ratio for a watershed might be assigned to hydroelectric development. But the benefits of dams for power generation turn out to be ephemeral over a time horizon of 100 to 300 years, because by that time the dam has silted up and is a giant white elephant. However, a

carefully managed forest industry in the site that would have been sub-
merged by the dam will produce trees in perpetuity. Under this type of
very long-term projection, breeder reactors appear to be a very much
more attractive source of power than dams.

To this point, it has been assumed that conflicts of interest can be
resolved strictly by an appeal to economic arguments and criteria: that
strategy which maximizes the benefit/cost ratio over a very long time
horizon is the best strategy. Is this true? In fact, the conflicts between
different competing interests in watershed management are so difficult
to resolve, because of the complexity of the issues under contention,
that a much deeper analysis is required. In fact, to set up an adequate
basis for determining how to resolve conflicts over resource management
policies, we must move outside the sphere of pure economic theory into
the area of political science, and consider the appropriate goals for a
democratic society. Maass (1962, 1966) has considered this general
topic in the context of water resources use, and his analyses are strongly
recommended to interested students.

Maass points out that conflicting demands must be resolved by reference to common standards, not by pure power, or power constrained
by "the rules of the game." Then what should be adopted as common
standards? It would appear that conflicts should be resolved by evaluating different strategies that might be used to achieve four goals of a
democratic society, in such a fashion that pursuit of any one of the goals
does not impair pursuit of the others. We will first present the four goals,
and then show the line of action indicated in a number of specific
instances by referring to them.

1   It is self-evident that each citizen of a democratic society is entitled to a
safe, attractive, pleasant, and healthy environment. These words involve
aesthetics. Since people's aesthetics differ, goal 1 implies that a great
diversity of decisions must be reached in the "population" of watersheds.

2   Each citizen should be free to better his economic lot by the investment
of time, energy, talent, or capital, and earn a reasonable return on such
investment. This goal promotes competition and industriousness, which
are demonstrably useful as motivators to men to excel. It will be noted
that a broad interpretation of this goal is compatible with a variety of
economic systems.

3   It seems self-evident that there should be more government regulation
of profit-seeking activity, so that speculators or sharp business practices do not exploit other citizens, leading to an inequitable distribution
of income.

4   It appears self-evident that humanity should conduct its affairs so as to
ensure that all peoples can exist for as long a time as possible on this
planet. That is, an important constraint on strategy selection is that irre-
versible or destructive activities with long-term consequences should be
systematically avoided. This goal has considerably more far-reaching
implications than may appear at first. Granted, for example, that the
planet is running out of crude oil and natural gas and readily accessible,
cheap uranium ores at an exponential rate (Chap. 15), what significance

does this assumption have for strategies having to do with methods of energy use? Energy inputs into ecosystems determine the carrying capacity of the habitat for biomass. Thus, the current great increase in fossil fuel use on the planet is building up the carrying capacity of the planet for people. However, when the fossil fuel runs out, as it will sooner than most people expect, there will be a great drop in the carrying capacity of the planet for people. If this energy shortage cannot be made up with nuclear energy, then there will be a great reduction in the number of people on earth.

This line of argument can be applied to river basin use. The argument for building more dams and hydroelectric power generating stations is that we are running into energy shortages, and consequently should increase power generation capacity. But there is another way of looking at this issue. Power generation and population growth constitute a positive feedback loop, in which the generation of additional power makes possible additional human population growth, and then the additional human population growth creates a need for still more power generation.

The other approach is to consciously use power shortages as a negative feedback loop that will tend to diminish the rate of human population growth. Clearly, this negative feedback loop could not control human population growth by itself, but it is one of a number of such loops, all of which acting together could make children so expensive that the population growth rate would diminish sharply. Otherwise, if the response to power shortages is to build more hydroelectric dams and generating plants, the unstable and precarious human population predicament which is being set up by our increasing dependence on fossil fuel is made even worse. That is, if both energy subsidy to the human ecosystem from fossil fuels and hydroelectric plants increase exponentially, thus allowing population to increase, then the fossil fuels run out and impoundments gradually silt up, as they must; thus humanity is hit with two terrible and gigantic problems simultaneously, rather than one. If there is a conscious effort to hold back on building dams, then a positive feedback loop, which intensifies a problem, is converted to a negative feedback loop, which will introduce a dampening effect into the system fueling population growth rates.

The point of this argument is that if we accept the premise that humanity should plan on living on this planet as long as possible in an approximately stable state, free of major disasters, we must also accept important constraints on the way we make decisions about the second goal, making money.

EXAMPLES OF CONFLICT RESOLUTION

Real estate developers sometimes buy raw wilderness land at very low prices, put in roads, lay out lot boundaries, and then resell the lots at immense profits. The two most noteworthy points about this activity are the magnitude of the profit and the consequences to the watershed. For example, it is typical for the developer to sell the lots at a price per acre ten times his purchase price. A typical consequence to watersheds is siltation in river basins, due to careless road construction, which wipes out half of all the fish present. Also, habitat of waterfowl, upland game birds, or big game is destroyed and populations are almost annihilated. The recreational values of streams and lakes can be ruined by eutrophication due to primitive sewage disposal systems, which simply run the raw sewage, untreated, into the water.

What are the implications of the four stated goals for such activity? Clearly, inequitable income distribution is being promoted by allowing developers to make huge profits while passing on the costs of restoring environmental quality to the taxpayers. Government can play a role in internalizing costs and benefits in this situation (adjusting the distribution of costs and benefits, so that the developers pay more of the costs associated with the benefits they reap). This can be done by means of an unearned value tax, under which the government receives a proportion of the difference between the price the developer paid for his land and the price at which he sells it to the public. The government can then use that unearned value tax to pay for restoration and maintenance of the environment after development is complete.

But the tax would have a deeper significance. The reason why there is so much overdevelopment of wilderness real estate is that the potential profits are very large in comparison with prospective costs. Again, we have an element in the human ecosystem characterized by absence of a regulating negative feedback loop. However, the unearned value tax is a means of bringing prospective costs more into line with prospective benefits, and diminishing the incentive for development unless there is a real demand.

Another consequence of trying to maximize the attainment of all four goals jointly is that great pressures will be created for multiple use of land. The third goal militates against one special-interest group that gains profit from a resource without considering the net benefits to other groups and interests within society. This means, for example, that forestry practices should be adjusted to minimize the loss of the recreational values of a forest as a consequence of logging operations. There will be some lowering of the profit to lumbering industry, but this is more than compensated for by increased benefits to campers, hunters, and naturalists. As the tensions of modern society become greater, there will be increasing recognition of the value of wilderness areas as a therapeutic as well as aesthetic resource.

In general, we are suggesting that benefit/cost analysis is a useful tool, but all the hidden costs and benefits must be figured into the analysis, and it will typically be necessary to impose constraints on the procedure for selecting watershed management strategy because of noneconomic benefits.

MAKING THE POLITICAL SYSTEM WORK

Many economists have pointed out the importance of separating the intelligence or data-gathering and interpretation function, the regulatory function, and the action function in government and other organizations. Maass also makes this point. Galbraith (1967) has pointed out the types of symbiotic relations between government and industry that have come to characterize what he calls the new industrial state. Too often, we have a highway industry and highway departments with a community of interests, forest industries and forestry departments with a similar community of interests, fishing industries and fisheries departments, and oil and other energy industries which have mutually useful working relations. Can one imagine a highway agency in government deciding that the highway industry should slow down on its rate of building highways? Or can the Atomic Energy Commission, an action agency, really be regulated effectively by itself? The point is that in order to make our political system work effectively, and ensure the most rational, objective possible system of selecting optimal management strategies for resources, including watersheds, we must separate administratively the functions of collecting and interpreting data, action, and regulation. Each of these three functions must be under entirely separate control, so that collection and interpretation of data and regulation are negative feedback control loops regulating the action agencies, rather than being easily overruled adjuncts to the action agencies.

CONCLUSION The multiplicity of conflicting and competing interests in watershed management reveal the true nature of modern society in a microcosm. The inherent complexity of all such situations makes it extremely difficult to decide rationally and objectively which of many alternative resource management strategies is best. The only way to do this is to make use of two tools jointly. First, it is necessary to make the most sophisticated possible application of benefit/cost analysis, computing the net present value of each of the alternative strategies, and using simulation in computers to assess costs and benefits each year as far into the future as possible. Second, it is necessary to impose constraints on the process of selecting strategies, on the basis of explicit statements about the appropriate political goals of a democratic society.

This chapter illustrates an interesting application of Principle 13 to society. If diversity breeds stability, then it is worthwhile for government to regulate the rate at which different political and economic interests acquire power and wealth so that no one grows too fast or at the expense of the others. Much of the unwise use of land and resources we now see is due to inadequate regulation by government. Thus, it was easy to urbanize wilderness because all the advantages were on the side of the developer. Government regulation is necessary in such instances, because the free play of the market will not prevent excessive development. This is because of time lags in economic systems: many lots may have been laid out, and wilderness bulldozed, before it is clear that demand for wilderness homes in a given area has been enormously overestimated by a developer.

SUGGESTIONS FOR INDIVIDUAL AND GROUP PROJECTS Can you think of another ecosystem besides watersheds that has conflicts of interest? Invent realistic legislative devices to resolve these conflicts in accordance with some reasonable set of rules.

Is there any evidence that bad management practices are having a deleterious effect on watersheds you visit on field trips? Is there any evidence of changes in soil?

REFERENCES Bormann, F. H., G. E. Likens, D. W. Fisher, and R. S. Pierce: Nutrient Loss Accelerated by Clear-cutting of a Forest Ecosystem, *Science*, **159**:882–884 (1968).

Galbraith, J. K.: "The New Industrial State," Houghton Mifflin, Boston, 1967.

Levno, A., and J. Rothacher: Increases in Maximum Stream Temperature after Logging in Old-growth Douglas-fir Watersheds, *U.S. Forest Serv. Res. Not.*, **65**:1–12 (1968).

Maass, A.: System Design and the Political Process: a General Statement, in A. Maass, M. M. Hufschmidt, R. Portman, H. A. Thomas, Jr., S. A. Marglin, and G. M. Fair (eds.), "Design of Water Resource Systems," chap. 15, pp. 565–604, Harvard, Cambridge, Mass., 1962.

————: Benefit-cost Analysis: Its Relevance to Public Investment Decisions, in A. Kneese and S. C. Smith (eds.), "Water Research," pp. 311–328, Johns Hopkins, Baltimore, 1966.

10 grassland and range systems: examples of reciprocal interactions between plants and animals

Grassland and range systems are worth special attention because they represent well-documented examples of circular causal systems. By this, we mean an environmental system in which variable A affects variable B, and then variable B has a reciprocal effect on variable A. This type of phenomenon is characteristic of range systems. Different species of range animals, and even different age groups within the same species, differentially select certain range plants in preference to others which are available for use as food. This means that by shifting the species or age composition of the population of grazing animals, we shift the feeding pressure from one group of plant species to another group of plant species. But this in turn results in a shift in the species composition of plant species, which means a shift in the species or age composition of the population of range animals for which the range will be most suitable.

Some of the essential features of range systems may be arranged in a highly idealized and oversimplified flow chart, as in Fig. 10-1. Although actual ranges may contain many different animal and plant species, with different age groups for each animal species, the essential features of the reciprocal circular causal range system are brought out by this diagram, in which only two animal species and two plant species are depicted. Man harvests both animal species, but the harvesting intensity may be different for the two species. That is, each year man may remove x percent of all the pounds of meat produced by one of the two animal species and y percent of the pounds of meat produced by another species. The two animal species both trample the ground, but the impact per pound of animal may be different for different species. Trampling has an effect on the suitability of the environment for plant species, but different plant species may not be affected to the same extent by a given degree of trampling. Man can increase the concentration of soil nutrients by fertilization of range, but there is a reciprocal causal relation between soil nutrients and the various plant species. All plant species are positively affected by increase in concentration of essential nutrients, but different species of plants have different sets of nutrient requirements. Also, the effect of plants on the concentration of nutrients in the soil varies from plant species to species.

A number of critical variables for the dynamics of range plant-animal systems are not included in this diagram. For example, fire has important effects on the nature of the dynamic balance between different plant species, and hence between different animal species.

The remainder of this chapter will be occupied with the presentation of evidence in support of the generalizations just itemized.

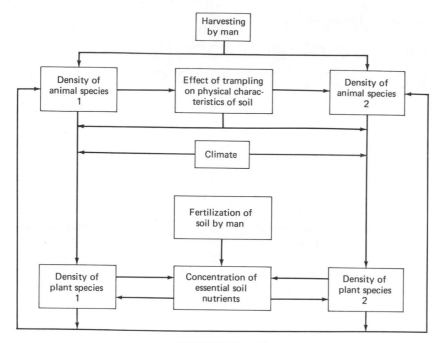

FIGURE 10-1 Flow chart of the causal relations among the different key variables operating in range systems.

THE STEPS IN THE ARGUMENT

In order to document the main thesis incorporated in Fig. 10-1, it is necessary to provide evidence 'in support of each step in the following sequence:

1 Various species of animals living on the same range have different patterns of preference for the plant species available to the animals on that range.

2 Because of these food preferences, the animals can, by selective grazing, shift the species composition of the plant species on the range. This does not necessarily mean that some plant species are driven to extinction; rather, it means that some species are made significantly less abundant because of this selective feeding.

3 Because of the shift in relative abundance of different plant species, there is a resultant shift in the relative abundance of different animal species.

step 1

This has been demonstrated for range systems all over the world. For example, Talbot (1966) states that in east and central Africa each species of ungulate appears to have a yearlong preferred diet different from and complementary to the other ungulate species. The more striking differences are illustrated by giraffes, which eat mostly trees, rhinoceroses, which eat mostly brush, and wildebeests, which eat grass almost exclusively. However, there are much more subtle differences than this, which can be illustrated by the relations of three species to red oats grass. Some ungulates eat none of this particular grass species, but it is the principal item in the diet of wildebeests, topis, and zebras in western Kenya and Tanganyika. The wildebeests select the fresh leaves of this grass until they grow to about 4 inches long, but they avoid stalks, seed heads, and dry red oats. Zebras eat the leaves of red oats more than 4 inches long, and stalks, heads, and dry red oats grass. Topis show a marked preference for dry red oats grass. Consequently, while all three species crop red oats grass, they fill different niches, like the *Drosphila* species discussed in Chap. 4. Thus range systems illustrate several of the principles previously introduced: Principles 15 (zones of tolerance), 26 (competitive displacement), and 27 (coexistence implying different niches).

A very elaborate study by Van Dyne and Heady (1965) on sheep and cattle diets on California range revealed the same pattern of subtle differences in diet, with the further complications of species-species shifts in diet through the season. The general pattern is indicated by the following table:

	grasses as proportion of total diet		forbs as proportion of total diet	
	cattle	sheep	cattle	sheep
Early summer	61	50	25	32
Middle summer	59	54	21	30
Late summer	52	57	35	27

Further, when the diets of the two species are analyzed in terms of the frequency of different plant species in the diet relative to their frequency on the range, differences are found in both the rank order of preferences and the intensity of preferences.

step 2   In Australia, the effect of sheep on the species composition of range is particularly well documented and understood. Rainfall is not reliable throughout the season. Consequently, sheep are particularly hard on those plants whose main stress period occurs during vegetative growth outside the rainy period. Plant species for which the main stress period occurs within the rainy period are less deleteriously affected by intensive sheep grazing. Consequently, heavy sheep grazing throughout the year shifts the species composition of the range plants in the direction of species with a shorter vegetative growth period, and a shallower root depth. This means that a result of pasture degradation due to overgrazing is a reduction in the period during which forage is being produced (O. B. Williams, 1961). Williams illustrates this process with the data in Table 10-1.

Clearly, there is a trend toward a dominant pasture plant that is short-rooted and has a short vegetative growth season.

step 3   The consequence of overgrazing of pastures by sheep in Australia has been to deprive the red kangaroo of its traditional habitat. However, another subsequent result has been the invasion of the grassland by hill

TABLE 10-1 Vegetative growth period and depth of rooting in a series of degraded pastures in southern New South Wales

	dominant plant	vegetative growth period, months	root depth, inches
Good condition	Old-man saltbush	10–12	72
	Bladder saltbush	6–8	36
	Wallaby grass	3–6	20
Deteriorated	Barley grass	3–4	10

SOURCE: Williams (1961).

kangaroos from the surrounding hills; the deterioration of the grassland by the sheep turned it into a suitable habitat for the hill kangaroo (Ealey and Suijdendorp, 1959; Main, Shield, and Waring, 1959).

The consequences are grave, from a practical point of view, when overgrazing is allowed and pasture deteriorates: in a degraded pasture the production, in weight, of green material after a rain may be only 23 percent of that in a pasture in good condition. This, of course, has an important effect on the carrying capacity of the range for sheep or cattle. Consequently, by attaching importance to the short-term profits from rangeland, one can destroy it over the long term.

FIRE AND RANGELAND

Fire may be an important factor in maintaining rangeland. Hulbert (1969) found experimentally that the major short-term effect of fire in Kansas bluestem prairie is caused by removal of litter, rather than direct heat or fire-induced nutrient changes. The effect of the litter removal, in turn, is to cause a lower soil moisture content throughout the summer. This is because litter reduces evaporation more than it intercepts precipitation, and also because the lesser plant production on nonburned areas is equivalent to a lessened transpiration compared to plots with no litter. Consequently, the significance of fire for trees is that it causes a lowered water supply for tree seedlings, as well as causing fire damage to the young seedlings. Consequently, fire prevents invasion of prairie by trees; it is not a disturbing factor in prairie succession, but rather a necessary factor in the maintenance of natural prairie.

THE EFFECT OF SHIFTS IN THE AGE DISTRIBUTION OF GRAZERS ON SPECIES COMPOSITION OF RANGE PLANTS

It was mentioned at the beginning of this chapter that not only changes in the species composition of range animals could affect the species composition of range plants, but shifts in the age composition within a single species of range animal also could do this. An interesting example of this has been studied by Rawes (1961) in the sheep country of northwest England. In this area a constant problem is the spread of matgrass high up on the hill grazing areas. This grass normally indicates poor fertility soils, and because of the gradual decline in the fertility of the hills in this area, the acreage of grassland dominated by this species is increasing. The increase is at the expense of grasslands dominated by better-quality sheep food. Chemical analyses of the production of different nutrient components of the matgrass, compared with the better grasses it is replacing, show that in pounds per acre the matgrass is much lower in virtually every constituent.

In the past, matgrass has been kept in check by older sheep. They ate the grass during the winter, and also uprooted whole plants, drifts of which could be seen 50 years ago, filling dikes and collecting against stone walls. However, the age composition of the grazing sheep populations has changed because of consumer demand for lamb and smaller joints of mutton. This means that the grazing population has a higher incidence of ewes and lambs. These animals have higher demands for minerals than the older males, and consequently they graze the most nutritious plant species selectively and avoid the matgrass, which is low in mineral content. The result is that matgrass not only spreads into the overgrazed surrounding grassland but becomes an unhealthy plant itself; it covers the surface of the ground with dead fibrous leaves that decay slowly and consequently stifle the growth of other plant species that are more nutritious.

In short, the sheep population being managed by a human population in equilibrium with its environment would have a stable age distribution and a stable sex composition. However, because the demands being placed by man on his environment are unreasonably high, the sheep population is constantly being maintained with the age and sex composition of a rapidly growing population, with a higher proportion of females and lambs than would be found in a stable population. The consequence is that the fertility of the grazing country is gradually declining.

This illustrates an interesting variant of Principle 13. When man overharvests and distorts the sheep age distribution, he operates as the analog of a physical factor introducing environmental instability. This in turn leads to lowered age and sex diversity in the sheep population (underrepresentation of old males). The final consequence is to introduce an instability factor for the plant population, which is made to undergo succession from one stable state to a less desirable stable state.

The remedy for this situation would appear to be fertilization. For example, Love and Williams (1956) have shown that by applying phosphate fertilizer to range, the forage production could be increased threefold and the protein production could be increased sixfold, compared with untreated land. Rawes (1961) applied fertilizers, but in spite of an increase in the production by range plants, he was not prepared to conclude that the effect was lasting or that the response was great.

However, there is a deeper issue. If maintenance of range becomes dependent on fertilizer application, as maintenance of crop production is already, then man will have increased his dependence on a fossil fuel energy subsidy to agriculture. As will be shown in Chap. 15, the lifetime of crude oil and natural gas on this planet will probably be much less than most people believe, owing to the tremendous annual increase in demand, and it is not nearly so certain as most people think that nuclear fuels or coal can constitute a complete substitute. Consequently, by increasing our dependence on a crude oil and natural gas energy subsidy to agriculture, we are placing ourselves in an increasingly precarious position. That is, the carrying capacity of the planet for people is being sharply increased beyond that which can be maintained indefinitely by incoming solar radiation, and there is no guarantee that we can maintain this very high carrying capacity. Indeed, we show no indication of wishing to maintain the carrying capacity at current levels but are, in fact, constantly increasing it sharply. Thus, we run the risk of converting the carrying capacity of the planet for people from the 1 to 3 billion which can be supported by the sun to between 10 and 15 billion, with the energy subsidy, only to discover, when we run out of crude oil and gas, that we now have 10 to 15 billion people on a planet which can support a population of only 1 to 3 billion on a long-term sustained-yield basis. What then?

SUGGESTIONS FOR INDIVIDUAL AND GROUP PROJECTS

Gather enough quantitative data on range use in your state or region to apply numbers to the causal pathways in the flow chart of Fig. 10-1. Express all variables in the same units: kilocalories per square meter per year.

REFERENCES

Ealey, E. H. M., and H. Suijdendorp: Pasture Management and the Euro Problem in the North-west, *J. Agr. W. Australia*, (3)**8**:273–286 (1959).

Hulbert, L. C.: Fire and Litter Effects in Undisturbed Bluestem Prairie in Kansas, *Ecology*, **50**:874–877 (1969).

Love, R. M., and W. A. Williams: Rangeland Development by Manipulation of the Soil-Plant-Animal Complex in the Difficult Environments of a Mediterranean-type Climate, *Proc. 7th Intern. Grassland Cong.*, pp. 509–517, 1956.

Main, A. R., J. W. Shield, and H. Waring: On Recent Studies on Marsupial Ecology, *Monogr. Biologicae*, **8**:315–368 (1959).

Rawes, M.: The Problem of *Nardus* and Its Productivity in Relation to Sheep Grazing at Moor House, Westmorland, *J. Brit. Grassland Soc.*, **16**:190–193 (1961).

Talbot, L. M.: Wild Animals as a Source of Food, *Bureau Sport Fish. Wildlife Spec. Sci. Rept: Wildlife* no. 98, pp. 1–16, 1966.

Van Dyne, G. M., and H. F. Heady: Botanical Composition of Sheep and Cattle Diets on a Mature Annual Range, *Hilgardia*, **36**:465–492 (1965).

Williams, O. B.: Principles Underlying the Improvement of Dryland Country, *Wool Technol. Sheep Breeding*, **8**:51–58 (1961).

suggested additional readings

Lewis, J. K.: Range Management Viewed in the Ecosystem Framework, in G. M. Van Dyne (ed.)., "The Ecosystem Concept in Natural Resource Management," pp. 97–187, Academic, New York, 1969.

Odum, H. T.: "Environment, Power and Society," Wiley, New York, 1971.

11 agricultural systems: the problem of replacing optimizing strategies of nature by optimizing strategies of man

We will begin this chapter with a simple, naïve statement of the aims of agricultural strategy. Then we will examine this statement for possible complications and reconsideration against the background of ecological theory developed in previous chapters. This will lead to a restatement of the aims of agricultural strategy. At that point, we will have set the stage for a penetrating reexamination of agricultural strategies, in terms of the fundamental significance and implications of various strategy options, for both the short and the long term. The discussion in this chapter will lead to an assessment of what mankind can reasonably expect from agriculture.

In the simplest possible terms, agricultural strategy has the following four aims:

1 Maximize production per unit of land area, in a way consistent with the next three aims
2 Operate at a maximal profit
3 Minimize year-to-year instability in production
4 Operate so as to prevent long-term degradation of the productive capacity of the agricultural system

A moment's reflection will show that all four of these desiderata involve a number of complications. What do we mean by "production?" Does this refer simply to pounds of food produced, or the energy content of the food in kilocalories per acre, or the protein content of the food, vitamins, minerals, or some optimal mix of dietary ingredients? How should such a dietary mix be determined? What do we mean by "operate at a maximal profit?" Do we mean that the operation should be efficient in terms of money, or do we mean that the operation should yield the highest (as opposed to most efficient) net return on capital and time invested, or energy, or both? What is the optimal strategy if efficient operation in terms of money is not produced by the same strategy which produces efficient operation in terms of energy? Does that matter? Does it make a difference whether we consider the short- or the long-term optimization? What kinds of factors induce instability? How much instability can be tolerated, and how do we go about answering that question? How do we measure long-term degradation of an agricultural system? Does avoidance of degradation mean that no change in the physical chemical characteristics of the soil is tolerable or does it mean that some changes are tolerable and other kinds of changes are not?

As a prelude to this discussion, it is worthwhile to compare man's agricultural strategy with the natural community.

The overall strategy objective of agriculture is to replace a natural community, which may well have had high diversity and stability, but low humanly useful productivity per unit biomass, by a nondiverse but more productive system without sacrificing stability. "Productivity" in this context refers to production rate of plant or animal tissue per unit time.

Nature seems to select for communities with the highest possible biomass per unit area for given incident solar radiation and available soil nutrients. Chapter 3 shows that succession at a given site involves a gradual increase in biomass at that site, as herbs give way to shrubs and shrubs give way to trees, so that the volume of vegetation-filled space over a given land area becomes greater and greater. Since the energy input in natural communities is limited to that from the sun, this means that in a fixed area there is gradual succession toward that community with the highest possible ratio of biomass per unit of incident solar radiation. That is, succession goes in the direction of lowest possible P/B.

Man, however, wants the opposite situation: the highest possible P/B. He achieves this by using species and strains of plants and animals in which the highest possible proportion of the biomass is useful: short-stalked grain with big heads, livestock with a high ratio of lean meat to fat and bones. Also, he sets up monocultures which allow for simplicity, and hence energetic efficiency in farm operations, and subjects these to a variety of techniques which artificially increase productivity. Irrigation, fertilizers, herbicides, fungicides, tractors, and farm amalgamation (to achieve economies of scale in farm operations) are a few of the techniques used.

By such means, desired high productivity is obtained, but at certain costs, which we shall examine in this chapter. The lowered species diversity and increased intraspecific genetic homogeneity produced by selection of strains imply a highly unstable system, with little inherent ability to adjust to environmental changes. Farm practices are becoming fantastically energy-hungry, and the high rate of energy flux, associated with a high rate of soil mineral turnover, can produce long-term destructive processes in the soil. Given the variety of attributes we seek in an optimal agricultural strategy, and the great variety of techniques available for seeking such optima, it seems worthwhile to take a very basic look at the fundamental nature of agricultural strategy selection. Indeed, as we shall see, there is a real question whether an optimal agricultural system can be put into worldwide practice, given the enormous energy required to do this on a sustained basis and the current prognosis for a worldwide energy shortage. This in turn raises the question whether the present world human population can be supported indefinitely. There are reasons to believe that only a smaller worldwide human population density can be supported with reasonable stability in the necessary agricultural production.

From the preceding discussion, it appears that a restatement of the aims of agricultural strategy is required.

The overall agricultural strategy objective is to produce the maximal amount of an optimal mix of calories, protein, vitamins, minerals, and essential amino acids per acre with the available energy, subject to the constraints that both money and energy must be used as efficiently as possible, and practices must minimize long-term instability in production and prevent long-term degradation of the productive system.

We will now consider the types of strategy options available to the farmer to help him meet this aim.

AVAILABLE KINDS OF AGRICULTURAL STRATEGY OPTIONS

Many strategy options are open to the farmer. We will begin by listing twelve of the most important of these; the first six concern broad questions of the strategy of farm operation; the last six concern more specific questions of the strategy of crop management. No attempt is made here to prove that these options are significant; this is merely a list to give the student an overview of the types of options that are available; documentation follows the list.

1   A basic question for all farmers is almost never considered by them: Do I raise domestic strains of plants or animals, or do I attempt to crop native strains that have always been here, even before farming? Or, perhaps, do I crop wild animals imported from another country? For example, do I crop bison or Shorthorns, Herefords or antelope?

2   How much of the forest should I cut on my land? Is it important to keep a lot of forest on my land, for shelterbelts, firewood, fence railings, erosion suppression, and evening out the rate of water flow through the year, or is my forest land more valuable if converted to cropland or pasture?

3   How much meat should I produce, in comparison with the amount of plant material I produce? Beef sells for more per pound, but it takes a lot more land to produce a pound of beef.

4   Should I get rid of all my draft animals and replace them with tractors? Tractors generate more horsepower than horses, per pound, but they consume gasoline, cost more, and don't produce fertilizer.

5   Should I irrigate, and if so, how should the system be laid out, and how much of my property should be irrigated?

6   How big should my farm be?

7   What kind of crops should I grow—cereals, vegetables, or what?

8   How densely should my seed be planted?

9   Should I grow only one type of plant in each field or orchard, or should I grow mixed crops in each field?

10   How much pesticide should I use?

11   Should I use clean or dirty culture in my orchards? That is, how much care should I take to eliminate weeds and shrubs from under my orchard trees, and shrubs, weeds, and other trees from the edge of the orchard?

12   Granted that different crops may show different increases in growth rate in response to fertilizer application, how should this knowledge modify my plans about which crops to grow and which fertilizers to use?

Unfortunately, in recent years an important thirteenth option has become all too obvious for many farmers in North America: Should I sell my farm for tract housing or industrial sites? During the last few years, holding on to farmland for purposes of farming has been one of the most unattractive financial investments in America. In fact, it has been only half as rewarding as investing in a randomly selected set of common stock of equivalent value. Selling farmland for real estate, however, has been one of the more attractive financial moves in our society. This has probably been a temporary situation, made possible by a short-lived period of food surpluses in North America.

EVALUATION OF AGRICULTURAL STRATEGY ALTERNATIVES

Whenever settlers went out from Europe to colonize other parts of the world, two broad strategy options for agriculture were available to them. The situation was almost never understood in these terms, however, and the existence of one of the options was rarely perceived. The two options were, first, to eliminate wild plants and animals and replace them with imported strains and species from Europe, and, second, to cultivate the plants and animals found in the new lands. The former option was typically perceived as being without question the appropriate strategy. Thus, America, from the Alleghenies to the Rockies, was converted from a land of 80 million buffalo (in 1800) to a land of Shorthorn, Aberdeen Angus, and Hereford cattle today.

Ecology is a fairly recent science, and measurements on energy flow in ecosystems date chiefly from the period after 1940. The early settlers could not have known that the natural vegetation and animal life might have been adapted to make better use of incoming solar radiation than species or strains imported from some other place by man.

The most serious challenge to the conventional wisdom has come from a large body of measurements made in various countries, but particularly in Africa. It should be noted at the outset that in making comparisons between meat production by wild animals and meat production by domestic animals, we must be careful to take account of any fossil fuel energy subsidy to the domestic animals, in the form of fertilizer or care.

Tables 11-1, 11-2, and 11-3 bring out the essential differences. The first of these shows that wild ungulates can exist at higher population densities than domestic livestock. Table 11-2 shows that for many wild ungulates, the live weight gain per day per animal is higher than for domestic sheep and cattle. Table 11-3 shows that there is a higher percentage of lean meat on the cold carcass of the wild game than on the carcass of cattle.

Thus, it seems clear that wild animals may be a very sensible way to raise meat on rangeland. Further, they produce a high proportion of lean meat; the current understanding about the relation between animal fat in the human diet and coronary artery disease can raise real questions about the medical advisability of having a high proportion of fat in a carcass to be used for human food.

These data indicate that much more experimentation would be justifiable to compare the actual poundage of lean meat produced per acre per year from game compared with domestic stock. It may well be that wild game represents a superior way to use the land.

Forests are of great value to agriculture for their manipulation of the water supply and the microclimate around fields. The evidence is assembled in Fig. 8-1A, B, and C. But forests adjacent to agricultural land serve various other purposes that are often not appreciated. For example, wild flowers in and at the edge of the forest serve as a reservoir of

TABLE 11-1 Yearlong ungulate biomass data from east Africa and elsewhere

approximate year-long biomass, pounds per square mile	approximate size of area, square miles	animals	range type	location
70,000–100,000	1,000–2,000	Wild ungulates	Savanna	East Africa
30,000	2,000–3,000	Wild ungulates	Bush	East Africa
21,300–32,000*	. . .	Cattle	Managed savanna (European ranches)	
26,700*	. . .	Domestic livestock	Average of virgin long and short grass	Western U.S.
14,000–20,000	. . .	Bison and associated wild ungulates	Prairie	U.S.
19,700*	1,126,500	Domestic livestock	Average of all virgin ranges	Western U.S.
11,200–16,000	. . .	Domestic livestock	Savanna (tribal grazing land)	East Africa
10,600	. . .	Black-tailed deer	Managed chaparral	California
6,900	. . .	Black-tailed deer	Average of oak woodland plus chaparral	Pacific Coast North America
5,800	2	White-tailed deer	Woodland proper density	Michigan
3,550	4,373	Red deer	Deer forest	Scotland
1,360	88,080	Mule deer (5 races)	Average all ranges	California
1,300	. . .	Mule deer	Average all ranges	Arizona

SOURCE: Talbot and Talbot (1963).
* Based on animal units equivalent to one 1,000-pound steer.

TABLE 11-2 Liveweight gains per day, domestic livestock and wild ungulates on east African rangeland

species	liveweight gain per day		period, months	average adult weight		approximate age when adult weight reached, months
	lb	kg		lb	kg	
Thomson's gazelle	0.13	0.06	10	41–	18.6–	18
	0.08	0.04	15	53	24.0	
Impala	0.26	0.12	10	101–	45.8–	
	0.20	0.09	18	131	59.4	24
Grant's gazelle	0.26	0.12	10	101–	45.8–	
	0.22	0.10	18	146	66.2	24
Topi	0.44	0.20	12	252–	114.3–	
	0.34	0.15	24	292	132.4	30
Kongoni	0.50	0.23	12	270–	122.5–	
	0.39	0.18	24	332	150.6	30
Wildebeest	0.52	0.24	12			
	0.44	0.20	24	360–	163.3–	
	0.41	0.19	30	460	208.7	45
Eland	0.54	0.24	48	625–	283.5–	
				830	376.5	?
Domestic sheep	0.12	0.05	18	44–	20.0–	
				100	45.4	18
Domestic cattle	0.30	0.14	. . .	350–	158.8–	
				1,000	453.6	60

SOURCE: Talbot (1963).

TABLE 11-3 Carcass composition of domestic cattle compared with east African wild animals

species	lean (percent cold carcass)	fat (percent cold carcass)
Cattle (sample 1)	56.3 ± 5.4	28.4 ± 7.3
Cattle (sample 2)	60.9 ± 6.0	19.3 ± 7.9
Wildebeest	76.6 ± 2.2	5.4 ± 2.7
Thomson's gazelle	80.4 ± 1.6	1.6 ± 1.3
Kob	81.3 ± 2.2	3.1 ± 1.5

SOURCE: McCullogh and Talbot (1965).

insect and other parasites and predators that serve to keep agricultural crop pests in check. Indeed, the fecundity and longevity of certain parasites of crop pests depend on whether they have fed on certain wild flowers (Leius, 1961). If the pests on a crop build up in numbers to an excessive degree, the enemies of the pest can migrate in from wild flowers and control them (Bierne, 1963). In general, the presence of woodlots or forests adjacent to agricultural land implies increased species diversity in the agricultural biome, and resultant higher species population stability. Also, forests serve as an additional source of income and add stability to the farmer's economic condition, by giving him an alternative source of funds during periods when poor market conditions make it inadvisable to sell other products over the short term.

A basic question of strategy for farmers concerns the choice of crop they should be producing. It has already been indicated that the efficiency with which energy is transferred from trophic level to trophic level within ecosystems is very low, and this is true in agricultural systems as well.

H. T. Odum devotes considerable attention to these energy conversion efficiencies in agriculture; he finds that they are very low. The percentage efficiency of crops in using incoming solar radiation is only about 0.02 or 0.03 without fertilization, and 0.12 to 0.25 with fertilization. The efficiency of livestock, in turn, in converting range and grain into meat varies widely, from 1 percent of the available food energy on unfertilized range to 40 percent of the available food energy in a cattle feedlot, where the cattle expend no energy in walking to obtain food. In the case of the 1 percent efficiency, however, the farmer requires 100 times as much land to produce a kilocalorie of meat per year as he does to provide a kilocalorie of grain. Clearly, this makes sense only if the higher price for meat per unit weight makes it possible, or if he cannot grow grain on the land being used for the pasture or range.[1]

As we might expect, then, the tremendous energy inefficiency in converting plants to cattle is reflected in prices. On February 11, 1971, the wholesale cash price for 100 pounds of wheat in the United States was $2.62, whereas it was $33.15 for steers. This difference underestimates the magnitude of the conversion inefficiency, because the $2.62 represents a larger land area than the $33.15, much of the beef weight being produced in small cattle feedlots.

[1] H. T. Odum, "Environment, Power and Society," New York, copyright 1971 by John Wiley & Sons, Inc.

In order to decide what he should be growing to make the most profit, the farmer must balance off the higher price for meat against this great loss in energy throughout the food chain, together with the operating costs associated with grain or beef.

The question whether draft animals should be replaced by tractors is more complicated than it appears at first, and there may be an optimal short-term answer that is different from the optimal long-term answer. The catch in the argument is that the tractor gives the farmer much greater power at the price of an immense fossil fuel energy subsidy: now he has to buy fuel, and also the fertilizer to replace the manure from the draft animals. The added power makes it worth it, unless the fossil fuel becomes so scarce that gasoline and diesel fuel are too expensive for the farmer. At that point, he might want to go back to draft animals, but then another problem shows up. On some of the acreage being used to grow food for people he now must grow food for draft animals. This transformation all over the country (any country) sharply reduces the carrying capacity of the country for people. This is what would happen if fossil fuel ran out and nuclear fuel was inadequate to replace it.

One way to demonstrate how tremendously yields are influenced by increased available horsepower, in the form of tractors, is by making comparisons between countries. This is done in Fig. 11-1, where we see that an increase in total available horsepower per hectare from 0.1 to 2.0 is associated with an increase in yield of major crops from about 700 to about 5,500 kilograms per hectare. Thus, a twentyfold increase in power is associated with a roughly eightfold increase in yield. Thus, while yield is quite sensitive to power input, it takes a great deal of power to get the increased yield. Clearly, world agriculture has become very dependent on a fossil fuel energy subsidy. Once again, we have an example of Principle 3.

Perhaps the central problem with respect to irrigation is disposing of the salts that are produced when water evaporates in an arid or semi-arid region, leaving behind a much more concentrated solution. It is possible to lay out a drainage system which does this, but it takes skill, ef-

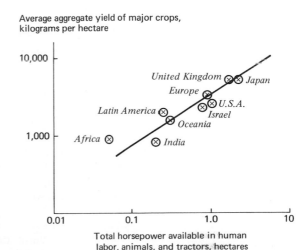

FIGURE 11-1 The relation between fossil fuel energy subsidy to agriculture, in the form of power applied to agriculture, and crop production. (*Replotted from data in fig. 2 of Giles, 1967.*)

fort, and organization to maintain it. Where the system of flushing out these salts breaks down, the soil fertility declines because of excessive salt concentrations, and the resource base of an entire nation can be significantly diminished, as in Iraq, once the center of the most advanced civilization on earth, that of Babylon.

This is an illustration of Principle 15: the salt concentration moved the land to the edge of the zone of tolerance for crops.

Profitable farming and farm size are closely related. Particularly if a cash crop has a relatively low value per unit weight, as in the case of wheat, a very large farm, of three or more sections (there are 640 acres per section), may be required for profitable operation. It follows that the size of the farm required for profitable operation depends on the profit per unit and the value per unit of the crop.

The kinds of crops that will do best in any given habitat will be determined by the climate, the soil type, temperature, and water availability, in line with the theory of Holdridge presented in Chap. 8. Apples, for example, require a cold climate which would kill citrus, and bananas require a very hot moist climate. Further, particular strains of a cereal, such as wheat, are adapted to grow in particular bands of latitude.

For each set of habitat characteristics, there is a particular planting density which is optimal for each type of crop, as we would expect from Principle 4. When the plants are young, competitive pressure is slight or nonexistent, and yield in grams per unit area is proportional to the density of the seeds planted. However, as the plants get older, competition for nutrients and light becomes progressively more important at excessively dense plantings. This means that at planting densities above the optimal level there is no further increase in yield, although this does not show up until the plants are well through the growing season. Figure 11-2 is a typical set of data illustrating this intraspecific competition in a crop plant.

Answering the question whether crop plants should be grown separately or in mixture is complicated. In general, the total yield is close to what one would expect from the yields of the two species grown separately. Indeed, the combined yield can be calculated from the separate yields (Baeumer and de Wit, 1968). In such instances, there is no increased total production to provide a motive for growing two plant species together.

There may be other motives for growing two crop species together, however, provided that the benefits are not absorbed by the additional handling costs of separating the two species after harvesting. By growing two species together, we increase the extent to which the combined canopy can trap incident light energy (Rhodes, 1968, 1969; Singh and Misra, 1969). Another motive for mixed cultivation is to minimize the probability of buildup of insect populations to high densities. The application of chemical controls for insect pests may not be necessary if there is a sufficiently diverse array of plant species in a crop (NRC-NAS, 1969). This illustrates Principles 13 (stability and diversity) and 23 (the distribution of individuals in a population, in this case pests, affects their energy efficiency).

Another quite different reason for growing two plant species together is that different plants have different zones of tolerance. In

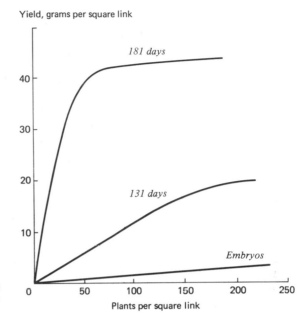

FIGURE 11-2 Dependence of yield of white clover on plant density and time since planting. (*Donald*, 1963.)

southern California, near Palm Springs, a desert environment, there is a tremendous amount of incident solar radiation per day, and a great deal of evapotranspiration. Crops are grown under irrigation, in two layers. The top layer, date palms, can withstand high light, heat, and drying. The understory, oranges, is shaded by the date palms. Consequently, the temperature, light, and drying are less than they would be in the absence of the date palms.

The best amount of pesticide to use is the smallest amount necessary. This minimizes the ratio of gross overhead to gross profit, but it also makes sense on biological grounds. Pesticides reduce the population density of the target organism quickly (at least at first), but typically they have a number of subsequent effects that may more than nullify the immediate positive effects. First, the sharp immediate reduction in pest densities reduces intraspecific competition pressure in the pest population, so that rates of survival, growth, and reproduction in the surviving individuals are elevated. This means that in the generation following the sprayed generation, there may be as many or more pests than there were at the time of spraying.

Also, if a pesticide reduces both the pest, an herbivore, and its enemies, which are carnivores, by the same percentage of their densities, then the absolute effect on the carnivores will probably be much greater, because they were present at much lower densities than the pest in the first place. This follows from the low energy conversion efficiencies from one level to another of trophic pyramids, already discussed in this chapter (Principle 2).

A third mechanism which enhances the survival probability of the remaining pests after pesticide application is resistance: the pesticide tends to eliminate all those individuals except those preadapted to survive it. This means that the most intense kind of selection pressure is being applied to the pest population: we are selecting for those individuals that survive the pesticide (Principle 6). Such individuals are likely to be inordinately vigorous in other respects, such as egg laying, and this has indeed been found to be the case (Kuenen, 1958). For the three reasons just mentioned, a typical consequence of using pesticides is that it becomes necessary to keep using them year after year. It is surprising how infrequently people ask why this is necessary, in view of the meaning of the word *pesticide*. If it is at all possible, pesticides should be used in conjunction with another method of controlling pests or should be replaced altogether by a different method (Chap. 12).

It has been discovered that where rainfall is great enough to permit it, "dirty" culture in orchards makes it possible to raise fruit more profitably without using any pesticide at all, because a large enough resident population of parasitic and predacious insects can be supported by the variety of plant species present to ensure control of any insect pests that might tend to build up in density (Pickett, 1949; Pickett and Patterson, 1953). Quantitative evidence on the effect of parasites and predators on orchard pests is found in the symposium edited by Le Roux (1963).

Pesticides can knock pest populations down quickly, but they lack the desirable long-term effects of other methods of control. Other methods of control may not be able to knock a pest population down quickly. Consequently, the concept of integrated control has developed, in which several different kinds of insect pest control are used together, each method being selected in part because of its compatibility with the others (van den Bosch and Stern, 1962).

Not all crops respond the same way to a particular fertilizer at specific levels of application, and the profitability of different crop-fertilizer combinations may vary considerably from field to field. Indeed, the responses of different plant species to a given fertilizer program may be so different that fertilizer treatment may be used as the means of altering the species composition of a plant community in a desired direction. For example, Robertson and Nicholson (1961) studied the effect of different combinations of calcium and phosphorus on various plants in hill pastures of northern Scotland. Calcium was applied as ground limestone, 0, 4, or 30 hundredweight per acre, and the phosphorus as superphosphate, 0, 50, or 150 pounds per acre; all nine possible combinations were tried. Fescue grasses and white clover increased in response to calcium, or phosphorus, or the combination. Matgrass, on the other hand, decreased in response to either fertilizer, or the combination. Other plants showed more complicated responses, increasing in response to one fertilizer, but decreasing in response to the other, and in some plant species there seemed to be unusual interaction effects.

WHAT CAN WE EXPECT FROM AGRICULTURE IN THE FUTURE?

There are two means by which we can estimate how great agricultural yields might be in the future. One means is to consider the maximum theoretically possible yields. The second is to find out the maximum yields that have already been obtained anywhere in the world, for each crop, during absolutely optimal growing seasons and using the best available agricultural technology. These would then be taken as a reasonable upper limit to worldwide *average* yield for that crop and would provide a much more conservative estimate than reasoning from a theoretical maximum limit.

We will consider three different estimates of theoretical maximum crop yields, by de Wit (1967), Kleiber (1961), and Howard Odum (1971). Table 11-4, based on data from de Wit, derives a worldwide mean estimate of 6.4×10^7 kilocalories of energy produced in crops per year. Kleiber estimated that under optimal conditions, 5 percent of the radiant energy from the sun appears as chemical energy in the crop. Suppose we take as a fairly average worldwide incident solar radiation 1.6×10^{10} kilocalories per year, the figure for California, per hectare. This would represent production of 8×10^8 kilocalories per hectare per year in grain crops, and following de Wit, if a quarter of this is usable for human consumption, we have the figure 2×10^8. This is about three times de Wit's figure, but some of the difference is attributable to using California as the source of the mean incident radiant solar energy figure. Much of the world's grain is grown considerably farther from the equator.

Howard Odum also has made a very elaborate analysis of the ability of the world to produce food, in which he relates the food yield, in kilocalories, from a unit of land surface to the auxiliary fossil fuel energy

TABLE 11-4 Computation of the maximum possible worldwide mean production of kilocalories per hectare per year*

latitude, °N	land surface, hectares × 10⁸	percentage agricultural land	total agricultural land, hectares × 10⁸	theoretical maximum energy production, kilocalories per hectares × 10⁶	total × 10¹⁴
70	8	52	4.2	12	50
60	14	38	5.3	21	112
50	16	18	2.9	59	170
40	15	13	2.0	91	178
30	17	11	1.9	113	211
20	13	10	1.3	124	161
10	10	10	1.0	124	124
0	14	10	1.4	116	162
−10	7	10	0.7	117	82
−20	9	10	0.9	123	111
−30	7	10	0.7	121	85
−40	1	14	0.1	89	13
−50	1	53	0.5	12	6
Total	132		22.9		1,465

SOURCE: Columns 1, 2, 3, and 5 from de Wit (1967).

* The conversion from kilograms of carbohydrate to kilocalories was made by using the following information from de Wit. Reasoning from known figures for photosynthetic efficiency, the light-scattering coefficient, the leaf area index, the leaf display, the light intensity, and the direction of the incoming light, he computes the potential photosynthesis of a crop surface in the Netherlands of 50,000 kilograms of carbohydrate per hectare. He assumes that half of this will not be obtained because of respiration losses and nonclosure of the crop surface. Only half of the remainder is usable for human consumption. This remaining 12,500 kilograms of carbohydrate yields about 50×10^6 kilocalories of energy. Thus, we have the conversion: 50×10^3 kilograms of carbohydrate per hectare represents 50×10^6 kilocalories per hectare, or 10×10^3 kilograms represents 10×10^6 kilocalories. For each line of de Wit's original table, carbohydrate production was converted to kilocalories by this method.

The world mean theoretical maximum crop production in terms of kilocalories per hectare comes from

$$\frac{1465 \times 10^{14} \text{ kilocalories of energy for all the world's arable land}}{22.9 \times 10^8 \text{ hectares, the world's total arable land}}$$

or 6.4×10^7 kilocalories per hectare.

in work flows added into the agricultural system in the form of technology, over and above the incident solar radiation. Thus he gives a range of production estimates, from agriculture without an energy subsidy to the highest attainable yields, with algae culture, and immense inputs of fossil fuel energy. He estimates the gross photosynthesis line at 5×10^8 kilocalories per hectare per year, but gives 10^6 to 10^8 as a reasonable range of production values, for various amounts of auxiliary energy. The high end of this range, 10^8, is intermediate between the estimates of 6.4×10^7 by de Wit and 2×10^8 by Kleiber.[1]

It is instructive to compare these theoretical estimates with the highest yields of grain actually obtained by those countries which have the most energy-intensive and technologically sophisticated agriculture. The highest wheat yield on record for any country was 4,810 kilograms per hectare in Denmark in 1968. The highest rice yield was 7,190 kilograms per hectare in Australia in 1968. Using conversions of 3600 kilocalories per kilogram for both of these, we obtain actual energy productions by crops of 1.7×10^7 and 2.6×10^7 kilocalories per hectare per year, 27 percent and 41 percent respectively, of de Wit's estimate. The figures 4,810 and 7,190 are national averages. Wheat yields of about 9,000 kilograms per hectare have been obtained on experimental fields with proper disease control (de Wit, personal communication concerning work of De Vos, Holland).

Clearly, a very high level of agricultural technology will be required to attain anything close to the theoretical maximum yields on a worldwide scale.

Figure 11-1 illustrates the sensitivity of agricultural yield to inputs on farm machinery. Yield is equally sensitive to all other inputs of agricultural technology. Ennis et al. (1967) have examined the relation between average yield of principal crops and intensity of fertilizer application in India, the United States, and Japan. To get from the Indian 1958 yield of about 700 kilograms per hectare to the Japanese 1963 yield of about 5,500 kilograms per hectare would require an increase in fertilizer application from about 1.5 to about 280 kilograms per hectare. Thus, it takes an amazing $\frac{280}{1.5}$ or 186-fold increase in fertilizer to produce a mere $\frac{5,500}{700}$ or eightfold increase in yield. This illustrates the dramatic increase in energy input to agriculture that will be required if the world's hungry are to be fed.

THE GREEN REVOLUTION IN AGRICULTURE

There has been a great deal of enthusiasm in the popular press about the Green Revolution. This refers to the exciting and admirable program supported by the Ford and Rockefeller Foundations under which new high-yield varieties of grains have been developed and introduced on a large scale into the Philippines, Mexico, India, Pakistan, and Turkey. This breakthrough has been a necessary contribution to solution of the population-resource crisis in underdeveloped countries. It is not sufficient, however, to solve the problem — revolutions in birth control, economic and political organization, power supplies, and education are needed also.

[1] H. T. Odum, "Environment, Power and Society," New York, copyright 1971 by John Wiley & Sons, Inc.

It is very important that the achievements of the Green Revolution be seen objectively, and in perspective, rather than as a panacea that will let us relax and ease off on aid to underdeveloped countries.

Paddock (1970) has indicated some of the additional factors we must be alert to in evaluating the current status of underdeveloped countries.

First, we must be careful not to assign credit to new varieties for increased yields that are actually due to improved weather. The years from 1967 to 1970 have been good weather years for crops in southern Asia, whereas 1965 and 1966 were bad. Thus, much of the apparent improvement in yield was not due to new strains. According to Principle 3, there are five categories of resources, including matter (such as rainfall) and energy (such as temperature) that can be expected to affect yield.

Second, national yields in India were increased from 1951 to 1961 by putting new land into production. But that game can be played only until all new land is in production.

Third, continuing increases in yield will depend on a massive increase in fertilization, irrigation, and use of farm machinery. All this will require a great deal of capital in nations that are desperately poor.

Fourth, the Green Revolution confronts us with a genetic danger. If exceedingly large acreages are planted out to a small number of new "miracle strains" of wheat, rice, and maize, there is a terribly high risk of vulnerability to rusts or insect pests. In accordance with Principle 13, we should be concerned about use of diversity to promote stability.

Thus, although a great advance has been made, it gives us precious little reason for being sanguine about the future of the underdeveloped countries. Unless there are massive and continuing efforts at population control, economic growth, and food production, they have a truly bleak future ahead.

SUMMARY   Modern farming is a highly complex science, involving the selection of optimal strategies for replacing natural communities with man-supported communities which are productive, yet stable. Selection of the strategies which are best involves a knowledge of ecology and economics as well as more obvious topics in agriculture. Unfortunately, all farm operations have become tremendously dependent on an outside energy subsidy, and it is not clear that it will be possible to keep increasing the magnitude of the subsidy indefinitely; it may not be possible even to maintain it at the present level (see Chap. 15).

SUGGESTIONS FOR INDIVIDUAL AND GROUP PROJECTS   Obtain the latest Production Yearbooks of the Food and Agriculture Organization, United Nations. Make graphs of yields per hectare from 1950 to the present for the following crops:

Wheat in Mexico
Wheat in Denmark
Wheat in the Netherlands
Wheat in Japan
Wheat in the Soviet Union

Wheat in the United States
Wheat in India
Rice in India
Rice in Japan
Rice in Spain
Rice in the United States
Rice in Australia

Try to explain differences from year to year for each country, and differences in the trends from country to country in terms of agricultural revolutions and weather. Does it appear that yields can be increased indefinitely?

REFERENCES

Baeumer, K., and C. T. de Wit: Competitive Interference of Plant Species in Monocultures and Mixed Stands, *Neth. J. Agr. Sci.*, **16**:103–122 (1968).

Bierne, B. P.: Ecology in Biological Control, in E. J. Le Roux (ed.), Population Dynamics of Agricultural and Forest Insect Pests, *Mem. Entomol. Soc. Can.*, no. 32, pp. 7–10, 1963.

de Wit, C. T.: Photosynthesis: Its Relationship to Overpopulation, in A. San Pietro, F. Greer, and T. J. Army (eds.), "Harvesting the Sun," pp. 315–320, Academic, New York, 1967.

Donald, C. M.: Competition among Pasture Plants. I. Intraspecific Competition among Annual Pasture Plants, *Australian J. Agr. Res.*, **2**:355–376 (1963).

Ennis, W. B., Jr., L. L. Jansen, I. T. Ellis, and L. D. Newsom: Inputs for Pesticides, in "The World Food Problem," vol. 3, pp. 130–175, the White House, Washington, 1967.

Giles, G.: Agricultural Power and Equipment, in "The World Food Problem," vol. 3, pp. 175–208, the White House, Washington, 1967.

Kleiber, M.: "The Fire of Life," Wiley, New York, 1961.

Kuenen, D. J.: Influence of Sublethal Doses of DDT upon the Multiplication Rate of *Sitophilus granarius* (Coleopt. Curculionidae), *Entomol. Exptl. Appl.*, **1**:147–152 (1958).

Leius, K.: Influence of Food on Fecundity and Longevity of Adults of *Itoplectis conquisator* (Say) (Hymenoptera: Ichneumonidae), *Can. Entomologist*, **93**:771–780 (1961).

Le Roux, E. J. (ed.): Population Dynamics of Agricultural and Forest Insect Pests, *Mem. Entomol. Soc. Can.*, no. 32, 1963.

McCulloch, J. S. G., and L. M. Talbot: Comparison of Weight Estimation Methods for Wild Animals and Domestic Livestock, *J. Appl. Ecol.*, **2**:59–69 (1965).

National Research Council–National Academy of Sciences: "Insect Pest Management and Control," Subcommittee on Insect Pests, Committee on Plant and Animal Pests, Agricultural Board, National Research Council, Washington, Publication no. 1965, 1969.

Odum, H. T.: "Environment, Power and Society," Wiley, New York, 1971.

Paddock, W. C.: How Green Is the Green Revolution? *Bioscience*, **20**:897–902 (1970).

Pickett, A. D.: A Critique on Insect Chemical Control Methods, *Can. Entomologist*, **81**:67–76 (1949).

——— and N. A. Patterson: The Influence of Spray Programs on the Fauna of

Apple Orchards in Nova Scotia. IV. A Review, *Can. Entomologist*, **81**:472–478 (1953).

Rhodes, I.: Yield of Contrasting Ryegrass Varieties in Monoculture and Mixed Culture, *J. Brit. Grassland Soc.*, **23**:156–158 (1968).

———: The Yield, Canopy Structure and Light Interception of Two Rye-grass Varieties in Mixed Culture and Monoculture, *J. Brit. Grassland Soc.*, **24**:123–127 (1969).

Robertson, R. A., and I. A. Nicholson: The Response of Some Hill Pasture Types to Lime and Phosphate, *J. Brit. Grassland Soc.*, 117–125, 1961.

Singh, J. S., and R. Misra: Diversity, Dominance, Stability and Net Production in the Grasslands at Varanasi, India, *Can. J. Botany*, **47**:425–427 (1969).

Talbot, L. M.: Comparison of the Efficiency of Wild Animals and Domestic Livestock in Utilization of East African Range Lands, in *The Conference Papers*, pp. 329–335, the Arusha Conference, I.U.C.N. (N.S.), 1963.

——— and M. H. Talbot: The High Biomass of Wild Ungulates on East African Savanna, *Trans. 28th North American Wildlife and Natural Resources Conference*, pp. 465–476, 1963.

van den Bosch, R., and V. M. Stern: The Integration of Chemical and Biological Control of Arthropod Pests, *Ann. Rev. Entomol.* **7**:367–386 (1962).

12 biological control: the reestab-lishment of a tolerable steady state

In view of the immense amount of controversy about chemical pes-ticides, it is surprising how little discussion there is about alternative means of controlling pests. There are, in fact many means of doing this: (1) the use of parasites, predators, and diseases against a pest, (2) the use of sex attractants so that an entire area smells like the female pest, and the males can no longer find prospective mates by following smell, (3) chemosterilants, which prevent a pest from reproducing, rather than poisoning it, and (4) a wide variety of cultural control techniques.

However, the term *biological control* is typically applied only to the use of a parasite, predator, or disease, or a mix of many different species of these, to control a pest. Various combinations of species have been in-volved in such experiments. The original cases most frequently made use of insect parasites and predators to control insect pests. The parasites were various types of hymenopterous insects, such as ichneumon flies or braconid flies, or dipterous (true fly) parasites, such as sarcophagid or tachinid flies. Various types of predators have been used, such as ladybird beetles (coccinellids), and various species of fly larvae that are predacious on other insects. The idea was gradually applied to many other combinations, however, so that insects were used to control weeds, and fungi, bacteria, and viruses were used to control insects. Predacious snails have been investigated as possible control agents for other snails, and with man's growing interest in and concern with the ocean, some very exotic organisms are being studied. For example, the crown-of-thorns starfish is known to be attacked by certain species of shrimp and conch shells, and we may witness experiments to use these in control pro-grams on a commercial scale.

The fundamental idea of biological control is always the same: the balance of nature is disturbed either because a pest is accidentally im-ported into a part of the world where it has no controls, or because a pest builds up to unusual densities in its native habitat. In either case, the aim of control is to reestablish a natural balance by importing one or more species which are enemies of the pest in its original home, or enemies of close relatives of the pest in another region or country, in order to decrease the survival probability of the pest.

Some of the successes in biological control have been immediate and spectacular. In other cases, there was no immediate, obvious reduc-tion of the pest following introduction of the biological control agent, but a gradual decline accompanied by a gradual increase in the parasite's ef-fectiveness over a period of up to several decades, by the end of which the pest was no longer important. Unfortunately, by the time the pest is under control, people have forgotten that it was a pest, or that a parasite was introduced to control it. When benefits are immediate enough for the public to be able to associate cause and effect, they are often attributed to chemical pesticides although they may be more apparent than real, and very short-lived.

Biological control has always been of great interest to ecologists, because it is one of the few practical activities in which it has been possible to apply ecological theory, from population theory to the theory of community stability, and to see whether results match predictions.

SOME EXAMPLES OF BIOLOGICAL CONTROL

One of the most spectacular examples of biological control was the first major success. In 1887 the citrus industry in southern California was in danger of being wiped out by the cottony-cushion scale, which apparently had been accidentally introduced from Australia. Albert Koebele was sent to Australia in 1888 by the Division of Entomology, U.S. Department of Agriculture, to find natural enemies of the scale (Doutt, 1964). He found two species, one a tiny fly, *Cryptochaetum iceryae*, and one a ladybird beetle, *Rodolia cardinalis*. The fly larvae, interestingly enough, belong to a family which is predominantly phytophagous, but the genus *Cryptochaetum* is anomalous, containing eight species which are all internal parasites of Homoptera, the true bugs (the group of insects containing the scale insects) (Clausen, 1940). Koebele wrote to Charles Riley, then Chief of the Division of Entomology, that he had found the two species, the fly parasite within the scales, and the larvae of the ladybird beetle eating the eggs of the scales. Riley replied to Koebele that "the sending of the coccinellids is of course desirable but I think we have much more to hope from the *Cryptochaetum*" (Doutt, 1964).

Accordingly, Koebele sent about 12,000 of the *Cryptochaetum* to California, and 129 of the ladybird beetles. The latter were placed on a tree that was heavily infested with scales. Within about $2\frac{1}{2}$ months, nearly all the scales had been eaten up by the ladybird beetles, which were now allowed to spread to adjoining trees. Within two more months, 10,555 of the beetles had been sent to 228 orchards. Subsequently, another 385 of the beetles were sent to California from Australia. Within 11 months of the time that Koebele had been sent to Australia, the scale in California had been brought under control! The shipment of oranges from Los Angeles County jumped in 1 year from 700 to 2,000 carloads. The cost of the biological control project was $1,500 (Doutt, 1964).

This history reveals several general principles of biological control. First, if a parasite or predator is going to be successful as a control agent, it is in most cases likely that a *very small number will be required to be effective*. An immediate corollary is that successful biological control programs may be fantastically cheap in comparison with the benefits they yield.

Second, it is currently impossible to tell in advance which of a series of potential biological control species is likely to be most effective in controlling the pest; this is the significance of the Riley letter to Koebele. Either field testing or some type of highly sophisticated laboratory and insectary experimental program, linked with a sophisticated and currently unavailable body of theory, would be required to evaluate parasites and predators in advance.

Third, an essential idea basic to almost all successful control programs is importation. There is typically little point in trying to build up the densities of a parasite or predator already coexisting with the host or prey species, the pest which we wish to control. In the first place this is a tremendously and continually expensive undertaking, and in the second place the present densities of a parasite or a predator are the product of natural forces. Unless the expensive manipulation is continued, the probable outcome of any attempt to raise the density of the parasite or predator will be its ultimate or rapid return to the density at which it occurs naturally and at which it is incapable of controlling the pest. This follows from the notion of "carrying capacity," a corollary of Principle 4. Consequently, if the complex of parasitic and predatory species coexisting with a pest is incapable of controlling it, biological control must be sought by importing one or more new species of enemies from some other region or country. These are found by going to other parts of the world and discovering the enemies of the pest in those areas where it originates, or alternatively finding enemies of its closely related species.

Fourth, the probability of being successful with a biological control agent quickly is heightened if the parasite or predator is spread quickly in the entire range normally inhabited by the pest.

One of the most interesting chapters in the history of biological control is the campaign against the European spruce sawfly in Canada (McGugan and Coppel, 1962). Foresters first noticed this species severely defoliating spruce stands in the Gaspé Peninsula of Quebec in 1930. By 1935, 6,000 square miles were infested, and the affected area increased abruptly thereafter. The outbreak did not decline sharply until after 1942. The total loss of spruce has been estimated at 11,400,000 cords: prices change, but a crude estimate of the worth of this loss is at least 44 million dollars. The efforts made by the Canadian government to control this situation were truly heroic: a parasite production plant was opened in Belleville, Ontario, in 1936, and over 890 million parasites, comprising 27 different species, were released to control the pest. Many species of parasites were gathered in Europe, but the sawfly species was so rare there that many of the parasites were obtained from nine species of its close relatives in Europe.

The consequence of all this activity was that the sawfly numbers began to decline in 1938. It is not clear, however, exactly why the pest was controlled. Field studies showed that during the period of control by the parasites, a great many of the sawflies were dying because of a disease that was subsequently identified as a polyhedral virus. It is generally believed that this was not a native disease, but was fortuitously introduced from Europe, either with parasites or sawfly hosts, and was then spread to insects which were released in the field to control the pest. The virus spread rapidly, and this incident probably figured prominently in the development of a new type of biological control: the use of viruses, bacteria, and fungi.

A different type of situation, described by Huffaker and Kennett (1969), concerns the control of Klamath weed, or Saint-John's-wort. This weed had overrun and essentially replaced grass in many range areas of California. A number of insects were released to control it, including the chrysomelid beetle *Chrysolina quadrigemina*. This beetle gradually displaced the other three insect enemies of the weed as the principal controlling agent, and it was strikingly effective. In localities in which about 60 percent of the ground had been covered by the weed, it was eliminated by the beetle in between 3 and 6 years.

FACTORS DETERMINING THE SUCCESS OF A BIOLOGICAL CONTROL AGENT

All authorities agree that certain attributes must be possessed by a biological control agent in order to be successful. Most authorities stress the primary importance of searching ability. The species must be able to find whatever species or organism it is supposed to be controlling, or it will not be very useful. It must have a high enough reproductive rate to build up in numbers fast enough to prevent buildup in numbers of the pest. It must be able to operate effectively against the pest without unduly interfering with other members of its own species. This means, for example, that it should not lay eggs on a member of the pest population if enough eggs to ensure its destruction have already been laid on it. To be successful, a biological agent must be able to perform well in the climate in which it is released. Also, it must be insensitive to competition from other enemies of the pest that might be released, or already present in nature. That is, its physiological tolerance characteristics must match the characteristics of the environment. Before releasing a biological control agent we must be certain that it will confine its diet to the pest species and avoid eating other species of economic value. Finally, the biological control agent should have no special requirement for some resource that the environment cannot supply in sufficient quantity to maintain the agent's density. For example, suppose a particular parasite species can maintain itself as an adult only on the nectar or pollen from a particular plant, and suppose the density of that plant in the habitat is limited. Then the density of the parasite will be limited by the plant, not the abundance of the host, and it may never be able to build up to great enough density to limit the host effectively.

None of these are vague theoretical concepts. In fact, all these parameters can be measured, and have been measured, and are incorporated into modern theories of pest control. For example, one of several equations that have been proposed to account for the numbers of the pests that get attacked over a given time interval by parasites is

$$N_A = PK(1 - e^{-aN_0P^{1-b}})$$

where N_A = number of pests attacked by parasites

N_0 = number vulnerable to attack

P = parasite population density

a, b, and K = constants equivalent to the searching rate, the parasite intraspecific competition pressure, and the maximum possible egg-laying rate, respectively

Parenthetically, it should be noted that this is only a preliminary model which historically has been useful but now has been superseded by a slightly better model. Thus this is an object lesson for students, alerting them to the predictable appearance of successor models of progressively greater accuracy.

This model can be used in simulation studies to determine the effect of different values of a, b, and K, and other parameters on the economics of operating a forest (Watt, 1964). For example, we can ask ourselves how much money we would lose over a 30-year period because of dead trees which could not be salvaged and sold, assuming we used parasites with different sets of parameter values for these, and other parameters. This is the method of *simulation*, and it is coming to be used a great deal in resource management. That is, instead of trying to find out how to manage forests by conducting an enormous amount of experimentation, we analyze all available data on particular types of forests and build a mathematical model which can be programmed for an electronic computer, so that the computer can be made to simulate the behavior of the real forest. Then we can play all types of games to find out the consequences for the economics of operating the forest over a century, say, of using a very wide range of management strategies.

Table 12-1 is an example of the summary output from one run of such a game. Five variables were being studied:

PARCOM	b
SE	a, the searching rate
eggs	K, the maximum possible egg-laying rate
CT4	the control threshold for strategy 4, the release of parasites. This is the pest density at which parasites are first released in the field. The lower CT4 is, the earlier in an outbreak the parasites will be released; the higher it is, the later in the outbreak they will be released
PAGRAD	the parasite gradient coefficient, a measure of the rate at which the parasites disperse outwards from the site at which they are released

The table shows that as we would expect, we cut down on tree losses with lower values of CT4 and higher values of PAGRAD, eggs, and SE. However, because this equation was built on actual information from two species of parasites important as enemies of the spruce budworm, the output has a deeper meaning. It can tell us the range of values an actual parasite ought to have in order to be able to control the spruce budworm. Neither of the two important parasites does control it, and this table shows why: they do not have the right combination of values for the

TABLE 12-1 Effect of changing parasite parameters on effectiveness of pest control by parasites

cost of dead trees to and including 1947 in millions of dollars when PARCOM = 2.0

CT4	PAGRAD	SE = 0.0005			SE = 0.0020		
		50 eggs	100 eggs	200 eggs	50 eggs	100 eggs	200 eggs
15.0	0.01	580.1	534.5	424.4	453.4	213.7	0.0
	0.04	577.9	530.1	418.2	439.0	193.7	0.0
	0.08	576.8	528.7	415.2	433.8	182.7	0.0
25.0	0.08	584.4	544.8	451.2	477.5	264.3	25.2
50.0	0.08	614.8	613.6	606.0	614.8	613.6	603.4

SOURCE: Watt (1964).

appropriate variables. One of the parasites has a high enough searching rate, but too low an egg-laying rate, and the other has a high enough egg-laying rate, but too low a searching rate. Nevertheless, a table like this shows what type of parasite we should be looking for in order to achieve successful control.

Similar tables can be developed for each of the important types of control: pesticide sprays, aerial sprays of viruses, bacteria or fungi, sex attractants, or sterilization.

THE ECONOMICS OF BIOLOGICAL CONTROL

A number of estimates have been made of the benefit/cost ratio for biological control projects. DeBach (1964) estimates the savings to the agricultural industry in California during the period 1923–1959 from notably successful biological control projects. The total savings for the entire period by preventing crop losses, plus pest control costs, was 115.3 million dollars. Against this, DeBach charged the total budget for all biological control work in California for the entire period, including salaries: 4.3 million dollars. The net benefit, therefore, was 111 million dollars! Few enterprises are so profitable! This cost represents a tremendous subsidy to agriculture. Perhaps it would be fair to charge 50 percent of savings and use the money to fund further research.

BIOLOGICAL CONTROL AND THE THEORY OF COMMUNITY STABILITY

A controversy of considerable interest has developed among ecologists about the relation between the strategy of biological control and the theory of community stability discussed in Chaps. 2 and 3. One point of view is that instead of releasing many different species of parasites and predators, and letting the best one gradually win out in the field, only natural enemies whose effectiveness can be ascertained in advance, from laboratory tests, should be released (Pemberton and Willard, 1918; Turnbull and Chant, 1961; Watt, 1965; and Turnbull, 1967). The reason for this position is that interspecific competition between the different species of parasites and predators may operate so as to render the combination less effective in controlling a pest than one good species of parasite by itself. The type of mechanism that has been postulated for this is allelopathic competition directed by the larvae of poor parasite species A against the larvae of good parasite species B when both are living in the host. "Good" and "bad" mean good searcher with high egg-laying rate and a long egg-laying period and poor searcher with low egg-laying rate and a short egg-laying period.

The contrary point of view is that this is a thoroughly dangerous proposition: "Administrators and leaders in entomology in positions of decision in these matters tend to view the conclusions of these gentlemen as a warning that introductions should not be made except after the most elaborate study, and that even then only the absolutely best species should be introduced. This implies an endeavor for which funds will seldom or never suffice" (Huffaker and Kennett, 1969). This argument has to be taken very seriously. It will be noticed, for example, that the total budget for biological control in California over a 36-year period was 4.3 million dollars, or about $120,000 a year, during a period when biological control should have been expected to have terrific political support, because of a series of five major spectacular successes and a

SE = 0.0050

50 eggs	100 eggs	200 eggs
154.7	0.0	0.0
113.9	0.0	0.0
100.2	0.0	0.0
219.9	66.0	0.7
614.8	613.6	602.9

string of minor successes. Consequently, the argument that biological control has been, and perhaps always will be, underfunded is simply correct. Further, this controversy is made more confusing, and hence more interesting to the student, by the fact that there are rather clear-cut findings supporting both cases. For example, Willard and Mason (1937) showed that by increasing the number of parasites acting against the Mediterranean fruit fly in Hawaii, the total parasitization increased even though competition between the parasite species decreased the effectiveness of the parasite that would have been most effective if operating by itself. On the other hand, Force (1970) shows just the opposite: in laboratory studies the total parasitism is less effective when there are several species, including the one that would be best by itself, than when the best species is by itself. This was because either the fecundity of the superior parasite was depressed in the presence of the inferior parasites, or the larvae of the superior parasite could not compete effectively against the larvae of inferior parasites when both were in the same host. Students are urged to explore this controversy further, as an exercise in scientific logic, using the readings in the reference list and the suggested additional reading. They should be alerted, however, to the fact that the ultimate outcome of many scientific controversies is usually that considerable truth exists on both sides of the agrument. We certainly could use more critical data bearing on this matter.

Now what does all this have to do with the theory of community stability? Figure 3-13, with the associated discussion, is really the basis for a theory of biological control strategies. The student is strongly urged to study the argument of Southwood and Way (1970) if he is interested in this issue. However, the embarrassing fact remains that some of the most spectacular pests of all thrive in spite of a large number and variety of enemies. This is true of locusts and grasshoppers, mice, the spruce budworm, the gypsy moth, and many others. Why are these animals pests? How are they able to build up to such great numbers on occasion if in fact a large number of species at the carnivore level implies stability throughout the trophic pyramid, rather than stability at just the carnivore level? This would appear to be a question like which came first, the chicken or the egg. Do we have a large number of species of grasshopper enemies because grasshoppers can become so abundant, or do grasshoppers become so abundant because they have so many species of enemies that there is excessive stability at the carnivore level, owing to interference and competition between the different species of parasites and predators? The implication is that large numbers of enemies may be the cause of great grasshopper abundance, not the effect. The position of this author is that we do not know yet, and the question is left as one of a great number to occupy the imagination, initiative, and ingenuity in experimentation and analysis of subsequent generations of scientists.

This chapter illustrates the applicability of Principle 13 to biological control. A high level of diversity at the parasite (carnivore) trophic level promotes stability, but only at that level.

SUGGESTIONS FOR INDIVIDUAL AND GROUP PROJECTS

Try to obtain data from government documents, annual reports of laboratories and institutions, and published budgets showing the costs and benefits of different kinds of pest control. Now try to obtain data on the amounts of money spent on research and development for each type of control. Do the amounts of money spent on each type of control seem to be related to the economic attractiveness? (We would expect expenditures to be higher for those control measures with higher benefit/cost ratios.) If there is a discrepancy, how do you account for it?

One useful source of information is "Restoring the Quality of Our Environment," the White House, Washington, 1965.

REFERENCES   Clausen, C. P.: "Entomophagous Insects," McGraw-Hill, New York, 1940.

DeBach, P.: The Scope of Biological Control, in P. DeBach (ed.), "Biological Control of Insect Pests and Weeds," pp. 3–20, Reinhold, New York, 1964.

Doutt, R. L.: The Historical Development of Biological Control, in P. DeBach (ed.), "Biological Control OF Insect Pests and Weeds," pp. 21–42, New York, 1964.

Force, D. C.: Competition among Four Hymenopterous Parasites of an Endemic Insect Host, *Ann. Entomal. Soc. Am.*, **63**:1675–1688 (1970).

Huffaker, C. B., and C. E. Kennett: Some Aspects of Assessing Efficiency of Natural Enemies, *Can. Entomologist*, **101**:425–447 (1969).

McGugan, B. M., and H. C. Coppel: Biological Control of Forest Insects, 1910–1958, in "A Review of the Biological Control Attempts against Insects and Weeds in Canada," Technical Communication no. 2, pp. 35–216, Commonwealth Institute of Biological Control, Trinidad, Commonwealth Agricultural Bureau, Farnham Royal, Bucks, England, 1962.

Pemberton, C. E., and A. F. Willard: A Contribution to the Biology of Fruit-fly Parasites in Hawaii, *J. Agr. Res.*, **15**:419–465 (1918).

Southwood, T. R. E., and M. J. Way: Ecological Background to Pest Management, in R. L. Rabb and F. E. Guthrie (eds.), "Concepts of Pest Management, pp. 6–29, North Carolina State University, Raleigh, 1970.

Turnbull, A. L.: Population Dynamics of Exotic Insects, *Bull. Entomol. Soc. Am.*, **13**:333–337 (1967).

——— and D. A. Chant: The Practice and Theory of Biological Control on Insects in Canada, *Can. J. Zöol.*, **39**:697–753 (1961).

Watt, K. E. F.: The Use of Mathematics and Computers to Determine Optimal Strategy and Tactics for a Given Insect Pest Control Problem, *Can. Entomologist*, **96**:202–220 (1964).

———: Community Stability and the Strategy of Biological Control, *Can. Entomologist* **97**:887–895 (1965).

Willard, H. F., and A. C. Mason: Parasitization of the Mediterranean Fruit Fly in Hawaii, 1914–1933, *U.S. Dept. Agr. Circ*, 439, 1937.

suggested additional reading   Zwolfer, H.: The Structure of the Parasite Complexes of Some Lepidoptera, *Z. Angew. Entomol.*, **51**:346–357 (1963) (in German).

13 the environment and infectious disease: a special class of ecological regulating systems

TYPES OF DISEASES

"Disease" can be categorized in several broad classes. The nature of the causal pathways by which environmental factors affect disease differs from one class to another.

1   The most important class of diseases until the twentieth century were the infectious diseases, caused by bacteria, protozoa, worms, fungi, and viruses.

2   A group of diseases becoming very important for man recently are the pollution diseases; emphysema, bronchitis, and various kinds of cancer, particularly lung cancer, are all increased in incidence by exposure to specific environmental contaminants.

3   Another group of diseases more prominent now than in earlier times are the stress diseases, such as coronary disease, which can be related to the social environment of man.

4   Nutritional diseases are related to social and economic conditions, agricultural practices of man, and food-processing practices.

5   Allergies are clearly at least in part the product of environmental factors.

6   Developmental anomalies (teratologies) can be caused by environmental contaminants which affect the mother while she is carrying the fetus. One of the best-publicized recent examples is the drug thalidomide, which resulted in the birth of many babies with shortened and deformed limbs.

The second and third classes of disease are discussed in Chap. 15. In this chapter, we will be concerned with detailed examination of infectious diseases, which can cause mass mortality in such diverse types of organisms as American elm, grasshoppers, herring, and man.

Infectious diseases fall into several broad categories, depending on the number of types of organisms involved in the dynamics of epidemics. We follow the classification of May (1958), who referred to infectious diseases as two-factor, three-factor, and four-factor complexes. A two-factor complex involves only a pathogen and a host species; influenza, a viral disease affecting man, is an example. Three-factor complexes, such as malaria and yellow fever, involve a pathogen species, a vector (the mosquito), and man. Four-factor complexes, such as plague and typhus, are dynamic systems wherein an intermediate host, such as a rat or other small mammal, carries ectoparasites such as ticks, fleas, or mites, which are vectors for the pathogen, which may be a bacterium (plague) or a rickettsia (typhus), for example. The fourth factor is man, the final host, who is infected by a bite from an infected ectoparasite which has left a dead or sick intermediate host (the small mammal), which had come into association with man.

THE INITIATION OF EPIDEMIC AND EPIZOOTIC WAVES

Throughout the history of medicine, various theories have been invoked to account for the dynamic behavior of epidemic waves. Gradually, it has been recognized that a complex of factors interacting with each other are responsible for the temporal and spatial characteristics of epidemics. Specifically, the behavior of a wave of disease spreading out from an epicenter, or a number of epicenters, is due to the interaction of three groups of factors: (1) density-dependent factors, such as the density and distribution of the host (people may be packed into slum dwellings or they may live in single-family residences each situated in a 1-acre lot); (2) the biological attributes of specific strains of viruses and other pathogens, such as the virulence and lifetime of viral particles, and the biological attributes of vectors and hosts, such as their resistance to the disease; and (3) density-independent environmental factors, such as weather fluctuations and the state of sanitation in the community.

In any actual epidemic, all three of these groups of factors interact. However, for the sake of clarity in exposition, we will consider the three factors one after another.

DENSITY AND WAVES OF DISEASE

There is an extensive literature on both the theory and the observations relevant to the role of density in infectious diseases. We will first give an oversimplified explanation of the theory, and then indicate the type of observations on which it is based, using only influenza for illustrative material, in the interests of brevity.

The classical theory of epidemics (e.g., Kermack and McKendrick, 1927) is based on the notion that at any time during an epidemic, each person in the population can be assigned to one of three groups:

1 The number of individuals still unaffected (N_U)
2 The number of individuals who are infected and ill (N_I)
3 The number of individuals who have been removed from group 2 by recovery or death (N_R)

Further, it is assumed that the rates of change of the numbers in these three groups at any point in time can be simply explained in terms of two rate parameters, which we shall call I, for infectivity rate, and R, for removal rate. We define I as the probability that an unaffected individual will become sick if he encounters an infected individual, and R as the probability per unit time that an infected individual will recover or die. Now where dN_U/dt means "the rate of change in numbers of N_U per unit time" and the time units are vanishingly small, classical epidemiological theory holds that the dynamics of epidemics can be approximately described by the following three equations:

$$\frac{dN_U}{dt} = -IN_UN_I$$

$$\frac{dN_I}{dt} = IN_UN_I - RN_I$$

$$\frac{dN_R}{dt} = RN_I$$

The first equation means that the rate at which people get sick starts out small, when few infected people are in the population, then gets larger as this number increases, and finally gets smaller again because of a decline in N_v. Consequently, if we make a graph of the numbers sick as a function of time, we get a bell-shaped curve, as in Fig. 13-1, a graph for one of the most famous epidemics in history, the plague of London in 1665. The third equation means that the rate at which people are removed from the pool of people in the sick group depends on the number of people in that group. The second equation, for rate of change of numbers infected, is obtained by subtracting the removal rate from the rate at which people get sick.

Some immensely important consequences follow from these three simple equations. First, because the rate at which people get sick depends on the total number of unaffected people present, it follows, and can be demonstrated mathematically, that the proportion of the population that gets sick depends on the density of the population at the outset of an epidemic. This means that epidemics are density-dependent phenomena, and hence are able to regulate population numbers, just as starvation, predation, and parasitism do. Further, it implies that mankind has a very powerful motive for not becoming too dense. The more dense the population at the outset of an epidemic, the larger will be the absolute number of the people who get sick.

Also, these simple equations show that the dynamics of epidemics are strongly influenced by I and R. The first of these parameters measures a great many variables, such as the level of sanitation, the alertness of public health agencies, and the virulence of the pathogen relative to the resistance of the host. The second parameter measures a variety of variables also, including the level of public health care and the nutritional plane of the population before being infected.

Declared total deaths per week, thousands

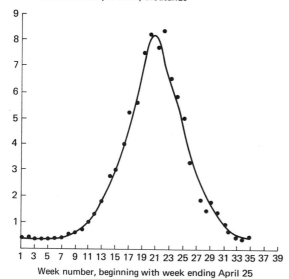

Week number, beginning with week ending April 25

FIGURE 13-1 The Black Death in London, 1665. (*Data from Creighton*, 1891.)

Now if this theory is correct, it ought to be possible to demonstrate that the proportion of the population that gets ill depends on the size and density of the population at risk. One of the more interesting papers on the relation between crowding and influenza is by Brand-Auraban (1959), who found the following relation between the percentage influenza sickness rate in Israeli schoolchildren and the numbers of persons per room in their homes:

numbers of persons per room	mean percentage sickness rate
0.0 to 1.0	39.9
1.1 to 2.0	44.3
2.1 to 3.0	52.4
3.1 to 4.0	58.9
4.1 to 5.0	67.3

Selivanov (1958) showed that 31 percent of the people living in towns became ill with influenza in Belorussia in 1957, but only 2 percent of village inhabitants succumbed. A comprehensive review of available evidence shows that crowding, as in school or military barracks, increases the probability of becoming ill with influenza (Jensen et al., 1958).

EPIDEMIC DYNAMICS AND THE BIOLOGICAL ATTRIBUTES OF PATHOGENS AND HOSTS

Unimportant epidemics of influenza occur frequently, but major epidemics occur at intervals, which may be as severe as that of 1918–1919, which killed about 21 million people throughout the world (Fig. 5-6B). The question why some epidemics are much more severe than others naturally arises.

Two facts appear clear on the basis of research up to the present. First, particularly severe epidemics are caused by the introduction into the human population of strains of influenza virus which are either entirely novel to the population or have not been encountered by the population for long periods. There is a possibility that the appearance of influenza viruses of unusual virulence in the human population is a cyclical phenomenon, with a period of about 40 years. Indeed, reference to such a cyclical phenomenon goes back in medical literature to the time of Hippocrates (Cooray, 1965). Second, evidence is accumulating to suggest that the reservoir for such unusually virulent strains of influenza, between major epidemics, is swine, and perhaps other mammals. One line of evidence to support this thesis comes from Philip and Lackman (1962). They tested inhabitants of all ages in two Alaskan villages that had influenza outbreaks in 1918, five Alaskan communities that did not have such outbreaks, and residents of two Idaho towns, for relative immunity to influenza A/swine and other types of influenza A viruses. In both the Alaskan villages that had influenza in 1918 and the Idaho towns, nearly all the persons born before 1920 had relatively more immunity to A/swine influenza. In contrast, relative immunity to A/swine influenza was low among residents of all ages in Alaskan villages that did not have 1918–1919 influenza epidemics. The findings suggest that the particular strain of influenza virus which caused the 1918–1919 influenza pandemic is related to the A/swine virus, and that the latter is the prototype of the former.

Many facts suggest that people develop specific immunity to strains of influenza they encounter in their lifetime, and further, very old people seem to have immunity to unusually virulent new strains. This suggests that particularly virulent strains may be closely related to strains that caused great havoc in previous cycles. For example, Mulder and Masurel (1958) reported that relatively higher immunity to the Asian influenza pandemic of 1957 occurred in people of 70 or thereabouts. Thus, the 1957 strain of virus may be related to the strain that produced the great influenza epidemic of 1889, 68 years earlier. McCord (1966) has made the interesting observation that in the terrible influenza epidemic which hit Camp Sherman, Ohio, in 1918, very few older relatives who came to visit dying soldiers became sick themselves. McCord inferred from this that perhaps these parents still had immunity acquired in 1889.

Vesenjak-Hirjan et al. (1966) have shown that the immunity to one strain of A2 influenza viruses is less likely to confer immunity to another if the two strains of virus are not closely related with respect to time and place of origin. The picture which emerges from all available virological and immunological studies on influenza is that the disease is always present in the human population, but the particular array of strains involved is constantly changing. As each strain appears, the human population gradually develops immunity. Immunity to one strain confers partial immunity to related strains, the degree of immunity being less, the less close the relation. Pandemics or severe epidemics occur at the appearance of a strain quite unrelated to strains to which the population has total or partial immunity. However, such novel strains appear only at long intervals, for example, 1889–1918–1957, so that when they do appear, only older members of the population will have any immunity, assuming that the particular infective strains occurred 30 or 40 years previously.

ORIGIN AND SPREAD OF PARTICULARLY VIRULENT INFLUENZA STRAINS

There is no doubt that major influenza pandemics do not represent the simultaneous appearance of different pathogenic strains in various parts of the world, but rather represent the wavelike spread of one viral strain throughout the world, from an origin in a particular geographic region. For example, the 1957 Asian influenza epidemic has been traced to a region along the road between Kutsing and Kweiyang, in the province of Kweichow in southwest China, where it appeared at the end of February, 1957 (Tang and Liang, 1957). Andrewes (1959) has summarized the staggering mass of data on subsequent international spread of the virus strain.

Three major conclusions emerge from his analysis. First, the spread did show a regular wavelike pattern, moving from China to Japan, then the Middle East, Africa, and South America, and finally the United Kingdom and the United States. Second, the virus epidemic was important in any country only during the winter season for that country. Thus, the epidemic was most important in Chile and Australia during July and August, but in the United Kingdom and the United States during the period September to December. Third, the late development of the 1957 epidemics in the cool Northern Hemisphere countries was positively not due to late arrival of the virulent strain of virus, because it was being dicovered in such countries as early as mid-June. Thus, there can be no doubt that some seasonal factor is implicated in influenza, but we must now determine which one.

Three hypotheses seem reasonable. First, the onset of cold weather in each country coincides with the return of children to school, and the attendant crowding in school promotes initiation and spread of epidemics. A second hypothesis relates to heating of homes at the onset of cold weather. Unless humidifiers are used to keep the relative humidity up, artificial heating of air produces a drop in relative humidity. As we shall see, a considerable body of evidence indicates that dry air provides an optimal environment for maintaining infectivity of influenza virus particles. A third hypothesis is that cold air creates body stress, which lowers resistance to disease, particularly upper respiratory infections. This argument creates problems, however, because as Andrewes (1959) has pointed out, if cold weather per se is the explanation for influenza epidemics, we are left with the problem of accounting for such epidemics in tropical countries.

In order to evaluate the importance of the three postulated causal pathways, we now need to consider in some detail the evidence relating to the effect of weather on pathogens and hosts.

THE EFFECT OF CHANGES IN WEATHER ON THE DYNAMICS OF EPIDEMICS

There are many different causal pathways by which weather can have an effect on the initiation or regulation of epidemic waves. In the case of two-factor complexes, weather can affect the infectivity and lifetime of the pathogen and the resistance of the host. In the case of three-factor complexes, the most obvious consequence of weather is its effects on ecology of the vector. Mosquitoes cannot survive and build up unless there is a considerable area under shallow, calm water. Weather can operate on any of the four factors in a disease like plague. Changes in weather affect not only the survival probability of the intermediate hosts, but also their distribution. This latter point is critical, because whether man becomes affected depends on whether he comes in proximity to the intermediate hosts. That probability is increased if the latter have an extension of their range. Weather determines the probability of survival of the ectoparasite vectors of plague.

The diverse nature of the transmission system for various diseases means that different weather factors are more important for one than for another. Temperature and humidity are important for influenza, rainfall is important for dengue fever, and humidity is important for plague.

To illustrate the point that weather can affect a disease by more than one causal pathway, we shall consider in some detail two such pathways for influenza.

RELATIVE HUMIDITY AND THE INFECTIVITY OF INFLUENZA VIRUS PARTICLES

It has been recognized for some time that relative humidity has an effect on the infectivity of pathogens (Baker and Twort, 1941; Robertson, 1943; De Ome, 1944). Therefore, it was reasonable for Loosli et al. (1943) to study the effect of relative humidity on the infectivity of influenza A virus, using mice. Groups of ten mice were introduced into a room with fixed relative humidity, for short periods, different numbers of hours after the virus had been released in the room in droplets, using an atomizer. The infectivity of the virus was measured as the percentage of the group whose lungs were revealed to have lesions after the short exposure. The results can be summarized as follows:

relative humidity, percent	number of hours after introduction of virus into room at which virus was still infective enough to produce lesions in 20 percent of mice
89	1
48	3
23	22

The infectivity of the virus persists much longer where the relative humidity is low. In general, there is now an accumulating body of evidence to indicate that each type of virus has an infectivity versus humidity curve which is characteristic of the particular strain of virus (Lester, 1948; Hemmes et al., 1960). Hemmes quotes his Utrecht thesis, in which he showed that within the range of typical indoor temperatures, viral infectivity is much more sensitive to humidity changes than to temperature changes.

Thus, one possible interpretation of the higher incidence of influenza in winter is that the cold outside temperatures force people to spend much time indoors, where the humidity is low and the infectivity rate of influenza viral particles is consequently high.

THE EFFECT OF WEATHER ON RESISTANCE TO PATHOGENS

Caroli and Pichotka (1954) studied the relation between the coagulation of rabbit blood and weather. They found that coagulation took less than normal time a few hours before passage of a storm front, then was markedly slower at or near the time of frontal passage, and then returned to normal after passage. Assmann (1962) has shown that meteorological factors can have an effect on the clottability of blood and the composition of serum albumin. Tromp (1963b) has discovered that dropping temperatures are associated with high blood sedimentation rates and low hemoglobin values, and that they can be followed by influenza outbreak.

The present state of our knowledge about the entire range of functions of blood plasma proteins is incomplete. However, we do know that the plasma proteins called globulins play a critical role in the body's mechanisms of disease resistance, and it is known that some of the plasma proteins are affected by weather. It is reasonable to fill in the missing links in the argument and assume, tentatively, the weather has an effect on those particular plasma proteins which are critical for antibody formation, either directly, or indirectly through physical or chemical effects on other plasma proteins.

Tromp (1963a) made another interesting point. Capillaries at the body surface constrict when chilled; this reaction is one of the mechanisms by which blood flow from the body core to the surface is minimized when the body is in a cold environment. It is an important mechanism in preventing body heat loss. However, the vasoconstriction of capillaries might have other implications. Cooling also increases viscosity and lowers the rates of all chemical reactions near the body surface. Perhaps, then, cooling of capillaries, particularly in the lungs, impedes operation of the various mechanisms by which agents in the blood (leucocytes, antibodies) combat disease. Clearly, this problem area suggests a number of ideas for experimental programs to test various hypotheses.

One very interesting possibility merits consideration. What is the effect of alternation of environmental conditions on the organism? This question is not relevant in the summer, because when people go outdoors, they are not going to an environment that differs greatly from the one they just left. In the winter, however, indoor and outdoor environments are very different, and the constant alternation of conditions may play a role in the generation of influenza epidemics. This notion provides us with something to look for in those typically wintertime epidemics which sometimes show up anomalously in the spring or summer. In such cases, was the preceding weather very variable?

WEATHER AND VARIATIONS IN THE INCIDENCE OF UPPER RESPIRATORY INFECTIONS

Rybinskaia (1957) reviewed data on Russian influenza epidemics going back to 1772, and found only one instance of an epidemic occurring in months other than November to March. The exception was the 1918–1919 pandemic, and he notes that in the countries of origin, it began in winter months. He shows that for the winters of 1952 to 1955, the incidence of influenza invariably rose in Kiev after sharp drops in the daily minimum air temperature. The most compelling evidence he presents implicating temperature, however, comes from schools that trained builders specializing in carpentry, stonemasonry, and painting. In the training program, masons were mostly out of doors, painters worked mostly in heated interiors, and the carpenters worked in and out of doors. Among the masons, 47.7 percent had clinically evident forms of influenza during epidemics; only 23 percent of the painters had a similar degree of influenza, and the percentage for the carpenters was intermediate. This evidence seems to rule out either low humidity or wide-amplitude daily temperature fluctuations as the key factor producing influenza; it rather implicates low outdoor temperatures. Tateno et al. (1963) in Japan provide corroborating evidence: influenza epidemics were usually seen when both relative humidity and temperature were at their lowest for the year. Gorgiev et al. (1959) made an observation noticed by workers in many countries: the A2 influenza virus was present long before the epidemic started, and the final trigger seems to have been related to the fall opening of school and a drop in temperature.

A number of highly sophisticated statistical studies have been conducted on the relation between the incidence of upper respiratory diseases and weather. Gafafer (1931), Pacy (1958), Loudon (1964), Boyd (1960), Holland et al. (1961), and Lidwell et al. (1965) all show that temperature was the important variable in determining the incidence of respiratory infections. Various sophisticated analytical tricks were used by these authors. Many different weather variables, time intervals, and time lags were used to discover what variable is most important in determining the incidence of respiratory illness; wherever a separation could be made, temperature turned out to be more important than any other meteorological variable. In all cases studied, colder weather was correlated with increased incidence of disease.

The Gafafer study produced a conclusion analogous to that for the forest insect study mentioned in Chap. 5. In the case of both the independent weather variables and the dependent variable (i.e., incidence of colds) the variable Gafafer used in the statistical analysis was not the raw datum, but the positive or negative deviations around a long-term cycling mean. Also, he analyzed data from cold and warm months separately. The results of this study showed, first, that by far the most important variable determining incidence of colds in each of the cold and warm parts of the year was the average temperature. Second, the correlation coefficients were considerably higher for the warm than for the cold period. In all cases, the correlation coefficients were negative; that is, the colder the weather than what was expected for the week, the higher the incidence of colds.

Thus this study, like that of the forest insect populations, shows that the greatest effect of weather on a biological system does not occur when the weather conditions are most apparently deleterious, but rather when they are most unlike the expected normal pattern for a given time of year. Thus, a temperature drop in the summer period increases the incidence of colds more than the same drop in the winter, just as warm winter weather is more likely than a very cold period in winter to be lethal to pupae that would normally overwinter in a very cold place. Thus, our overall conclusion is that biological systems evolve so as to be adapted, at each time of year, to the usual weather conditions at that time of year. Weather that is abnormal for a particular time of year is deleterious, even though the conditions which are deleterious then might not be at all deleterious at some other time of year.

It should be noted that although none of the six statistical studies just mentioned deals with influenza (they all deal with colds or other respiratory infections), the types of phenomena for which they obtain statistical evidence have in fact been observed in influenza (Tromp, 1963c) Specifically, a sudden drop in midsummer temperatures has been observed to trigger influenza outbreak, and the causal mechanism appears to have been discovered in the biophysical changes in the properties of the blood proteins, following the weather change and coincident with the influenza outbreak.

In each of the studies conducted on influenza or other upper respiratory viral infections, the length of the time lag discovered from the time of a sharp drop in temperature to the onset of disease was as short as the design of the research could detect. In general, these studies show that infection follows temperature drop with a lag of only a few days. However, inspection of these papers suggests that at least two different phenomena are being studied, which are characterized by different time lags. Boyd found higher correlation coefficients with temperature of the previous week than with temperature of the same week. On the other hand, Lidwell, Morgan, and Williams found evidence of some process which has its peak effect 2 days after operation of the relevant cause (cold temperatures). This alerts us to the possibility that cold temperatures, or drops in temperature, may operate on respiratory diseases through a number of different causal pathways characterized by time lags of different lengths.

Reviewing the literature, we are alerted to the interesting possibility that there may be a type of temperature stress operating in connection with influenza epidemics which has a time lag of 6 to 20 months. It has been observed repeatedly that major epidemics of influenza flag their arrival many months in advance by minor increases in incidence, followed by disappearance until the major wave appears. Thus, there is reason to believe that whatever phenomenon caused the epidemic to appear had been set in motion by some causal pathway operating long before the epidemic occurred. For example, Andrewes (1958) observed, "Again, it has been repeatedly noted that a small local outbreak of influenza in early summer has completely subsided, to be followed by a much bigger epidemic due to an identical type of virus in the following autumn. In other words, there is evidence that influenza virus may lie latent in a particular area, to be activated when, with colder weather, things are more favorable for an epidemic. In general we have to admit we do not know where influenza lies concealed between epidemics." This quotation raises the logical possibility that temperature stress one cold season may be involved in seeding a virus strain which may then produce an epidemic the subsequent cold season in response to temperature stress that season.

To explore this issue, winter temperatures and influenza mortality rates were compared for England and Wales during the 40-year period 1890 to 1929. Influenza rates for year t were compared with the sum I_t of mean monthly temperatures for December of $t-1$ and January, February, and March of t. The magnitude of the drop from I_{t-1} to I_t was examined for its possible relation to influenza incidence in year t.

The most severe influenza outbreak occurred in 1918, a year after the largest such temperature drop in the period. The second largest outbreak coincided with the fourth largest drop in the period. The third largest outbreak coincided with the second largest drop in the period.

Thus, it seems there may be a relation between an unusually large influenza outbreak and a winter that is sharply colder than the preceding winter.

The student should be clear as to the hypothesis being proposed here. It is not being postulated that winter-to-winter temperature differences are the only factor determining changes in incidence of influenza. Rather, the postulate is that changes in influenza incidence are under control of a group of factors which interact with each other: changes in the genetic nature of the virus through mutation and selection, changes in the weather and resultant characteristics of host and pathogen, development of immunity in the host to particular strains of virus, and density-dependent factors.

In general, the most microscopic possible examination of records on weather and influenza outbreaks is advocated, for a number of reasons. It may teach us about either or both of the resistance properties of the host, and the infectivity and lifetime of the virus. Further, development of empirical rules lays the basis for construction of the formal theory that can be used to predict waves of disease with respect to timing, severity, and location. Finally, empirical work of this type opens up the possibility of building a bridge between epidemiologists who practice statistical analysis of data and mathematical biologists and epidemiologists who have developed a formidable body of theory, to a large extent reasoning from a priori general principles and a small number of semiquantitative general observations. The interested student should consult Bailey (1957) for an introduction to this theoretical literature.

THE ROLE OF EPICENTERS IN INFLUENZA PANDEMICS

It is well known that influenza does not appear everywhere simultaneously, but rather appears first at certain points on the earth's surface from which it spreads outwards (Isaacs and Andrewes, 1951; Baroian, 1958). Perhaps influenza outbreaks originate at particular places because of climatic events there, and then spread outwards by migration of infective carriers. Such dispersal, which is known to occur for many infectious diseases, places a formidable obstacle in the path of any simple method of analysis. Such wavelike dispersal means that events at one point on earth at a given time may be the result of earlier events *at another place*. It behooves us to discover where such places are and what characteristics make them epicenters for disease pandemics. If we can do this, then future pandemics can at least be predicted by monitoring weather conditions at the epicenters.

Isaacs and Andrewes describe the spread of influenza in the 1950–1951 outbreak. It appeared first in Scandinavia, then spread to Holland, the British Isles, and then to many other countries. The sequence, month by month, was as follows:

End of October	Southern Denmark
End of November	Sweden, Norway
Late December	Holland, Newcastle, Aberdeen, Edinburgh, Northern Ireland
January	Iceland and spreading through England

Is southern Scandinavia a reasonable site for an influenza epicenter, on the basis of climatic characteristics? Could any climatically unusual events have triggered the 1950–1951 influenza pandemic? With respect to the first question, inspection and analysis of weather tables show that Copenhagen has by far the highest standard deviation, relative to its mean, of any of the cities the author examined for mean monthly temperatures for the coldest month of the year. With respect to the second question, the winter of 1949–1950 was characterized by an unusually mild October, November, and December in Copenhagen, followed by a sharp break in January.

All authorities agree that the Asian influenza epidemic of 1957 appeared in northern China simultaneously in a number of locations around March, 1957. Mainland Chinese weather records are not available, but It is noteworthy that there was a sharp deterioration of the weather at Tokyo from November to December, 1956.

**SUMMARY OF
FINDINGS ON THE
EFFECT OF
TEMPERATURE
STRESS ON
INFLUENZA
EPIDEMICS**

A reasonable interpretation of all the literature on the temperature-influenza relation is as follows. There appears to be some increase in temperature stress which commits a population to an influenza epidemic months before the epidemic occurs. After this, the precise timing of the epidemic is determined by microlevel phenomena operating over a period of days or weeks. Complications of various kinds render precise description difficult: there is evidence for variation in the length of the time lags and for cumulative effects, and there are suggestions that the critical factor is not the severity of the stress but its unexpectedness. Unexpectedness can be expressed either as deviation from a trend line or as deviation from the corresponding previous temperature index. Evidently, there is need for much more research to determine the best means of measuring and describing this stress.

If the general interpretation of this chapter proves to be correct, the implications will be most interesting. One of the most fascinating problems is how it makes a difference to the body what last January was like, when since last January it has experienced very much warmer weather in a sequence of intervening months. Perhaps it makes a difference only to the virus? Perhaps the virus goes into a special overwintering state? This would seem an especially inviting hypothesis, for it could explain how Asian flu, a markedly *new* strain of flu, could suddenly appear in several places in northern China. However, as indicated by various bodies of data, this "physiological memory" appears to be a real phenomenon.

It appears that the body is controlled by a large number of negative feedback control loops, which constantly adjust to an expected set of environmental conditions in the immediate future. The more such conditions depart from what the body is programmed to withstand, the more it is stressed, and the greater is the likelihood that it will succumb to infectious disease. Further, it appears possible that the mechanisms by which the body is programmed to adjust to a certain set of conditions may be complex, involving response to temperature differences of successive winters, time lags, cumulative effects, and information storage.

**WEATHER
FLUCTUATIONS
AND A
FOUR-FACTOR
COMPLEX. PLAGUE**

The possible number of main effects and interactions is very large when we consider the effect of weather on a four-factor complex such as typhus or plague. To illustrate, we have chosen plague, for a number of reasons. First, there can be little doubt about the importance of this disease. The first well-reported plague pandemic began in the reign of the Emperor Justinian and spread over the inhabited parts of the earth; it is generally believed to have killed about a hundred million people in that period, the sixth century. Since the world's total population at that time was only about 150 million people, the plague was an almost unimaginable catastrophe. The second major wave of plague, the fourteenth-century Black Death, killed between one-half and two-thirds of the British population, and the estimate of 25 million, or one-quarter of the European population who died because of it, is now considered by some authorities to be too conservative. Plague has killed great numbers in the third major wave, in the early twentieth century, and it is still with us. Whenever society begins to disintegrate, plague erupts, sooner or later. Thus, it has been reported that there were 56 deaths in Vietnam due to bubonic plague in the first 11 weeks of 1968. If modern society should begin to crumble for some of the reasons discussed in Chap. 15, then we can be reasonably sure that plague, a very ancient enemy of our species, will return to bring our numbers back down to the carrying capacity of the planet. We all have plenty of motive to learn about plague.

A second motive for studying plague is that there is a vast scientific literature on every aspect of the subject: one recent review covering 5 years of new research discussed 450 projects (Pollitzer, 1960). Thus, it is possible to gain considerable insight into the dynamics of the disease.

Third, several sets of quantitative data make it possible to check various hypotheses about the possible role of weather in affecting plague epidemics and epizootics. Because it has been thought for a long time that climate is implicated in plague pandemics, a considerable body of literature has developed presenting epidemiological data, laboratory findings, and conjectures bearing on this postulated set of causal pathways.

Finally, plague is without doubt one of the most complicated phenomena ever studied in any branch of science, and it therefore represents a real challenge to any type of analysis. There are many reasons for the complexity, including a multitude of species of intermediate hosts and vectors; various complicated and interacting factors related to the ecology of Homo sapiens, from degree of crowding to procedures for storage and distribution of grain; and various causal pathways by which weather can affect each step in the causal chain leading to infection in man. Consequently, by revealing the main features of how such a complicated system works, we reveal by example some of the important features of complicated ecological systems. In general, it will be shown that there are very large differences in the relative importance of different independent variables. Thus, while there are very large numbers of independent variables involved in such complicated systems, it is possible to account for most of the behavior of such systems in terms of only a small number of them. The term *key factor* has been used to describe those variables whose importance in systems dynamics overrides or dominates that of many other independent variables (Morris, 1959).

In this discussion, we shall try to minimize presentation of background information on plague research, because this has been dealt with comprehensively in an excellent recent series of monographs by Pollitzer (1951, 1952a, 1952b, 1953, 1960 are most relevant to the present discussion). Instead, we will focus our attention on those aspects of plague which are most related to climatic effects, together with other information necessary to ensure that we are giving a balanced and undistorted treatment of the role of climate.

The four biological components involved in plague are the pathogen, a bacterium, *Pasteurella pestis*; insect vectors, fleas ectoparasitic on small mammals, particularly the rat flea, *Xenopsylla cheopis*; a wide range of mammalian intermediate hosts, including 200 species, of which the black rat *Rattus rattus* is by far the most important; and man, the final host.

One point about plague must be clarified before beginning any discussion of the role of weather as a possible regulating factor on its incidence. The disease occurs in two forms which are so different that weather must operate on them via different causal pathways. Bubonic plague, sometimes referred to as the zootic form of the disease, is caused by the bite of insect vectors infected with *Pasteurella pestis*. The principal insect vector is the rat flea. The other variant of the disease is pneumonic plague (the demic phase) and is caused by spread of the infection from man to man through inhalation of the pathogen expelled into the air in the form of fine droplets when an infected carrier coughs. In the case of bubonic plague, spread depends to some extent on the effect of weather on the ecology of the vectors and intermediate hosts. In the case of pneumonic plague, the principal causal pathways for weather effects are of the type discussed in connection with influenza: effects on the pathogen, indirectly on the host through crowding at certain seasons and microclimatic effects in buildings, and directly through weather effects on the resistance and susceptibility of the host. To complicate matters further, pneumonic plague originates with bubonic plague outbreaks, but it can afterwards exist independently, spreading to locations where there are no infected rodents or rat fleas. To make matters still more complicated, sometimes, although rarely, bubonic plague may be produced from pneumonic plague, because a human flea will bite a pneumonic plague carrier, become infected, and then bite a noninfected human, thus transmitting the plague bacillus and producing the bubonic form of the disease.

It is clear, then, that in attempting to relate plague incidence to weather, we must take care that we know which form of the disease is operative at any point in time, and hence which causal pathway is relevant as a means by which weather can operate. Further, the epicenter phenomenon can clearly be of great importance in plague, because weather may trigger an outbreak at one point or area on the earth's surface, after which migration of travelers can spread the pneumonic form to new locations or transmit the disease by carrying rat fleas.

The historical record leaves no doubt whatever that plague pandemics have been spread about by travelers. Various historical clues about the origin of Black Death in the fourteenth century have been pieced together by many workers, sometimes through stumbling by chance over critically important fragments of information found by archaeologists on tombstones, or in long-lost Latin manuscripts. Politzer (1951) happened upon a reference to the work of a Russian archaeologist who found tombstones near the lake Issyk Kul (about 43°N, 77°E) showing that people had died there of plague in 1338–1339. Previously, Creighton (1891) had reported the discovery of a very old Latin manuscript of Gabriel de Mussis, describing the transfer of plague from the Tartar hordes to Genoese merchants at Kaffa (about 45°N, 37°E, on Kerch Strait in the Crimea). This occurred during the siege of Kaffa in 1345 and 1346, and plague was carried with the Genoese merchants by ship back to Genoa (44°N, 9°E), about 1346; within a matter of days plague began to radiate rapidly. By November, 1348, it was in London, and by 1357 it had spread to the towns of England, and most of the then known civilized world, including Europe, and Middle East, and Africa.

This is a clear-cut instance in which effects spread over a very wide area were the ultimate result of a cause at an earlier time and very different place, and may have originated in the effect of climate on the population dynamics of wild rodents, and the interaction of these phenomena with the parasite-host dynamics of the rodent-flea system. Thus, in attempting to build mathematical models to account for season-to-season changes in the incidence of bubonic or pneumonic plague, we may have to trace back through a very long cause-effect pathway indeed.

Several points about the preceding narrative are noteworthy. First, the area around Issyk Kul, particularly to the west of the lake, is suspected of being an ancestral epicenter for plague. Second, it will be noted that although the Black Death spread over a vast distance from east to west, its main effects were considerably more confined with respect to a north-south gradient. This may have implications for the thermal and moisture tolerances of the principal vectors and hosts in the traditional transmission and reservoir system for plague. Third, it may be that wide-amplitude swings from year to year in temperature and moisture regimes in arid or semiarid areas are highly important in regulating the danger of outbreak in plague. (Issyk Kul is at the edge of a vast arid or semiarid area.) The plague center of the United States is similarly located in arid cold desert country: the plains of Colorado to northern Arizona. Fourth, great population migrations associated with wars and merchant shipping were clearly implicated in the fourteenth century Black Death. Further, although bubonic plague was important in triggering the vast pandemic, the pneumonic phase subsequently became of central importance in the wildfire spread of the disease. Thus, we can have a phase switch in the midst of an epidemic which has influence on the rate of spread and the nature of the causal pathways which are important in regulating the system.

We shall review the literature on possible effects of climate and weather on plague by considering, in turn, how the pathogens, vectors and intermediate and final hosts can be affected.

Before considering matters of detail, however, some general impressions that result from review of the plague literature seem in order, and useful. First, there appears to have been a definite trend in research on plague paralleling almost all research in this century, away from holistic studies to reductionistic studies dealing in great depth with very small fragments of whole processes of whole systems. It is certainly important to gain detailed understanding of the components that combine to produce a large-scale process, but it is equally important to continue holistic studies. The conspicuous lack of such studies in the modern literature may be a contributing factor to the striking want of balance in the overall research effort. Many of the plague research pioneers were convinced that climate was the ultimate regulator of plague; yet in a recent extremely comprehensive survey of plague, at most a handful out of 450 papers reviewed regard climate as being of central importance in the research reported (Pollitzer, 1960).

Indeed, Pollitzer, who probably had as comprehensive a grasp of the literature on plague as any person now living, notes (1953) that research of great potential importance on the relation between climate and plague has not been followed up, as indicated in this excerpt from his review: "Rogers found it possible to use such meteorological observations as a basis for forecasting plague epidemics in certain parts of India and obtained, as stated by Wu Lien-teh (1936), satisfactory results during the period 1930–1932. However, no further advantage seems to have been taken of this method." The state of affairs is rather amazing. In view of another generalization that can be made about the plague literature, that is, its controversial and inconclusive state on many absolutely central issues, it is hard to imagine why no use has been made of striking correlations between various weather factors and plague, for these provide clues as to which biological mechanisms really are the key factors in regulating incidence of the disease.

As an example of the somewhat unsatisfactory state of the literature, it seems difficult to obtain an answer acceptable to all authors to why bubonic plague switches to pneumonic plague. What is the mechanism responsible for this all-important switch? Weather may very well be the critical factor, colder weather increasing the likelihood of switch to a demic phase, but this notion is certainly not accepted by all authors (Pollitzer, 1953).

Three more general impressions emerge from a review of the plague literature. First, it is probably true to say that more is known about the effect of climate as a regulator of bubonic plague than about climate as a regulator of pneumonic plague, probably because the former has been more important recently than the latter, and hence has been more suited to study using modern techniques. Second, in research on pneumonic plague, more thought seems to have been given to the effect of climate on the susceptibility of the host than to the effect of climate on the virulence of the pathogen. Third, in research on bubonic plague, there seems to have evolved a more certain understanding of the effect of climate on the vectors than on the relations between epidemics in man and population dynamics of the intermediate hosts. This may well be because the role of climate in governing the vector-host system is probably the most complex part of the causal pathway by which climate operates on plague epidemics. This is true because of the very large number of intermediate host species, the additional complexities created by relations between intra- and interspecific groups of hosts with different degrees of resistance to plague, and a multiplicity of causal pathways relating climate to food of the intermediate hosts and their population growth, decline, and dispersal.

With these general impressions as a background, we now examine in more detail the general status of information about the various links in the plague system.

One very old notion about *Pasteurella pestis* is that the pathogen is more virulent in overcrowded and ill-ventilated houses, because it has been shown that carbon dioxide has a favorable effect on the pathogen (Marsh, 1901; Fraser, 1901; Jennings, 1903). However, as Pollitzer (1953, p. 160) notes, contradictory observations by different workers with regard to the effects of different temperatures and humidities on the pathogen-host system in pneumonic plague make it very difficult to draw any generalizations that will stand up against close scrutiny.

More is known about the effect of weather on the flea vectors. Whereas there has been controversy about the effect of a dry atmosphere on adult fleas, there has been no controversy about the effect of drought on larval fleas: it is inimical to their development (Buxton, 1938; Mellanby, 1933; Sharif, 1948). Also, there is an intermediate range of humidities that is optimal for larval fleas; excessively high humidities are as harmful as excessively low humidities (Sharif, 1949). In general, fleas, like many other animals, thrive best at intermediate temperatures and humidities (see Pollitzer, 1952b, for recent review). However, this subject is complicated enormously by the fact that the climatic optima will be different for the same species living in different climatic regions, or different flea species living in the same or in different climatic regions. Further, changes in climate may have several different kinds of effects on fleas. Besides governing survival, climate will have an effect on fecundity and dispersal. In particular, rat fleas seem to move off their rat hosts temporarily in rainy weather (Hirst, 1927).

In general, there is such a bewildering variety of different effects that climate can have on the whole weather-vector-host system, that this variety itself dictates the approach that one must take to understanding plague dynamics in nature. Instead of determining from laboratory studies what causal mechanisms operate, and then looking for all of these in nature, it seems more efficient use of research effort to examine laboratory studies to find out the type of causal pathways that might be important in nature, and then to look at the actual data from the field completely objectively, using statistical analysis to determine which causal pathways and variables in fact play key factor roles. Then, laboratory studies can be used to interpret the findings from the field. Thus analysis of variance is used as the final arbitrator of the relative importance of various factors in the field, not a priori conceptions carried over from laboratory studies.

In the case of the rodent intermediate hosts, as well as the flea vectors, weather might function through various causal pathways. For example, the suslik, a prairie-dog-like, ground-squirrel-like wild rodent, is an important intermediate host in southeastern Russia; its young disperse from the parental burrows in June. Now it seems reasonable to postulate that the area covered in this dispersal might be affected by climate, because the more favorable the climate is for growth of the food plants, the smaller would be the territories required by the susliks in order to obtain adequate food. Also, the poorer the climate is for suslik food, the more one would expect population mixing of susliks while they go on long-distance foraging searches, producing chance contacts with other susliks. However, another complication arises, in connection with mating. The average distance a suslik will have to travel in order to find a mate will be in part determined by the suslik population density, and this will in turn be dependent on the effect of past climate on the availability of suslik food.

In short, climate can have an effect on three factors that will have an effect on the dynamics of plague in suslik populations: the mean distance traveled to find a mate; the number of chance encounters with other individuals per unit area traversed, and hence the probability of contact with another infected individual; and population growth rate, which determines the ratio of susceptible to immune individuals in the population, this ratio being higher when there is a high proportion of young individuals in the population. On balance, drought will favor the first mechanism for spreading the disease, but it will operate against the second two mechanisms that promote a rise in incidence of the plague. Therefore, we might expect that wet periods in those arid regions that are epicenters for plague will promote dangerous buildup of the disease, provided it is not too wet and too cold.

A number of causal pathways by which weather can operate on plague in the hosts are postulated by Pollitzer (1953, pp. 154–162). He notes that it is a well-established fact that travelers who fall ill with bubonic plague either before leaving an infected locality or en route are particularly likely to develop secondary lung involvement. Pollitzer postulates that a possible cause may be cold or rainy weather during the journey. He mentions that adverse weather may be one of the possible conditions regulating the rise of pneumonic plague outbreaks. In view of the previously mentioned controversy over exactly what these factors might be, it seems best to let analyses of the data speak for themselves.

LONG-TERM PLAGUE CYCLES RELATIVE TO LONG-TERM CLIMATIC CYCLES

It seems worthwhile to determine whether any clues about the causation of plague pandemics can be found in historical climatology. It will be remembered that there have been three major and well-documented plague pandemics in history: Justinian's plague of the mid-sixth century, the Black Death of the mid-fourteenth century, and the modern pandemic that began in the last decade of the nineteenth century. The question that now comes up is: Can we explain the timing of these plagues in terms of unusual climatic conditions at the epicenters?

It is important, first, to ensure that we know where the epicenters were, as accurately as possible. According to Pollitzer (1951), the weight of evidence points to Ethiopia, Issyk Kul, and the Central Asian plateau of southern Russia, Mongolia, and northwestern China as the epicenters for the first, second, and third pandemics, respectively. Thus, it is worthwhile to search the records from historical meteorology to find out what we can at the critical time-place combinations. Unfortunately, it is difficult to describe a specific time and place for the origin of the third pandemic, except to say that by 1855 it was well established in the Yunnan province in southwestern China, from which it gradually spread via Hong Kong to the rest of the Eastern world. However, the ultimate epicenter for the third pandemic was the Central Asian plateau, probably in the eighteenth or seventeenth century. Thus, the second and third pandemics probably originated in the same general area.

Despite the fact that meteorological observations go back only to 1698, and are very sketchy before this century, it has been possible, using a great variety of indices and data-pooling techniques, to deduce the climate for parts of the world back to about A.D. 800 (Lamb, 1963). Unfortunately, the Central Asian plateau is one of the areas for which very few data are available. Lamb shows, however, that near 50°N, 50°E, it was quite wet in the summers at about A.D. 1300. After this, the geographical isolines of high wetness gradually migrated west so that peak wetness over the period 1070 to 1450 for 50°N, 0°W occurred between 1340 and 1380. This raises the intriguing possibility that the Black Death migrated from east to west because it was following isolines of high summer wetness which allowed for great buildup in rat flea populations. We have additional information from Buchinsky's (1963) research on the aridity of the southern Ukraine, the areas just west of the Central Asian plateau with which we are concerned. He has constructed from various old records on rains, floods, storm, thunderstorms, and locust raids, a drought index for each century in the arid southern Ukraine. The higher the index, the greater the amount of drought. Thus, the thirteenth century in the southern Ukraine was very wet, and the fourteenth was very dry. Extrapolating from Lamb's charts, it seems that the thirteenth-century wetness extended into the early part of the fourteenth. Thus, the Black Death may have been the consequence of a long series of wet years east of the southern Ukraine:

century	drought index	century	drought index	century	drought index
Tenth	20	Thirteenth	8	Sixteenth	27
Eleventh	67	Fourteenth	36	Seventeenth	35
Twelfth	26	Fifteenth	22	Eighteenth	35

Unfortunately, this analysis is highly speculative and is based on shaky evidence. It is mentioned, however, to demonstrate to students that it is possible to reconstruct meteorological records by extrapolating backwards from correlations between instrumental records and other indices in the modern period, and then to use the reconstructed meteorological indices to interpret the long-term history of diseases. There should be enough old records available to reconstruct the weather history in the Central Asian plateau and thus give us a basis for predicting how plague is likely to behave in the future.

STATISTICAL ANALYSIS

An important batch of data for testing various ideas about the dynamics of plague comes from India, particularly the Punjab, in the northwest of India. As recently as 1947, there were 78,937 plague deaths in all of India, and 51,455 deaths in just one province (Uttar Pradesh) (Seal, 1960). By comparison, the annual total number of plague cases in all of Egypt has not exceeded 1,702 in the twentieth century (this number occurred in 1916) (Hussein, 1955). Plague in the Punjab has been serious since 1900, has shown extremely wide-amplitude fluctuations, and has been very carefully studied throughout the period by epidemiologists, who have assembled all the data necessary for thorough analysis. Further, two long runs of data are available: one from 1900 to 1924 (Rogers, 1928) and another from 1939 to 1957 (Seal, 1960). Thus, we can develop and test relations on the early body of data and test their predictive usefulness on the later body of data.

We will begin by outlining the findings of Rogers (1928) with respect to the causal pathways that regulate plague in the Punjab.

To gain a general view of how the disease changes seasonally in relation to climate, Table 13-1 shows for each month, for the period 1900 to 1924, in the Punjab, the mean monthly number of plague deaths per 10,000, the monthly saturation deficiency in inches, the monthly rainfall in inches, and the mean monthly temperature. These figures, which were read off Rogers's (1928) chart 2 with a plastic rule, are reasonable in terms of other data with which they have been compared (e.g., New Delhi data in the Smithsonian Miscellaneous Collections, vol. 105). This table shows that plague is most severe in March, April, and May, with a very sharp decline in June and July. This decline follows a large increase in saturation deficiency of the air.

Previously in this chapter, however, we have been alerted to the importance of lag effects, and it will be noticed that just before the great decline in incidence of plague each year in the Punjab, during the period February to June, there is a great increase in the saturation deficiency of the air. It has already been noted that high temperatures and low humidities are inimical to the survival of young fleas. Therefore, it seems reasonable to postulate that by April the monthly saturation deficiency has become high enough to have a catastrophic effect on the flea population, and this results in a precipitous drop in the number of plague cases.

Another way of bringing out the great effect of saturation deficiency on plague is to examine the data on year-to-year changes in incidence of the disease as related to year-to-year changes in climate. Rogers (1928) has already done this, by assembling weather records and plague incidence records, and making graphs showing how both sets of variables change from year to year. Table 13-2 is based on his chart 3. By inspection of his charts for various weather factors and plague, all plotted against year, it appeared that the most important factor determining annual changes in the incidence of plague was the saturation

TABLE 13-1 Average month-to-month change in plague incidence and climate in the Punjab

month	monthly plague deaths per 10,000	monthly saturation deficiency, inches	monthly rainfall, inches	mean monthly temperature, °F
January	2.4	0.17	1.0	55
February	4.0	0.18	1.2	59
March	10.3	0.32	1.0	72
April	17.9	0.58	0.5	80
May	14.7	0.88	1.0	89
June	2.9	0.89	1.6	93
July	0.4	0.56	6.4	90
August	0.2	0.44	5.6	88
September	0.3	0.48	2.2	87
October	0.4	0.49	0.7	78
November	0.7	0.31	0.1	67
December	1.2	0.22	0.7	56

SOURCE: Data from Rogers (1928).

deficiency during the previous July to September period. This is a time of the year when saturation deficiencies are in an intermediate part of their annual range, and it appeared that high saturation deficiencies in this period were extremely inimical to the development of plague the following year, whereas low saturation deficiencies promoted it. Preliminary plotting showed that there was a great deal of scatter in the plot of plague deaths per mille against saturation deficiency for July to September the previous year, apparently due to the effect of the number of plague deaths the previous year. Hence, in Table 13-2 a third column is added, giving plague deaths each year as a proportion or multiple of the plague deaths of the previous year. When this ratio is plotted against saturation deficiency the previous July to September on semilogarithmic graph paper, Fig. 13-2 is obtained. The scatter on this graph is considerably less than in the comparable plot of plague deaths against saturation deficiency the previous July, August, and September.

Figure 13-2 is the graph of an equation of form

$$Y = b_0 - b_1 X$$

where Y is the natural logarithm of P_y/P_{y-1}; P_y is the number of plague deaths per mille in year y in the Punjab, and X is the saturation deficiency in inches of mercury in the Punjab during the months July,

TABLE 13-2 Plague in the Punjab, in India

year y	plague deaths per thousand P_y	saturation deficiency July-September in year y	P_y/P_{y-1}
1900	0.03	0.41	
1901	0.84	0.52	28.0
1902	8.78	0.48	10.5
1903	9.60	0.42	1.1
1904	19.82	0.58	2.1
1905	16.65	0.59	0.84
1906	4.56	0.45	0.27
1907	30.27	0.63	6.64
1908	1.53	0.33	0.051
1909	1.77	0.40	1.15
1910	6.74	0.48	3.81
1911	8.89	0.75	1.32
1912	1.54	0.53	0.17
1913	0.92	0.48	0.60
1914	3.31	0.41	3.59
1915	11.48	0.71	3.47
1916	0.17	0.36	0.015
1917	0.45	0.32	2.64
1918	4.94	0.69	11.0
1919	0.57	0.46	0.115
1920	0.32	0.66	0.56
1921	0.13	0.55	0.41
1922	0.38	0.55	2.92
1923	2.44	0.45	6.42
1924	12.24	0.44	5.02

SOURCE: Rogers (1928).

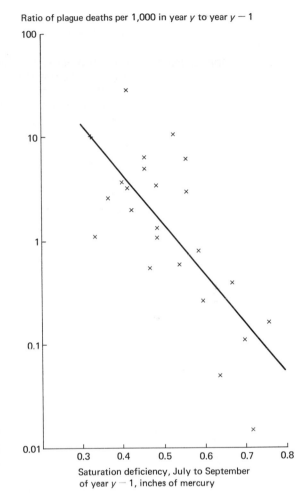

Ratio of plague deaths per 1,000 in year y to year $y-1$

FIGURE 13-2 The relation between the rate of increase in plague deaths from year $y-1$ to year y, and the saturation deficiency (drying capacity of the air) in year $y-1$ from July to September, inclusive. (*Data from Rogers*, 1928.)

Saturation deficiency, July to September of year $y-1$, inches of mercury

August, and September in year $y-1$. The line of best fit has been calculated and is plotted on Fig. 13-2. This line fits the equation

$$Y = 5.96 - 11.22X \tag{13-1}$$

which accounts for 54 percent of the variance in Y. The regression has an F ratio of 26.00, which with 22 degrees of freedom in the denominator is significant at the 99.95 percent level. That is, there is less than one chance in 2,000 that this regression would have been obtained by chance alone. Clearly, saturation deficiency the previous year has a tremendous effect on plague incidence the following year in the Punjab. No other climatic factor had anything like this effect on the system.

This striking relation between plague deaths and preceding humidity could not be confirmed in data from Seal (1960) on a resurgence of plague in India between 1943 and 1950. A reasonable explanation for this second twentieth-century Indian plague epidemic is that it was produced by breakdown of the Indian food storage and distribution system, associated with the war in the Orient. Rats, consequently, were free to move in and out of masses of stored grain, presumably leaving infected fleas therein.

CONCLUSION Influenza appears to be influenced by the density of the human population and by weather fluctuations. Plague seems to be influenced by social disintegration and weather fluctuations. In both cases, there is clear evidence that the diseases spread rapidly. There are several important messages for modern technological society in these conclusions.

The Paddock brothers (1967) have discussed the possibility that the world will become so short of food that the United States should give food only to those countries that have a chance of surviving. This would leave such countries as India and Egypt as centers of terrible poverty and social disintegration. The assumption on which this proposal, which was probably not meant to be taken seriously, was based, is that such countries can in effect be quarantined off, so that any major catastrophes that occurred there would not drag down the rest of the world with them. This chapter indicates that if major epidemics of infectious disease broke out there, it would be a historically novel situation if the results could be prevented from spreading to many other nations.

Second, this chapter makes clear the role of weather in infectious disease: it is unlikely that all the observed correlations are the product of coincidence, or some other factor. This is important for us at a time when man is putting so much pollution into the atmosphere, and the rate of production of air pollutants is increasing sharply because of the spread of industrialization (Chap. 15). Chapter 5 demonstrated that climate is affected by volcanoes, and Chap. 15 makes clear that pollution is beginning to have a volcanolike effect on global weather. This chapter has been intended to demonstrate beyond reasonable doubt that when man intentionally or inadvertently affects weather, he has a host of effects of a type that would never occur to most people, from effects on larval flea vectors of bubonic plague to physiological mechanisms regulating the resistance of man to upper respiratory infections. Truly, in playing with the weather, we are like a group of children playing with a box of dynamite.

Third, this chapter was designed to show that there are in fact major social costs of constantly increasing human density. This high density is bought at a price, and as densities go higher and higher, the price can become unacceptably steep.

Chapter 15 will indicate that there is real danger of social disintegration in the world because of man's overextending the planetary carrying capacity for our species and exhausting supplies of fossil fuel and other essential resources. Chapter 13 is intended to show that there could be an exceptionally steep price for such insanity.

SUGGESTIONS FOR INDIVIDUAL AND GROUP PROJECTS Collect data on the incidence of infectious disease in the cities of your state over the last several decades from your state department of public health. Which diseases are showing important increases or decreases from year to year? How do you explain these trends? Do you notice large year-to-year fluctuations in incidence above and below the long-term trend? Use any other data you can obtain to account for these fluctuations with the methods of this chapter.

REFERENCES Andrewes, C. H.: Epidemiology of Influenza, *Bull. World Health Organ.*, **8**:595–612 (1953).

——: The Epidemiology of Epidemic Influenza, *Roy. Soc. Promotion Health*, **78**:533–536 (1958).

——: Asian Influenza, a Challenge to Epidemiology, in M. Pollard (ed.), "Perspectives in Virology," pp. 184–196, Wiley, New York, 1959.

Assmann, D.: Zur Frage des Umwelteinflusses auf den physikalischchemischen Zustand des Blutes und die praktische Bedeuting in der Klinik, in S. W. Tromp (ed.), "Biometeorology," *Proc. 2d Intern. Bioclimatol. Congr., London, 1960,* pp. 510–519, Pergamon Press, Oxford, 1962.

Bailey, N. T. J.: "The Mathematical Theory of Epidemics," Hafner, New York, 1957.

Baker, A. H., and C. C. Twort: Effect of Humidity of Air on Disinfection Capacity of Mechanically Atomized and Heat-volatilized Germ aerosols., *J. Hyg.*, **41**:117 (1941).

Baroian, O. V.: The Spread of the World Influenza Pandemic in 1957, *J. Microbiol. Epidemiol. Immunobiol.*, **29**:933–946 (1958).

Boyd, J. T.: Meteorological Conditions and Mortality, *Proc. Roy. Soc. Med.*, **53**:107 (1960).

Brand-Auraban, A.: Some Observations on the Epidemic of Asian Influenza in Jerusalem, *Intern. Rec. Med.*, **172**:101–107 (1959).

Buchinsky, I. E.: Climatic Fluctuations in the Arid Zone of the Ukraine, in "Changes of Climate. Arid Zone Research, 20, UNESCO," pp. 91–95, Paris, 1963

Buckland, F. E., and D. A. J. Tyrrell: Loss of Infectivity on Drying Various Viruses, *Nature*, **195**:1063–1064 (1962).

Buxton, P. A.: Quantitative Studies on the Biology of *Xenopsylla cheopis* (Siphonaptera), *Indian J. Med. Res.*, (2)**26**:505–530 (1938).

Caroli, G., and J. Pichotka: Weitere Untersuchungen zur Beziehung zwischen Blutgerinnung and Wetter, *Arch. Meteorol. Geophys. Bioklimatol*, (B)**5**:403–412 (1954).

Cooray, M. P. M.: Epidemics in the Course of History, *Ceylon Med. J.*, **10**:88–96 (1965).

Creighton, C.: "A History of Epidemics in Britain from AD 664 to the Extinction of Plague," Cambridge University Press, Cambridge, 1891.

De Ome, K. B.: Effect of Temperature, Humidity, and Glycol Vapor on Viability of Air-borne Bacteria, *Am. J. Hyg.*, **40**:239 (1944).

Deutschman, Z.: Trend of Influenza Mortality during the Period 1920–1951, *Bull. World Health Organ.*, **8**:633–645 (1953).

Fraser: "Indian Plague Commission. Report 1898–1899," London, 5, appendix III, 1901, p. 482.

Gafafer, W. M.: Upper Respiratory Disease (Common Cold) and the Weather. Baltimore, 1928–1930, *Am. J. Hyg.*, **13**:771–780 (1931).

Gorgiev, T. B., V. G. Krasnova, B. M. Estrin, I. M. Yartsera, and R. R. Popova: Epidemiological Characteristics of the 1959 Influenza Outbreak in Dnepropetrovsk, *Probl. Virol.*, **4**:16–21 (1959).

Hemmes, J. H., K. C. Winkler, and S. M. Kool: Virus Survival as a Seasonal Factor in Influenza and Poliomyelitis, *Nature*, **188**:430–431 (1960).

——, ——, and ——: Virus Survival as a Seasonal Factor in Influenza and Poliomyelitis, *Antonic van Leeuwenhaek, J. Microbiol. Serol.*, **28**:221–233 (1962).

Hirst, L. F.: Rat-flea Surveys and Their Use as a Guide to Plague Preventive Measures, *Trans. Roy. Soc. Trop. Med. Hyg.*, **21**:87–104 (1927).

Holland, W. W., C. C. Species, and J. M. G. Wilson: Influence of the Weather on Respiratory and Heart Disease, *Lancet*, **2**:338–341 (1961).

Hussein, A. G.: Changes in the Epidemiology of Plague in Egypt, 1899–1951, *Bull. World Health Organ.*, **13**:27–48 (1955).

Isaacs, A., and C. H. Andrewes: The Spread of Influenza. Evidence from 1950–1951, *Brit. Med. J.*, **2**:921–927 (1951).

Jennings, W. E.: "A Manual of Plague," London, 1903.

Jensen, K. E., F. L. Dunn, and R. Q. Robinson: Influenza, 1957, A Variant and the Pandemic, *Progr. Med. Virol.*, **1**:165–209 (1958).

Kermack, W. O., and A. G. McKendrick: A Contribution to the Mathematical Theory of Epidemics, *Proc. Roy. Soc. (London)*, **(A)115**:700–721 (1927).

Lamb, H. H.: On the Nature of Certain Climatic Epochs Which Differed from the Modern (1900–1939) Normal, in "Changes of Climate. Arid Zone Research, 20, UNESCO," pp. 125–150, Paris, 1963.

Lester, W.: Influence of Relative Humidity on Infectivity of Airborne Influenza, a Virus, *J. Exptl. Med.*, **88**:361 (1948).

Lidwell, O. M., R. W. Morgan, and R. E. O. Williams: The Epidemiology of the Common Cold. IV. The Effect of Weather, *J. Hyg.*, 427–439 (1966).

Loosli, C. G., H. M. Lemon, O. H. Robertson, and E. Appel: Experimental Air-borne Influenza Infection. I. Influence of Humidity on Survival of Virus in Air, *Proc. Soc. Exptl. Biol.*, **53**:205–206 (1943).

Loudon, R. G.: Weather and Cough, *Am. Rev. Respirat. Diseases*, **89**:352–359 (1964).

McCord, C. O.: The Purple Death. Some Things Remembered about the Influenza Epidemic of 1918 at One Army Camp, *J. Occupational Med.*, **8**:593–598 (1966).

Marsh: "Indian Plague Commission. Report 1898–1899," London, 1901, 3, p. 73, 5, appendix III, p. 480.

May, J. M. (ed.): "The Ecology of Human Disease," M.D. Publications, New York, 1958.

Mellanby, K.: The Influence of Temperature and Humidity on the Population of *Xenopsylla cheopis*, *Bull. Entomol. Res.*, **24**:197–202 (1933).

Morris, R. F.: Single-factor Analysis in Population Dynamics, *Ecology*, **40**:580–588 (1959).

Mulder, J., and N. Masurel: Pre-epidemic Antibody against 1957 Strain of Asian Influenza in Serum of Older People Living in the Netherlands, *Lancet*, **1**:810–814 (1958).

Pacy, H.: Relationships between Cough and Climate in a Smogless New South Wales Coastal District, *Med. J. Australia*, **2**:194–196 (1958).

Paddock, W., and P. Paddock: "Famine—1975. America's Decision, Who Will Survive?" Little, Brown, Boston, 1967.

Philip, R. N., and D. B. Lackman: Observations on the Present Distribution of Influenza A with Swine Antibodies among Alaskan Natives Relative to the Occurrence of Influenza in 1918–1919, *Am. J. Hyg.*, **75**:322–334 (1962).

Pollitzer, R.: Plague Studies. 1. A Summary of the History and a Survey of the Present Distribution of the Disease, *Bull. World Health Organ.*, **4**:475–533 (1951).

———: Plague Studies. 6. Hosts of the Infection, *Bull. World Health Organ.*, **8**:381–465 (1952a).

———: Plague Studies. 7. Insect Vectors, *Bull. World Health Organ.*, **7**:231–342 (1952b).

———: Plague Studies. 9. Epidemiology, *Bull. World Health Organ.*, **9**:131–170 (1953).

———: A Review of Recent Literature on Plague, *Bull. World Health Organ.*, **23**:313–400 (1960).

Robertson, O. H.: Air-borne Infection, *Science*, **97**:495–502 (1943).

Rogers, L.: The Yearly Variations in Plague in India in Relation to Climate: Forecasting Epidemics, *Proc. Roy. Soc. Brit.*, **103**:42–72 (1928).

Rybinskaia, L. N.: A Study of the Correlation between Meteorological Factors and the Incidence of Influenza and Upper Respiratory Catarrhs, *Probl. Virol.*, **2**:37–43 (1957).

Seal, S. C.: Epidemiological Studies of Plague in India, *Bull. World Health Organ.*, **23**:282–292 (1960).

Selivanov, Ya. M.: Peculiarities of the Spread of Influenza in Belorussia in 1957, *Vopr. Virusol.*, **5**:51–53 (1958) (in Russian).

Sharif, M.: The Water Relations of the Larva of *Xenopsylla cheopis* (Siphonaptera), *Parasitology*, **39**:148–155 (1948).

———: *Cambridge Phil. Trans.*, b, 233, 581 (1949).

Tang, Fei-Fan, and Yung-Ken Liang: "Antigenic Studies of Influenza Virus Isolated from the 1957 Epidemic in China," Third International Meeting of Biological Standards, Opatija, 1957.

Tateno, I., S. Suzuki, S. Nakamura, O. Kiamoto, M. Togashi, and T. Sato: Etiology and Seasonal Incidence of Acute Nonbacterial Respiratory Infections in Adults with Emphasis on Influenza, *Japan. J. Exptl. Med.*, **33**:159–168 (1963).

Tromp, S. W. (ed.): "Medical Biometeorology," Elsevier, Amsterdam, 1963a.

———: Human Biometeorology, *Intern. J. Biometeorol.*, **7**:145–158 (1963b).

———: The Influence of Weather and Climate on Urinary Volume, pH, 17-Ketosteroids, Hexosamines, Cl, K, Na and Urea, *Intern. J. Biometeorol.*, **7**:58–73 (1963c).

Vesenjak-Hirjan, J., E. Egri-Hecimovic, and A. Hrabar: Influenza in Croatia, 1950–1965, *Bull. World Health Organ.*, **35**:273–277 (1966).

Waddy, B. B.: Climate and Respiratory Infections, *Lancet*, **263**:674–677 (1952).

Wu, Lien teh: "Historical Aspects: Epidemiological Factors" — Chaps. 1 and 10 in Lien-teh Wu, W. H. Chun, R. Pollitzer, and C. Y. Wu, "Plague: A Manual for Medical and Public Health Workers," Shanghai, 1936.

suggested additional reading

Zldanov, V. M., V. D. Solovev, and F. G. Epshtein: "The Study of Influenza," U.S. Department of Health, Education and Welfare, Washington, 1960 (510 pp).

14 urban, regional, and national planning in the light of ecological principles

Some readers may find it strange that a chapter on planning is included in a book on ecology. However, the relation between ecology and planning is real and important, because many important phenomena that should be considered in planning have an ecological cause. Thus, for example, many of the most important characteristics of money and energy flow in a city or a region have their basis in demographic factors: the density and growth rate of a population, its age distribution, and the extent to which it is regularly or contagiously distributed in space. Also, many of the principles discussed in Chap. 2 reveal themselves in cities. The city is a physical system, as well as a social and economic system; therefore, it should not surprise us to find that cities have an effect on the physical environment and the weather, and that this effect is modified by the way in which cities are designed (Principles 1 and 2). Time, space, matter, energy, and information are all categories of resources (Principle 3), and this is true whether we are concerned with community dynamics in a coral reef, or the pollution problem, or space-time saturation of transportation systems. All resources are subject to the saturation-depletion principle (Principle 4), which is very important in urban, regional, and national planning. Biological communities seem to evolve toward increased efficiency of energy use (Principle 10), and it is important to consider whether human systems can ignore this example and retain any kind of thermodynamic, economic, or social stability. Mature systems exploit immature systems (Principle 11), and it is important to consider how this applies to tax and capital flow from city to city. The relation between diversity and stability in biology (Principle 13) seems to be related to the economic health of cities (Jacobs, 1969).

THE CITY: OVERVIEW

What determines the location of cities? Huntington (1945) has noted that most of the world's great cities fall within a rather narrow range of climatic conditions. We would expect this from Principle 15 (zones of tolerance). Also, proximity to physical features favorable to transportation, such as great natural harbors, rivers, or mountain passes, is important. Dasmann (1965) has pointed out another very important fact related to city location: if one makes a soil map of a region, it appears that cities began as small villages in the center of the very best agricultural soils in a region, and gradually spread to cover those soils. The Los Angeles metropolitan area is a case in point. There is nothing surprising in this, in a sense. One would not expect a farm village to be situated 25 miles from the farm soil it depended on for its livelihood.

The irrational element is introduced when the farming village grows outwards over the best agricultural soils in a region. The fundamental principles illustrated by this are Principles 11 (mature systems exploit immature systems) and 17 (succession). Because farms produce foods and use equipment, they create an optimal environment for business growth. But then the business center attracts larger and larger business. The city grows, and spreads out over the farmland so that the farms which initiated the whole process are destroyed.

It has been noted in Chap. 4 that most biological systems have marvelous self-regulatory, or homeostatic, capabilities. By contrast, the city is in most cases a system temporarily out of control. If a city were like a plant or animal population, and became too large for its resource base, then it would suffer from a decline in nutrient availability. This would lead to a decline in survival and reproductive rates, a decline, therefore, in numbers, and the city would come back into balance with the resources on which it depended.

However, cities behave in very different fashion. If a city grows beyond its resource base, then land speculation at the urban fringe, combined with rezoning and elimination of farmers who cannot afford to pay urban tax rates for agricultural land, allows the city to grow further. The resources needed for the city are obtained by expending fossil fuels to haul in food and other supplies from increasingly remote locations. The availability of fossil fuel energy temporarily removes the controls which operate in the case of the natural system.

As the income of a family rises, they can afford to move farther and farther out from the urban core to progressively more attractive locations. Thus, any motive to maintain the attractiveness of the urban core is gone, with resultant blight of the core. Associated with this blight there may be a depletion of cultural amenities. Art galleries, opera and concert halls, museums, and other amenities of the type that one normally expects to find in large urban areas are underrepresented. It is noteworthy that such urban blight at the core is least likely to occur in cities on a small island (Manhattan) or peninsula (San Francisco), where the limited amount of available space keeps real estate values high at the core. Urban sprawl destroys the city as well as the countryside, because it relaxes the pressures that would normally create high densities and resultant amenities at the core, which normally give a city its character and charm.

Also, urban sprawl decreases the energy efficiency of a city enormously, as the mean distance required to traverse the distance between functionally related points (home and work; home and stores) increases for no useful reason. City life has many associated costs related to taxes, crime, pollution, and energy; these costs are justified only if the benefits more than compensate for them. If urban sprawl operates to prevent the development of the benefits not normally found in small cities, then the whole motive for development of large cities has been destroyed.

Consequently, important questions of national strategy with respect to cities are raised. Is it preferable to house a population of 10 million people in one large city of 10 million, ten cities of 1 million, or some other combination?

Finally, it is noteworthy that cities can decline and die, as well as grow. It is important to consider the mechanisms by which cities die. The argument that a great fire can destroy a city is invalid, for London, San

Francisco, and Chicago were all rebuilt after catastrophic fires. Also, cities have been almost totally destroyed by war and have been rebuilt (Hamburg, Coventry, Volgograd, Berlin). Thus, when a city dies, fire and war are not a sufficient explanation. If the resource base for a city is still adequate, any cause of destruction, if temporary, will still allow the city to be rebuilt. One possible explanation for the demise of great ancient cities now represented only by ruins is the destruction of a resource base (Persepolis, Palmyra, Babylon). In the case of the great cities of the valley of the Euphrates and Tigris in what is now Iraq, we know how the destruction of the resource base occurred. The irrigation system deteriorated owing to accumulation of the salts which should have been flushed out of the soil. The soil is now too saline to allow the once luxuriant crop growth of Babylonian times.

A city is healthy if it is characterized by a number of dynamic balances. There must be favorable proportions of young people and old people, rich and poor, skilled and unskilled, and healthy and sick. Unusually low values for any of these ratios are symptoms that the city is in trouble. Further, there must be functional balances in the city, between its ability to generate capital for its needs and its requirements for capital to deal with the needs of the people. There must be a dynamic balance among the proportions of the total land area devoted to different kinds of land use. If too high a proportion of the total land area is devoted to transportation, supply/demand ratios for generation and absorption of tax monies are affected. Some cities have lost too high a proportion of their ground area to the automobile, and seaports tend to lose too high a proportion of their land adjacent to the water to old, rotting, little-used warehouses.

It is possible to construct measures of the efficiency of energy use in urban transportation systems analogous to those that describe the efficiency of animal systems. One such measure might be the reciprocal of the energy cost to transport one man to and from work minus the metabolic energy of the man saved because he does not have to walk to work, and also saved metabolic energy because he has more time free to do other things, that would otherwise have been spent in walking.

By using energy as efficiently as possible, we allow a higher proportion of the world population to enjoy a high standard of living. Otherwise, for example, developed countries will have almost exhausted crude oil stocks before Nigeria and India can use them at a high rate per capita.

URBAN PLANNING: THE PHYSICAL ENVIRONMENT

Cities have very great effects on their physical environment. Duckworth and Sandberg (1954) have noted that there is a long history of observations on the "urban heat island effect" — that cities are hotter than the surrounding countryside. They made several measures of this difference in the San Francisco Bay area cities. One of the most straightforwardly derived measures is what they call the urban temperature differential, or the difference between the temperature at the hottest point in the city and that at the coldest point. The maximum temperature is typically located in the most densely built-up area, or the center of the urban heat island, and the minimum temperature in the urban region typically appears at some point on peripheral open lands.

The magnitude of the urban temperature differential depends on the magnitude of the urban center. Table 14-1 illustrates the difference in the size of this differential for three different urban centers in the same megapolitan region measured by the authors. The urban heat island effect is clearly much larger in larger cities than in smaller cities. What causes the urban heat island effect?

We can get clues to the nature of this phenomenon from three sources: a program of ground measurements of city temperatures at many different times and places in cities, aerial photography of cities with films sensitive to different and known parts of the energy spectrum, and computer simulation models. Samples of the results from the first two of these types of research are depicted in Plate 8. The photograph is of San Francisco taken in the middle of the day from the air using Kodak Ektachrome Infrared Aero Film with an orange filter. Superimposed on the photograph are isotherms for temperatures at 23:20 Pacific standard time on April 4, 1952, as taken by Duckworth and Sandberg and their associates. The temperatures were taken quickly by three cars with electrical resistance thermometers mounted at a 2-meter height to the front bumpers of the cars. It will be seen that the coldest part of the city about midnight is a large park, Golden Gate Park, and the hottest part is in the central business district.

To compare the isotherms with the color photograph, we must understand how the color infrared film responds to different wavelengths of the energy spectrum. This is explained by Fig. 14-1. The green vegetation in the park reflects infrared energy in the middle of the day, exposing the cyan (azure blue) layer of the film emulsion; the dyes left relatively unexposed are yellow and magenta, which combine to produce a red color in the developed picture. The reflectance of infrared energy, which is a short distance into the spectrum component carrying heat, is correlated with the cooling effect of healthy deciduous vegetation on the air, through the conversion of water to water vapor involved in transpiration. (However, it is not this mechanism which reflects back the infrared radiation.) In the hottest part of the day, the hottest parts of downtown areas are adjacent to low buildings with metal roofs, or automobile parking areas. Such areas emit radiation in all parts of the energy spectrum in the late morning and early afternoon, expose all layers of the emulsion, and consequently appear in the color plate as white to light cream-colored.

TABLE 14-1 The effect of city size on the urban temperature differential

city characteristic	San Francisco	San Jose	Palo Alto
Population	784,000	101,000	33,000
Incorporated land area, square miles	45.1	14.8	8.6
Population density, persons per square mile	17,383	6,824	3,837
Urban temperature differential, °F (representative evening values)	10–12	7–9	4–6

SOURCE: Duckworth and Sandberg (1954).

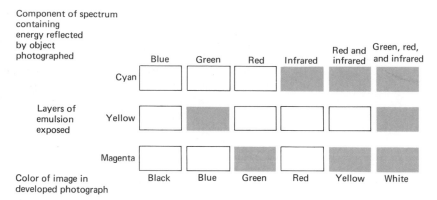

Component of spectrum
containing
energy reflected
by object
photographed

	Blue	Green	Red	Infrared	Red and infrared	Green, red, and infrared
Cyan				▓	▓	▓
Yellow		▓				▓
Magenta			▓		▓	▓
	Black	Blue	Green	Red	Yellow	White

Layers of
emulsion
exposed

Color of image in
developed photograph

FIGURE 14-1 Color formation by Kodak Ektachrome Infrared Aero Film. (*Based on chart and discussion by Knipling*, 1969.)

The photograph of Plate 8 was taken with an orange filter. Therefore, the blue and part of the green radiation reflected from the city was not allowed to expose the emulsion; all three dyes developed and combined in a subtractive mixture to form a black image. Green, red, and infrared radiation respectively expose the yellow, magenta, and cyan layers. In this chart an exposed layer is indicated by a solid-color rectangle. When any one layer is exposed, the dye in that layer does not develop, and the two remaining layers combine in a subtractive mixture to form the image color. When reflected energy from a surface exposes two layers, the image color becomes that of the single layer not exposed. This film has become an important new tool in ecology for two reasons. First, since it is normally used with a yellow or orange filter which blocks out ultraviolet energy, the bleaching effect of ultraviolet on normal films, which degrades resolution in aerial photography, is eliminated. Also, because this film is sensitive to infrared radiation, it may be used to discriminate between objects which appear very similar on the basis of radiation in the visible part of the energy spectrum, but differ with respect to radiation in the infrared part of the energy spectrum. The interested student can pursue this subject further in the book by Johnson et al. (1969) listed in the references at the end of the chapter.

Myrup (1969) has performed computer simulation studies which explore the effect of various factors on the heat exchange in cities. His conclusions can be summarized as follows.

1 Several different physical processes interact to determine the temperature of any part of a city at each time during the 24-hour cycle. In general, different processes tend to cancel each other's effects, so that the net effect of all factors operating jointly is less than the effect would be if only one were operating. Thus, while there is less evaporation from leaf surfaces near city centers, and this loss of the latent heat of vaporization, or refrigerator effect, tends to elevate temperatures, this elevation tends to be compensated for by the greater height of buildings which increases diffusion of heat upwards.

2 One of the most important factors that cut down heat in cities is increase in the proportion of the city surface where evaporation is occurring. An increase in the proportion evaporating from 0.0 to 0.5 cuts down the maximum temperature from 34.6 to 26.2°C in the simulation model output. The implications of this finding for urban planning are that parks and fountains have more than an aesthetic effect in cities: they also help to make the city livable during hot summers. This effect is very striking when one walks around cities on a hot summer afternoon: the temperature drop as one approaches parks and fountains is very large. Contrariwise, as one approaches large asphalt surfaces, the temperature increase is very noticeable.

3   Increasing building height decreases maximum temperatures about 6°C. Low buildings made of building materials which absorb a great deal of heat make for a very hot environment. This suggests one value of multi-story parking lots.

The urban heat island phenomenon is becoming more important for large cities, particularly those in semitropical or tropical climatic zones. For example, in the decade 1891–1900, the mean temperature for the entire period in Los Angeles was 16.7°C. By the decade 1951 to 1960, the mean temperature for the decade in Los Angeles was 18.4°C, even though the global mean temperature had begun dropping in 1950! Consequently, it is important to plan cities with the physical environment in mind: parks and fountains are not just a luxury. They have an important effect on the ability of people to live productive lives in a city.

URBAN PLANNING: THE EFFECT OF TRANSPORTATION SYSTEMS DESIGN ON USE OF SPACE, TIME, AND ENERGY

As cities grow in population and in size, the amount of traffic within them increases. In this section, we wish to make three points. First, as traffic systems approach complete saturation, small increases in additional traffic have large effects on the time cost of traveling in the system. Second, the square footage of roadway, freeway, or railway required per person depends on the mean trip distance per person each day, and that in turn depends on the area of the urban center. Finally, the efficiency with which matter, energy, space, and time are used in transportation systems is very different for different kinds of transportation systems, and at high population densities and population sizes, incorrect choice of mode of transportation for the urban area can bring mean trip velocities almost to zero miles per hour, and cause immense waste of energy and space.

Smeed (1968) reports that Wardrop found the following relation to describe the effect of numbers of cars using a road and v, the average speed of traffic in miles per hour:

$$\frac{\text{Number of cars using road per hour}}{\text{Width of road (in feet)}} = 68 - 0.13v^2$$

This relation describes road systems with the approximate frequency of intersections and widths found in London and many other towns. An implication of this equation is that at some high traffic flow density (car equivalents per hour per foot width of roadway) the journey speed drops to zero. Data from Thomson (1967) in Fig. 14-2, based on traffic speed measurements on main roads of central London, show how speed drops with greater density of traffic. It will be noticed that this curve does not drop as a straight line, but increasing car equivalents by 10 cars per hour per foot width of road causes a greater decrease in velocity when the number of cars is already high.

Another process prevents this supercongestion from being relieved, however, as cities become very large. That process is the gradual saturation of ground space in cities by cars, roads, parking lots, and so on, until the functional and economic activity of the city is impaired. This means that there is an effective upper limit to the extent that ground area within a city can be converted to automobile use. Thus, a city can get caught in a trap through overdependence on the automobile as the

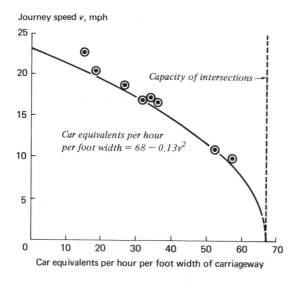

Journey speed v, mph

Car equivalents per hour
per foot width $= 68 - 0.13v^2$

Capacity of intersections →

Car equivalents per hour per foot width of carriageway

FIGURE 14-2 The effect of the number of vehicles attempting to pass through a traffic artery on mean vehicle speed: the impact of space as a limiting factor in the human ecosystem. (*From J. M. Thomson,* Speeds and Flows of Traffic in Central London, *Traffic Eng. and Control* 8:672–676 (1967). *Reprinted by permission.*)

sole means of passenger transportation: on the one hand, extreme congestion of all traffic arteries may require the construction of more freeways through the city, but on the other hand, construction of more freeways destroys the character of the city. This trap makes dramatically clear the role of space itself as a limiting resource in the affairs of man.

The trap is sprung because of another important condition: building freeways does not solve traffic congestion but, in a very real sense, *creates* it. This effect operates as follows.

As a city grows, the relation between freeways, population size, and urban area involves a set of reciprocal feedback systems. This may be described in stepwise fashion, thus:

1 As the population grows, traffic on existing roadways becomes congested, and so in order to relieve this congestion, more roadways are built.

2 Because more roadways have been built, the population size of the urban area can increase.

3 Because the population size increases, the city spreads out over the surrounding countryside, and the size of the urban area increases.

4 Because the size of the urban area increases, the mean trip distance per person increases. This effect can be understood without recourse to mathematics by considering what happens when a 10,000-person town grows to a city of 1 million. When a town has only 10,000 people, the distance to work cannot be very far for those that live and work in it, because the size of the town is limited. However, when the town becomes a city, it is probable that the average distance between home and place of work will become much greater. In fact, it has been discovered that mean trip distances are proportional to the square root of the area of cities (Smeed, 1967). Thus, as city areas increase, the space required

per person per trip increases, and, in turn, the need for square footage of roadway per person increases. For example, when the commuting population increases from 10,000 to 1 million, the area of roadway required per person increases by a factor of about 12 (Smeed, 1964).

5 Because the average amount of roadway needed per person increases as the city area grows, the congestion builds up rapidly, creating a need for more roadway or freeway construction.

6 A limit is finally imposed on this process, however, because the city is no longer a home if it is all paved over with asphalt for freeways. Thus, the ultimate consequence is a very large city with a high proportion of its ground area in freeways and roadways, with terribly congested traffic.

The preceding sequence describes the situation in which a city depends on cars as the sole means of passenger transportation. The real reason for the difficulty, however, is the large amount of space required per person to move people about in cars compared with the space required of other methods of transportation. The most dramatic effect of differences in mode of transportation is in the proportion of the ground area of cities sacrificed to transportation. In London, which makes very heavy use of subway trains and buses, only 15 percent of the ground area of the city is carriageway, whereas in Los Angeles the corresponding figure is 23 percent, because of almost exclusive reliance on cars (Smeed, 1967). It should be noted that these two figures apply only to the part of the city definable as central business district, but the difference applies throughout the two metropolitan areas.

The explanation for this difference can be seen in Table 14-2, which gives the carrying capacity of a carriageway when used in different ways. The amount of space required per person per hour to move passengers at a given speed can be computed from this table as follows. Suppose we take cars with one person (the driver) on 44-foot roads traveling at 15 miles per hour as the reference standard. This system moves passengers at the rate of 45 people per foot width per hour. To compare the space demands of this system and competing modes of passenger transportation, we express them as fractions or multiples, such as $45/x$, where x is the persons per foot width per hour moved by the other form of transportation. Thus, cars with four people each, including the driver, require $\frac{45}{179}$, or $\frac{1}{4}$, as much space per person; buses with 32 people require $\frac{45}{478}$, or $\frac{1}{11}$, as much space per person; and urban railway lines require $\frac{1}{65}$ as much space per person. Clearly, granted that space is a resource, cars are a very bad way to use the resource in cities.

The great space requirements for freeways in urban centers also destroy those urban centers. Thus nothing like the array of culture within a 1-mile radius of Piccadilly Circus in London is to be found anywhere in Los Angeles.

Transportation by car in cities is wasteful not only of space, but also of time and energy. Smeed (1964) has performed thorough operations research analyses of bus, car, and minicar travel in the London road system, including walking and waiting times in the case of buses; these

TABLE 14-2 The carrying capacity of a carriageway when used in different ways

	persons per foot width per hour	speed, mph
Urban street 24 ft wide, mixed traffic:		
Cars with driver only	25.5	15
	41	10
Cars with 1.5 persons	38	15
	62	10
Cars with 4 persons	102	15
	165	10
Buses with 32 persons	272	8.6*
	440	6.7*
Urban street 44 ft wide, mixed traffic:		
Cars with driver only	45	15
	63	10
Cars with 1.5 persons	67	15
	95	10
Cars with 4 persons	179	15
	254	10
Buses with 32 persons	478	8.6*
	676	6.7*
Urban mortorway (capacity per foot width is independent of width):		
Cars with driver only	125	40
Cars with 1.5 persons	187	40
Cars with 4 persons	500	40
Footway	800	2.5
Urban railway line	2,900	18
Suburban railway line	2,200	30

SOURCE: Smeed (1961).
* This has been calculated by assuming that a bus requires 3 minutes per mile for stops, deceleration, and acceleration.

TABLE 14-3 Effect on journey times of changes in mode of peak hour road traffic in London

	occupancy		journey time, minutes*			
			1-mile journey		5-mile journey	
	bus	car	with	without	with	without
1962 conditions	38	1.5	18	16	62	48
All by bus	38	. . .	18	18	50	46
All by car	. . .	2	x	x	x	x
	. . .	3	x	13	x	66
	. . .	4	14	6	72	29
All by minicar	. . .	2	. . .	9	. . .	46
	. . .	3	. . .	5	. . .	25
	. . .	4	. . .	4	. . .	22

SOURCE: Smeed (1964).
* Journey times by car exclude parking times. *With* denotes "with commercial vehicles." *Without* denotes "without commercial vehicles." 'x' denotes that the capacity is exceeded.

are summarized in Table 14-3. It will be noted that London could not accommodate its present commuting population if they all traveled one or two per car, and in standard-size cars they would beat bus times only by traveling in car pools with four people per car.

The relative energy costs of different modes of transportation can be seen from the annual report on highway statistics of the U.S. Federal Highway Administration. In 1967, the average mileage per gallon per vehicle in the United States was 14.08 in passenger cars and 5.36 in buses. Numerous statistical sources show that the average number of people traveling in a car is two or less, so that the passenger-miles per gallon in cars averages 28.16 or less. However, with only ten people in a bus, the passenger-miles per gallon would be 53.6. Thus, if properly loaded, buses make much more efficient use of fuel. Railways make still more efficient use of fuel. In 1967, railways in the United States used about 4×40^9 gallons of fuel, including the fuel used to produce electricity for electrically powered trains. With this fuel, the railways carried about 15.3×10^9 passenger-miles of passenger traffic, and yet passenger cars were only about 1 percent of all the cars operated by the railroads in 1967! (All railroad statistics are from the Interstate Commerce Commission.)

In summary, the United States has drifted into heavy dependence on that mode of urban passenger transportation which makes the worst possible use of space, time, energy, and matter, the automobile, which averages less than two people per car.

URBAN PLANNING: CRIME AND VIOLENCE, EDUCATION AND TAXATION

Crime, violence, education, and taxation are all linked because of their common dependence on the age structure of populations. Since for humans, as well as for other animals, all attributes of an individual change with its age, a change in the age structure of a population therefore changes the frequency distribution of all attributes of that population. This fact has many implications which are not generally recognized.

One of the most important of these concerns the effect of changes in population age distribution on the economic life of a city. Suppose the growth rate of a population increases so that a higher than normal proportion of the population is in young age groups. This means that the ratio of people in ages that consume education taxes to those in ages that produce education taxes is increased. What is not widely known is how sensitive the educational tax burden per taxpayer is to rather small changes in rate of population growth. It has been shown by computer simulation studies that this proportion varies as follows (Glass, Watt, and Foin, 1971):

percent per annum population growth rate	tax consumers per tax producer
0	0.33
1	0.45
2	0.59
3	0.74
10	3.53

This means that if a stable population (zero growth rate) increases its growth rate to 1 percent per year, the resultant change in population age distribution increases the tax burden per taxpayer to $100(\frac{0.45}{0.33})$, or about 135 percent of what it would have been with no population growth. Similarly, we can obtain the percentage increases in tax burden per taxpayer when a stable population increases at 2, 3, and 10 percent per annum growth rates. The figures are as follows:

percent per annum population growth rate	tax burden per taxpayer relative to tax burden under zero population growth
1	1.35 times as great
2	1.8 times as great
3	2.3 times as great
10	11 times as great

These results from computer simulation studies scarcely lend any support to the claim often heard from promoters that urban growth is a good thing because it broadens the tax base and lowers per capita tax costs.

However, to make matters worse, it is known that many types of crime are much more likely to be committed by young people. Therefore, in a growing population, as the growth rate becomes steeper, the proportion of young people will be higher and consequently the population-wide incidence of crime and the costs of police protection will be greater. Further, in a dense population there is more incentive to steal, and stress produces more aggressive and riotous behavior. The FBI Uniform Crime Reports for the United States show that larceny is more than three times as likely to occur, per capita, in people under 25, as in people of 25 to 34, and is about one-fifth as likely to occur in 35- to 44-year-olds as in people under 25. Thus, increasing population growth rates and the resultant increase in incidence of people under 25 would be expected to increase the incidence of larceny. If all crimes on the average show this same statistical pattern of higher incidence in young people, then we would expect to find high resultant police costs in rapidly growing urban areas.

Analysis of the data on United States cities indicates that it is difficult to develop a statistical relation between city growth rate and tax burden per capita for two reasons. First, there is a lag from the time a city begins growing rapidly to the time when the tax costs of the growth become apparent. For example, more public colleges are not needed for 18 years after a surge in birth rates. Second, the effect of city growth rate on tax burden is masked by the tremendous effect of city size on tax costs.

The data are summarized in Fig. 14-3. In the range of city sizes from 10,000 to 500,000 there are a large number of cities, and the data for both total taxes and police protection can be described by a smooth curve. Above that point, the number of cities becomes smaller, and there is a great deal of scatter, although the general trend is still up. This figure shows that the component of urban taxes per capita allotted to police protection is a more steeply rising function of city size than the total tax expenditure.

FIGURE 14-3 The effect of city population size on per capita cost of living in cities of different populations. The data used to construct this graph were obtained from the U.S. Bureau of the Census, 1960 Census and *Census of Governments: 1967*, vol. 4, no. 5; *Compendium of Government Finances*; and *City Government Finances in 1966–1967*. Expenses per person were obtained by dividing total expenses, and expenses for police protection in 1967 by the population for 1960. In all cases the figures used are for cities, not counties or Standard Metropolitan Statistical Areas. The first four points in each series are for a large group of cities in each of four population size ranges: 10,000 to 24,999; 25,000 to 49,999; 50,000 to 99,999; and 100,000 to 249,999. Above that size, the number of cities in each size range is smaller, and data points on individual cities have been plotted instead. There is a great deal of scatter in these points when data for all cities are plotted, because the effect of size of population alone is confounded by growth rate in population, population density per square mile, and other factors. However, the general trend is up for both lines.

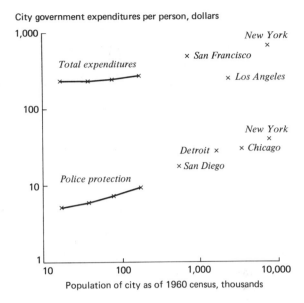

To summarize, these data indicate that the cost of crime per capita is a steeply rising function of city size, and the overall tax cost per capita is also a rising function of city size, but it rises less steeply. In short, these data do not support the argument that urban growth "broadens the tax base and reduces taxes per person." Rather, they indicate that it is costly to each taxpayer to have urban growth. Each taxpayer must answer the question for himself as to whether the economic and other benefits of living in a large city compensate for the increased costs. The general conclusion would appear to be that large national or regional populations are better accommodated in a large number of medium-sized cities than a small number of very large cities. This challenges the idea of some planners that humanity should plan for extremely large cities in the future.

It is not the intent here to argue against high taxes, which may often reflect an increase in the ratio of public to private sector spending, and hence an improvement in the quality of life. Rather, we seek to undercut a standard argument of developers who promote development so that they can profit, without regard to consequences for others.

DIVERSITY AND STABILITY IN CITIES

Chapter 2 pointed out that Principle 13, concerning the relation between diversity and stability, had implications for the economic organization of cities. Jane Jacobs (1969, chap. 3) has pointed out that cities like Manchester, dominated by a single industry (textiles in the nineteenth century), have a high level of efficiency because of economies of scale inherent in the mass production process. However, because such cities have a low diversity (information content) in their mix of industries, because of the domination of the city economy by a small number of types of activity, the city economy is vulnerable to economic stagnation if anything happens to the market for its principal products. This point has been dramatically demonstrated in 1970 by the problems confronting Seattle because of the domination of the Boeing Company in the city's economy at a time when the market for aircraft is declining sharply. Mrs. Jacobs has also noted the overriding importance for a city of having a large number of small companies in which a great deal of work is devoted to research or development of new products. The very nature of such work results in a high component of exploratory rather than routine activity. Consequently, the work has a low degree of efficiency, although, on the other hand, the existence of such activity means that the city has a high potential for evolution into new types of economic activity. The parallel with the natural world extends further. In high-diversity environments such as coral reefs or the rain forest, the species do not respond rapidly to sudden environmental demands for adaptation (Principle 12). Neither did the specialists at Boeing, many of whom became unemployed. The parallel is not perfect, because in the human situation, fluctuations in the physical environment are not an issue because of our use of fuel and air conditioning to even out the physical fluctuations in the external environment. Thus, diversity does not imply rigidity. However, the individuals are more functionally rigid if an urban area has a low level of corporate diversity.

Two major implications for city planning emerge from this discussion. First, cities should make every effort to have an optimal variety of industries. Second, sophisticated planning should look ahead to a time when fuel will be much more expensive than it was up to 1970.

CITY DESIGN AND FUEL AVAILABILITY

Chapter 3 noted that great distance between a site of demand and a site of supply implies a high transportation overhead. That many modern cities have such a low density, with farmland interspersed with fringe suburbs, is a result of the recent low cost of fossil fuel and the way it has been regarded as an inexhaustible resource. However, when resources of all types become more scarce, and the most cheaply accessible and highest-grade supplies are depleted, fuels of all types will become more expensive. At that time, the most economically efficient cities in terms of spatial organization will have great competitive advantage over those more spread-out cities that have been produced by the automobile age, which will soon be coming to an end because of gasoline shortage. Gradually, increases in the cost of transportation and shortages of agricultural land (Chap. 15) will force cities to be more compact, and the transportation overhead associated with all essential services will be minimized. Cities will tend to grow up more than out, following the path of evolution now associated with island and peninsula urban centers (Manhattan and San Francisco), rather than plains and plateau cities, such as Phoenix, Dallas, Houston, and Detroit, which are not bounded by water and consequently spread out rapidly into the surrounding flat land.

Because of lag times between time of change in city design and time of resultant economic benefits, government will have to underwrite the initial construction. At the time of construction it will not be profitable. Another strategy is to make it profitable by changing the tax structure.

REGIONAL PLANNING

It has been known for a long time that increase in distance between two points implies an increase in transportation cost. For example, von Thunen (1826) recognized that there was a relation between the prices of agricultural products and patterns of land use, such that there was an economically optimal way to distribute fields on a farm with respect to their distance from the farmhouse. Such research has developed until there is now an elaborate body of theory which deals with the economic geography of location of enterprises in space (Lösch, 1954; Isard, 1956; Chisholm, 1970). A central point of this theory is that the closer land is to high population densities, the higher the value of the land, and consequently the greater the production of crops per hectare adjacent to high population densities. This principle applies whether we are considering an individual farm, a city, a region, a country, or the world. Also, it implies that villages, towns, and cities will tend to be situated so as to minimize transportation costs.

Three tables will illustrate these points.

Table 14-4 illustrates the relation between the production per hectare and the distance from the farm center. Table 14-5 illustrates the relation between wheat production per hectare and population density in different countries. Table 14-6 is a hypothetical example showing how the transportation costs associated with different commodities can determine the economically optimal location for a village, town, or city.

The conclusion to be drawn from these tables is that regions should be planned so that transportation costs are kept as low as possible. This will be particularly true in the future as fuel becomes more scarce and consequently more expensive. Thus, there will be tremendous incentive to avoid the type of urban sprawl depicted in Fig. 3-9C.

Also, when or if energy becomes scarcer in the future, the cost of fertilizer will become exorbitant, and it will be important to produce agricultural products on soil which has the highest possible natural fertility. Thus, regional planning should avoid the type of situation now found, in which cities cover the very best agricultural soil in a region, and less productive soil is being used to produce crops at a great cost in irrigation

TABLE 14-4 Finland: relation of production per hectare and distance to farm plots

distance, kilometers	Wiiala gross output	Wiiala net output	Virri gross output	Virri net output	Suomela net output
0–0.1	100	100	100	100	100
0.5	92	78	89	67	83
1.0	84	56	80	50	68
1.5	77	34	73	40	56
2.0	69	13	67	33	46
3.0	. . .	. . .	57	25	32
4.0	. . .	. . .	50	20	
5.0	. . .	. . .	44	17	

SOURCE: Chisholm (1970).

water and fertilizer. We must develop a new outlook on regional planning, in which the importance of maximizing the efficient use of energy is recognized.

One interesting systems-level problem in regional planning is that because the economic return from land is higher when it is converted to subdivisions than when it is used to grow crops, each region plans to convert as much land as possible to subdivisions, leaving it to other regions to grow the crops. There is a trap in this strategy because all regions are planning the same thing. Who, then, volunteers to grow the crops? This is a situation in which the effect of supply relative to demand might not begin to act on agricultural commodity prices until it is too late. That is, land could be converted to subdivisions before it became apparent that it is really needed for agriculture. A further complication could be a rapid decline in yields per acre of land if increased upper atmospheric turbidity began to have a significant effect on the global weather regime, as explained in the next chapter.

Interregional economic competition could have some strange effects on the health of a nation. Suppose, for example, that both Florida and California quickly converted all their citrus land to tract housing, each assuming that the other would supply the citrus for the nation. The result might be no citrus. It is true that vitamin C can be made available in forms other than natural orange juice, but the substitutes might not be so attractive to children, whose health is affected by the level of vitamin C intake. Thus, conversion of citrus groves to tract housing by both Florida and California might have a very real effect on the nation. The point in this discussion is that what is good for a region is not necessarily the best strategy for the nation of which the region is a part. It may be necessary at times to prevent a region from eliminating an agricultural activity which is in the best interests of the nation.

NATIONAL PLANNING

If it is true, as suggested in Chap. 15, that the world will have less cheap fuel in the future, then each nation would do well to plan its mix of transportation systems so that the energy efficiency of the whole mix is as high as possible. In reality, this is not at all what has happened. For ex-

TABLE 14-5 Relation between wheat yields per hectare of wheat field and population densities in different countries

country	total area, thousands of hectares	1966 population, in thousands	population density, persons per hectare	1966 wheat yields 100 kilograms per hectare
Netherlands	3,615	12,455	3.45	40.4
Belgium	3,051	9,528	3.12	30.5
Japan	36,966	98,865	2.67	24.3
Switzerland	4,129	6,050	1.47	32.3
U.S.A.	936,322	196,920	0.21	17.7
Argentina	277,666	22,691	0.08	12.0
Australia	768,681	11,541	0.02	15.3
Canada	997,618	19,919	0.02	18.7

SOURCE: Data from Production Yearbook, vol. 21, Food and Agriculture Organization, United Nations, 1967.

ample, as shown in Fig. 6-6, jet planes actually have a lower mean portal-to-portal trip velocity than high-speed trains at distances under about 360 miles, because of the long trips to and from airports. In spite of this, however, the United States has drifted in the direction of using jet aircraft for medium-haul transportation by regional feeder air carriers, even though time of travelers and the energy required would be used more efficiently by very high-speed trains such as the Tokkaido line between Tokyo and Osaka in Japan. It is noteworthy that the long distances from airport to urban core typical of large urban centers, such as upwards of 40 minutes from Dulles Airport to Washington, D.C., are found also in most small communities. As fuel becomes more expensive, it will become more important to have more rational planning of the national transportation mix.

The future will very likely bring more rational national planning of food resources, as well as fuel resources. One consequence will be more attention to valuable offshore fisheries and shellfish resources. At present, many nations are pursuing policies highly destructive to offshore resources, even though they may be rich sources of food. It is increasingly clear that large quantities of pesticides and metals, originating in agricultural and manufacturing industries, are being washed into the sea by rivers and are causing great mortality of the highly productive continental shelf marine resources.

Figure 14-4 indicates the possibilities for culture of fish and shellfish at the edge of the ocean. The area depicted is the east coast of Molokai, one of the Hawaiian Islands. Much of the shoreline consists of interdigitated land and water which facilitate the construction of oceanside ponds to be used for intensive fish and shellfish culture. In fact, a very high proportion of the world's coastline lends itself to this type of activity: the coasts of Norway, Scotland, British Columbia, and New Jersey are examples.

Another matter that must be considered is the effect of major irrigation and hydroelectric power projects on fish and shellfish resources. For example, the Aswan High Dam on the Nile in Egypt has destroyed the fisheries at the mouth of the river by preventing the flow of nutrients to the Nile delta.

TABLE 14-6 Two hypothetical village locations x and y

A resource	D distance from village in kilometers		C units of cost per kilometers from village	D product of B × C	
	x	y		x	y
Water	0.1	0.5	10	1.0	5.0
Arable land	2.0	1.0	5	10.0	5.0
Grazing land	2.5	1.5	3	7.5	4.5
Fuel	2.5	2.0	3	7.5	6.0
Building materials	3.0	2.0	1	3.0	2.0
Total costs				29.0	22.5

SOURCE: Chisholm (1970).

FIGURE 14-4 The east coast of the island of Molokai, illustrating the great potential for intensive fish and shellfish culture in ponds at the edge of the ocean.

The largest question posed by ecology for national policy concerns national goals. What should they be? Are economic growth and economic efficiency the most appropriate goal for a country? Should survival, and consequently the avoidance of strategies which lead to internal and external conflict, be an important desideratum also? What about the importance of maximizing the efficiency of energy use, as well as economic efficiency?

Another important question concerns the relation between national demographic policy and economic growth. Kahn and Wiener (1967) have focused attention on the great growth in gross national product currently occurring in Japan. It is noteworthy that this growth has been achieved during a period in which there has been considerable success in minimizing the population growth rate. As pointed out earlier in this chapter, too high a rate of population growth rate can so distort the age structure of a population that any benefits from growth in economic product are absorbed by the capital costs of educating young people and providing various social services. Consequently, an excessive rate of population growth may effectively block even the most heroic efforts by a country to improve the lot of the people. By the same token, a country may be able to make rapid economic strides only if the rate of population growth is small enough relative to the resources of the country.

It is surprising how little attention has been focused on the relation between the cultural and economic health of a country and the population size relative to resources. It is clear that many countries, such as northern India, the Moslem world, and the entire Mediterranean basin group of nations, had vastly more cultural vitality in the past than at present, and it is interesting to speculate about the cause for the evident decline. Why? It is extremely interesting that most treatises on ancient history either never mention the words *forests* and *trees* or, if they do, mention fleetingly that the forests of Greece had been removed by the fifth century B.C., for example. In fact, the tremendous economic surge of the United States got its initial kick from fuel wood, and it was able to continue because of the fortuitous subsequent discoveries of coal, crude oil, and natural gas.

It is to be hoped that a time is coming when nation states will perceive their natural resources as an extremely important factor in determining their vitality, and will husband these resources, such as soil, water, fish, trees, and fuel, with the care they so evidently deserve.

SUGGESTIONS FOR INDIVIDUAL AND GROUP PROJECTS

Take temperatures at different times of the day in different parts of the city where you live. Select sites with different thermal characteristics (fountains, parking lots, your garden, a large park, a street in a residential area, a shopping center). What patterns do you observe in the hour-to-hour differences at a site, and the differences in daily trend between sites? How do you account for these patterns?

REFERENCES

Chisholm, M.: "Rural Settlement and Land Use: An Essay in Location," Aldine, Chicago, 1970.

Dasmann, R. F.: "The Destruction of California," Macmillan, New York 1965.

Duckworth, F. S., and J. S. Sandberg: The Effect of Cities upon Horizontal and Vertical Temperature Gradients, *Bull. Am. Meteorol. Soc.*, **35**:198–207 (1954)

Glass, N. R., K. E. F. Watt, and T. C. Foin: "Human Ecology and Educational Crises: One Aspect of the Social Cost of an Expanding Population," in S. F. Singer (ed.), "Is There an Optimum Level of Population?", pp. 205–218, McGraw-Hill, New York, 1971.

Huntington, E.: "Mainsprings of Civilization," Wiley, New York, 1945.

Isard, W.: "Location and Space Economy," MIT Press, Cambridge 1956.

Jacobs, J.: "The Economy of Cities," Random House, New York, 1969.

Johnson, P. L. (ed.): "Remote Sensing in Ecology," University of Georgia Press, Athens, 1969.

Kahn, H., and A. J. Wiener: "The Year 2000: A Framework for Speculation on the Next Thirty-three Years," Macmillan, New York, 1967.

Knipling, E. B.: Leaf Reflectance and Image Formation on Color Infrared Film, in P. L. Johnson (ed.), "Remote Sensing in Ecology," pp. 17–29, University of Georgia Press, Athens, 1969.

Lösch, A.: "The Economics of Location," p. 520, Yale University Press, New Haven, Conn., 1954.

Myrup, L. O.: A Numerical Model of the Urban Heat Island, *J. Appl. Meteorol.*, **8**:908–918 (1969).

Smeed, R. J.: "The Traffic Problems in Towns," Manchester Statistical Society, pp. 1–59, 1961.

———: The Traffic Problem in Towns, *Town Planning Rev.*, **35**:133–158 (1964).

———: The Road Capacity of City Centers, *Highway Res. Record*, **169**:22–29 (1967).

———: Traffic Studies and Urban Congestion, *J. Transport Economics and Policy*, **2**:1–29 (1968).

Thomson, J. M.: Speeds and Flows of Traffic in Central London, *Traffic Eng. and Control*, **8**:672–676 (1967).

Thunen, J. H. von: "Der isolierte Staat in Beziehung auf Landwirtschaft und Nationalokonomie," Rostock, 1826.

15 a global strategy for mankind

Whether the world has too many people is a subject of intense controversy now. In the interests of stimulating the student to examine the cases on both sides in detail, we will present fairly elaborate versions of the opposing arguments. To facilitate comparison of the data and logic used in support of the two cases, the arguments are each arranged in the following order: the demographic situation, the effects of population on social and economic factors, resources, the physical and chemical state of the planet, and overview.

THE OPTIMISTIC VIEW The environment-population optimists argue that there is no crisis with respect to world population density, and that most parts of the world can stand much larger populations than they have at present. Two important observations used in support of this argument are, first, that much of the world is underpopulated at the moment, most of the population being piled up in a few big cities, and second, that many parts of the world have had much larger populations in the past than at present. From the second observation it is argued that those parts, and others, could clearly support much larger populations in the future.

There is certainly a large body of evidence to indicate that many parts of the world have a fraction of their former population density. The Hebrew historian Flavius Josephus indicated in "Jewish Antiquities," a first-century history of the Jewish people, that the population of Jerusalem during the time of Nero (A.D. 37–68) was 2,720,000. The entire population of Israel in 1966 was only 2,629,000. There is a great deal of evidence that the entire Mediterranean basin, the Middle East, and Europe once supported enormous numbers compared with their populations today. For example, the book of Numbers (1:45–46) states that of the Israelites living in the Sinai Peninsula, those over 20 and able to go to war numbered 603,550. Suppose that only men were able to go to war, and 40 percent of the male population was of fighting age. Then the population must have been roughly $\frac{100}{20}$ times 603,550, or about 3 million. The area formerly occupied by these people is now virtually uninhabited.

Some readers may feel that such scattered impressions of large ancient populations in the literature are based on counting errors and are consequently not to be trusted. However, evidence which cannot be subject to any such errors exists in the form of vast ruins in many places where the population densities are now very low. Such sites which probably had much higher populations in the past include the valley of the Tigris and Euphrates, Palmyra in Syria, Nippur in Iraq, and Persepolis in Persia.

From such evidence it is possible in principle to argue that the world can support much larger populations than at present.

The incidence of many diseases in the world is dropping rapidly, as the plummeting death rates so eloquently attest, and this is the most compelling argument that high population densities do not imply a less salubrious environment. It can be argued that those few diseases for which the incidence of mortality is rising are not due to rising density, but rather a moral breakdown: smoking, drug use, and promiscuity.

People have always complained about taxes, which are not noticeably higher than they have been before. Increasing crime and violence are due to changes in social mores and moral change, not population density.

Clark[1] has summarized four types of economic arguments which have been put forth to support the notion that high population densities and growth rates are desirable.

The first argument is economies of scale. For example, in the nonagricultural sector of an economy, increasing the labor input leads to increased returns per unit of labor input. High population density decreases the cost per unit in training rare types of specialized professionals. A small population might not be able to train any numerical analysts, heart surgeons, or cereal breeders. High population densities permit development of many specialized types of equipment, which give a nation a competitive advantage over other nations. Only the United States, the Soviet Union, the United Kingdom, and France are important manufacturers of commercial jet aircraft. The higher the population density and the greater the use rate, the cheaper it is to operate large pieces of construction or equipment, such as harbors and transport systems. Economies of scale lead to development of specialized firms, which can lower costs. Densely populated countries require less capital per unit of product than less densely populated countries: from all such examples, it is inferred that the cost of living per capita diminishes as population density increases. But it should be noted that opposite effects may be present (social inefficiencies increase with increasing density).

The second argument is the high rate of increase. Where there is a rapid rate of population increase, an erroneously judged investment stands a good chance of being put to some alternative use. In a stationary economy, a mistake is more likely to be irretrievable. However, this argument can be turned around. If economies are growing rapidly, the failure to perceive an impending slackening off of a growth rate can have an astronomical penalty. This was incurred by several of the largest commercial airlines in 1970 and 1971. During the spring of 1971, four of the largest United States airlines lost an average of about 10 million dollars a month each because of overinvestment in new planes.

High rates of increase can generate "windfall profits." The ease of making money in a high growth situation minimizes risk, maximizes potential gain, and encourages technological innovation and investment gambles. An economy needs immense amounts of capital to gamble with to finance research and development of nuclear fusion reactors. Fusion reactors are desperately needed. Thus, it can be argued that high rates of population increase are a necessary precondition for solution of major social problems.

The third type of economic argument stems from observations on spatial location. Wages of workers are greater, on the average, where densities are high, because of economies of scale. But further, larger

[1] C. Clark, "Population Growth and Land Use," by permission of Macmillan, London and Basingstoke, 1967.

towns have a greater variety of manufacturing concerns than smaller towns, and hence are more economically self-sufficient. Thus, costs of living are lower because a lower proportion of all goods purchased in a large town is priced to include transportation overhead.

The fourth argument concerns the effect of population growth rates on personal freedom. The hold of serfdom weakened from the eleventh century onwards, in time of renewed population increase. The population decline of the late fourteenth and fifteenth centuries produced increasing rigidity in economic life. The probability of a person's attaining a high position in an organization is higher in an expanding than in a declining organization.

It is argued that there are no resource shortages at present, or in the foreseeable future, for two reasons. The first relates to food, and the second to energy. As for food, there will be a continuing surplus of grain in the United States for as far as we can see, it is argued, because there is a great deal of acreage which can be returned to grain production. Further, the most cursory examination of international agricultural production statistics indicates that agricultural production in most countries could be boosted enormously by more use of modern technology: fertilizers, irrigation, tractors, and new strains (Chap. 11). In turn, the ability of mankind to solve massive problems in agriculture or in any other area of activity is dependent merely on energy availability, and with the advent of the breeder reactor, energy will no longer be limited. As for marine resources, it can be argued that the productivity of the ocean can be increased enormously by using solar-powered engines to produce massive convection currents, so that upwelling of nutrients from the abysmal depths will increase nutrient concentrations near the surface.

Also, it is argued that the problems of smog, water pollution, and pesticides are all purely temporary, and that they will be solved by the advent of newer, more sophisticated technology and the unlimited cheap energy of the fusion reactors.

In summary, the argument of the technological optimists states that man, as a species, can solve quite literally any problem, given enough energy and technology, and that man's ingenuity will always find the needed energy sources and technological innovations.

Certain key elements in the above lines of reasoning are noteworthy. Hope rather than fact is basic to many assertions. Solutions are predicted which depend not on present technology, but on technology which has yet to be developed, and the time required to develop, and indeed invent, new technologies is typically ignored. (But time is also a limiting resource: Principle 3.) Data on the available supplies of resources, on which the new technologies depend, are seldom presented. Predictions involve a short look into the future, generally 30 years, not 50 years or longer. The possibility of novel types of problems due to interaction effects is ignored. The possibility that certain processes already started may be more serious than they appear to be, owing to a long lag time before effects show up, is also mostly ignored. Predictions are based on a remarkably optimistic view of the efficiency with which all large institutions can now make rational decisions. Finally, the notion that there are real thermodynamic constraints and resource limitations on the activities of man is simply denied. There is no acknowledgment of the existence of optima, asymptotes, or limits, as implied by Principle 4.

MAN AND GLOBAL RESOURCES: TECHNICAL DETAILS ON THE PESSIMISTIC VIEW

The counterargument challenges every component of the argument by the technological optimists. The counterargument says that mankind now faces several different kinds of problems due to high and rapidly increasing populations, social disintegration due to environmental and demographic dysfunctioning, public health effects of the same problems, resource depletion, institutional inefficiency related to organizational size, and physical and chemical degradation of the planet.

The counterargument contends that the technological optimists' viewpoint hinges on four key implicit or explicit assumptions, which are the product of a historical accident, rather than a broad view of the facts concerning our planet. The four assumptions are as follows:

1 Man as a species is incapable of destroying his civilization.
2 Resources are essentially unlimited and therefore need not be conserved (denial of Principle 4).
3 The knowledge and inventiveness of man as a species grows constantly, and the continued existence of the species is guaranteed thereby.
4 Therefore, human populations can increase for as far ahead as we can see.

One essential fact is ignored in these four assumptions. There are immense areas of the world where man has been a long time, and they are now denuded of forests, arid, and in general apparently burned out, compared with more recently inhabited areas. These overexploited areas include northern India and a large proportion of the Mediterranean basin: Portugal, Spain, southern Italy, Greece, Turkey, Lebanon, Israel, Egypt, North Africa, Cyprus, Sicily, Malta, and the Middle East. The magnitude of the difference between such areas and the newer, more recently exploited countries is illustrated by Fig. 15-1 and Plate 9. These large tracts suggest that man can in fact disrupt life on this planet, but this has not been clearly perceived by most modern thinkers because the centers of culture and civilization have shifted to new regions, where people are not constantly faced with the sight of ruined resources and lands. The historical evidence indicates strongly that the vast tracts which now appear ruined did not get that way because of climatic change, but were deforested and otherwise excessively and unwisely cultivated by man, with resultant long-term soil degradation. It is particularly noteworthy that certain regions of the world should not be exploited, because of the high likelihood that climate will produce deserts as indicated by the Holdridge system. Unfortunately, this is true in tropical and subtropical regions where the population growth rates will create great pressures for exploitation of marginal lands. Also, trees have a moderating effect on climate (Chap. 8).

Indeed, an overview of the history of mankind from an ecological point of view suggests that civilization has had a strongly nomadic characteristic. That is, man is able to develop civilization to a high level on a particular portion of the planet because of a useful resource base there.

A

B

C

FIGURE 15-1 Examples of the effects on land-scape due to intensive exploitation by man. **A** Soil removal on hillside associated with subdivision development in Trinidad. **B** Roadside erosion in California. **C** Overgrazing by goats in Guatemala.

However, in the course of developing the civilization, the readily attainable resource base (minerals, trees, soil, water) is depleted, degraded, and polluted, and for this reason the civilization loses much of the vitality it had in its prime. Then, as the place becomes a desert, or semidesert, the most adventurous and vigorous of the population have traditionally set off to find a new site to exploit. This process has now come to an end, of course, because on this planet there are no new places to exploit, and the energy resources are inadequate to send large enough groups of exiled colonizers to other planets, most of which probably have inhospitable environments in any case.

THE DEMOGRAPHIC SITUATION

In 1970 the world population of humans was large (about 3.6 billion), increasing rapidly (about 2 percent per annum), and had a highly distorted age structure (that is, about 40 percent or more of the population in a rapidly growing population is under 15 years of age; only about 26 percent or less of the population is under 15 in a slowly growing or stationary population) (Keyfitz, 1971). Assuming that the world population is too large already, a case we intend to prove in the remainder of this chapter, three important questions follow from the three demographic facts in the preceding sentence. First, what would be the required lifetime fertility in women in order for the population to achieve stability? Second, what is the likelihood that this stability will be achieved? Third, even if we do achieve it, how long will the demographic effects of the presently distorted age distribution persist?

To answer the first question, we can compute the number of children each woman can have if the population is to stop growing by working backwards from the present death rate. That is, if the population change each year is to be zero, then the addition to the population, number of births, must be exactly equal to the subtraction from the population, number of deaths (assuming no net migration). Thus, for example, the worldwide average birth and death rates per year per 1,000 population are now 34 and 15, respectively, and this difference is the reason for the population increase. To achieve zero growth the birth rate per 1,000 would have to drop immediately to 15.

To find out what this means in terms of complete lifetime fertility for each woman, we proceed as follows (Frejka, 1968). We assume that the birth rates of women of different ages remain the same relative to each other. That is, if the probability that a 35-year-old woman will give birth to a child in a given year is now only x times the probability that a 21-year-old woman will have a child in that year, we assume that x remains unchanged. Then if the birth rate for the entire population of women must be reduced $(34 - 15) \frac{100}{34} = 56$ percent in order to produce population stability, the age-specific birth rate for each age is reduced equally (an equal proportion, not an equal amount). From this, in turn, we can compute the allowable complete lifetime fertility schedule for each woman, or each family.

The magnitude of the reduction in children per woman allowable under this schedule is startling. For example, in the United States, where the death rate is now such that the birth rate would have to be reduced 45 percent, Frejka has shown that each family would have to limit its total number of children to about 1.2 for about the next 20 years, after which the mean number of children per family could increase gradually to about 2.8 children per family in the period 2030 to 2035! In most of the undeveloped countries, zero population growth can be achieved only if the average number of children per woman per lifetime drops below 1.00; for example, the required percentage reductions in the crude birth rate to attain stability are 65 for the United Arab Republic, 80 for Venezuela, 58 for India, 72 for Thailand, and 64 for Japan.

Clearly, nothing like this level of reduction is socially, politically, psychologically, or economically feasible. Birth rates in the United States, for example, have fallen from about 3.35 children per lifetime in 1962 to about 2.45 children per lifetime per woman in 1970. However, this has only reduced the United States population projection for the year 2000 from 356 million to 280 million (*Statistical Abstract of the United States*). Even if the lifetime birth rate for Unites States women falls immediately to 2.11, the year 2000 population will still be 266 million, about 30 percent more than the 1970 population of about 204 million. The reason why heroic efforts at reducing the birth rate will have little effect in stabilizing population in any country in the next few decades is that the population age distribution has become terribly distorted because of the post-World War II drop in mortality rates, which was not accompanied by a corresponding drop in birth rates. This means that the world is now teeming with young women in or approaching their most fecund years of life.

Davis (1967) has assessed the possibility that current national population policies will be able to deal effectively with the problem. He shows that in most countries, the targets that have been set for reduced birth rates would in fact allow for tremendous population increase. Most underdeveloped countries are aiming for reductions in the birth rate of 20 to 40 percent. While this is a gigantic social undertaking, it will not be adequate to stabilize population size. Davis reviewed the literature on polls taken to determine the number of children young women and married couples thought was the ideal family size. This number ranges between 2.7 and 4.3. Clearly, since this range applies throughout all polls in all countries, a tremendous worldwide revolution in thinking about population growth is mandatory.

From the very long-term point of view for mankind, a most frightening circumstance is the tremendous number of young people throughout the world who were produced by the post-World War II baby boom. No matter what happens now, the effects of this surge in year class strength will have very substantial long-term consequences. Frejka (1968) has done simulation studies to determine how long the effect of this pulse will last, assuming a constant total population to 2185. Even 215 years from now, the population will still not have reached a stable age distribution. The present great wave of young people becomes a great wave of 60- to 90-year-olds in 2025, which gives rise to another great wave which shows up as 50- to 90-year-olds in 2105. The social, political, and economic consequences of these waves are alarming. Some notion of the magnitude of the cause for alarm has been given in Chap. 14, in connection with such problems as the taxation and education consequences of distortions in the age distribution.

GLOBAL TRENDS IN RESOURCE DEMAND RELATIVE TO SUPPLY

We will consider the global status of four categories of resources: the oceans, land and food, minerals, and fuels. Their availability can be used to determine maximum possible limits on global population size.

Many writers tell us with great enthusiasm about the almost inexhaustible amount of food that can be obtained from the oceans. To put this remark in perspective, it is useful to indicate, first, how small the present world catch of ocean food is compared with the amount of food obtained from land; second, the current status of the major fishery resources; and third, the desertlike productivities of the ocean. In 1966, the total world production of cereal grains was 1,088 million metric tons, and the world ocean fish catch was 57 million metric tons. In 1966, world production of meat was about 70 million metric tons.

To understand better the significance of these numbers, it is necessary to know that although seafood is a rich source of animal protein, it is a poor source of carbohydrates. There is considerable variation in the caloric content of fish and shellfish, but 1200 kilocalories per kilogram of edible portion is close to a typical figure, whereas the corresponding energy content of cereals is about 3600 kilocalories per kilogram of edible proportion, about three times as much. Since production of seafood is about 5 percent of the worldwide production of cereals, by weight, the worldwide contribution to human energy needs by seafood is only about 2 percent of that of cereals.

Three kinds of arguments have a bearing on how much the seafood production can be increased. Clearly, examination of the fisheries statistics on species already being harvested intensively gives insight into the status of the stocks, and how much more exploitation pressure they are likely to be able to stand. Also, it is possible to reason from what is known about the intensity of the sun's energy reaching the surface of the ocean, and the efficiency of energy conversion through the various steps of the marine trophic pyramid, to estimates of the maximum possible harvest. Finally, it is possible, reasoning from ecological stability theory (Chaps. 2 and 3), to infer how great an intensity of exploitation unexploited fish stocks from the tropical seas could tolerate.

Table 15-1 shows the liveweight catch in thousands of metric tons of twenty of the major marine fisheries in 1958 and 1968. The most striking feature of this table is that most of the major marine stocks have shown slight increase in production, if any, within this decade, and many show declines in production. This would scarcely seem to provide

TABLE 15-1 Trends in production of major marine resources

	liveweight catch in thousands of metric tons	
	1958	1968
U.S. pink salmon	54.8	47.6
U.S. Pacific red salmon	30.8	25.1
Japanese marine pink salmon	91.4	47.0
English and Welsh plaice	34.0	40.1
U.S. Pacific halibut	21.7	11.7
Norway cod	373.4	358.5
Canada cod	287.9	320.4
Scotch haddock	90.5	89.2
Spanish hake	68.3	74.8
Japanese cod	60.0	109.0
Japanese croakers and drums	106.7	74.0
German redfish	123.6	99.0
Russian redfish	170.8	44.0
Japanese horse mackerel, jack mackerel, scads	324.3	358.0
Norway herring	607.6	704.8
Japanese albacore	63.2	70.0
Japanese yellowfin tuna	111.5	116.0

SOURCE: FAO, "Yearbook of Fisheries Statistics," vol. 26, 1969.

grounds for unbridled optimism about the magnitude of future oceanic harvests.

What are the prospects for a significant increase in the production of food from the sea? Ricker (1969) has reviewed recent evidence on the amount of living tissue produced in the world's oceans. All the following estimates should be accepted as preliminary, and as order of magnitude estimates only, owing to the still embryonic state of research. However, they do indicate the magnitude of the productive capacity. This table should be expected, for the great drops in production from trophic level to trophic level follow from Principle 2:

trophic level	production of organic matter in all oceans in millions of metric tons per year
Green plants (largely phytoplankton)	130,000
Primary consumers (largely animal plankton)	13,000
Secondary consumers (small fish)	2,000
Tertiary consumers	300
Quaternary consumers	45

The figures for the tertiary and quaternary consumers, which include most of the fish species of commercial interest, are remarkably close to recent world production statistics on ocean resources. Assuming that present commercial production is some mix of second-, third-, and fourth-level consumers, between 2.5 and 5 times the present harvest of fish seems to be the most we can theoretically expect to obtain. Thus, the greatest possible contribution ocean fisheries can make to the world food needs is about a quarter of the 1966 world cereal harvest by weight, or 8 percent of the 1966 cereal harvest by caloric content.

Some readers will be quick to point out that we do not need to confine our attention to fish, since it would make a far more efficient use of the solar radiation reaching the ocean to harvest plankton; that is, instead of harvesting tuna and salmon, we should harvest the tiny crustaceans or, better yet, the green algae. However, there are some problems with this proposal, which Ricker (1969) has pointed out. First, there is the expense (in energy, as well as money) of filtering or centrifuging algae out of seawater. Second, plankton is not a desirable food item for humans: the flavor is unattractive, the salt content is high, and some of the prominent species have siliceous skeletons which are indigestible. Finally, mass removal of phytoplankton from the oceans would so deplete the surface ocean waters of nutrients that fertilizers would be required.

One large resource which does attract people of temperate regions is the massive and species-rich fish fauna of the shallow tropical seas and reefs. How would this type of community hold up under intense exploitation? This question can be answered by application of Principles 9, 12, and 13, on the relation between diversity, stability, and productivity in ecosystems. A high level of community diversity is associated with a low productivity per unit biomass, even if there is a high productivity per unit area (Principle 9). Also, where there is low productivity per unit biomass and diversity is great, there seems to be a great deal of biological stability (Principle 13). However, species in stable physical habitats are unusually responsive to perturbations imposed by climate or man (Principle 12).

Do these notions have any relevance to how resilient different oceanic ecosystems might be when subjected to intensive exploitation? The highly species-rich fauna of the tropical coral seas live in a much more stable and salubrious physical environment than the fish fauna of the North Temperate regions. Many ecologists would argue that the more diverse ecosystems characteristic of the tropical coral seas have not had to adapt to wide-amplitude fluctuations in the physical environment. The evidence for this assertion is that tropical animals tend to have a lower fertility but longer life-span than their temperate counterparts. Intensive exploitation by man is most similar to drastic sudden deterioration in the physical environment. Thus, it is reasonable to argue that species which have not been forced to evolve homeostatic mechanisms to counter the effects of such environmental changes will be less likely to bounce back after intensive harvesting than those exploited by the commercial fisheries of North Temperate waters. It should be noted carefully that we do not have an adequate data base to make extrapolations like this with certainty of their being correct, but the assertions just made can be defended as being reasonable in the light of current ecological theory.

Thus, the extraordinarily diverse and rich-appearing oceanic tropical ecosystems may be chimerical, and might in fact disintegrate and drop in productivity precipitously under the influence of intensive harvesting. It is worth noting in passing that remarkably little is known about the population or community dynamics of the colorful fish fauna of these coral seas; investigating them would provide a remarkably instructive research system in the future.

We can now combine the information from Chap. 11, on maximum possible agricultural production, with new information on land and the information just presented on maximum possible ocean production, to obtain an estimate of the maximum possible human population that could be energetically supported with foreseeable food resources. The basic equation is

Maximum possible world population of humans

$$= \frac{\text{total annual possible harvest in kilocalories, from aquatic and terrestrial sources}}{\text{annual caloric requirement to support one average person in kilocalories}}$$

A human being of 70 kilograms body weight (about 154 pounds) has a basal metabolism of about 1700 kilocalories per day. A generous average international daily allowance is about 3000 kilocalories, or about 11×10^5 kilocalories per annum.

To meet this need, the maximum help we can expect from marine resources is about 350 million metric tons, at about 1200 kilocalories per kilogram. This represents about $350 \times 10^6 \times 10^3 \times 1200$, or 4.2×10^{14} kilocalories per year. Thus, with over five times the present marine production, only 3.82×10^8 people could have their energy requirements met, or a little over 10 percent of the *present* world population.

To compute the number of people that can be supported by food from land, a number of complications must be considered in the arithmetic. However, to give the student an overview of the argument before becoming involved with various subsidiary issues, we will begin by showing how, in principle, we can calculate the population that can be supported by food from the land.

Only five variables are needed:

N = the current world population (3.6 billion people)

X = the additional population that can be supported by food from the land

A_t = the total land area in the world suitable for growing crops

A_f = the land area required to feed one person

A_u = the total amount of agricultural land converted to other uses for each person added to the population

The complications arise because, first, land area has different meanings, depending on its latitude (the solar radiation received at the surface of the earth is different at different latitudes); also, there are problems about what is meant by agricultural land, and the inherent productivity of agricultural land. Then there are many problems associated with computing the land area required to support one person. As shown in Chap. 11, agricultural productivity is strongly dependent on the energy input from agricultural technology. Thus, we can use three different sets of estimates to compute maximum possible agricultural production: the theoretical limit, set by the thermodynamic efficiency of the photosynthetic process, which gives the highest estimate of production; a limit set by some estimate of the amount of energy that agricultural technology might be expected to impart to the growing process at some time in the future; and a reasonable estimate of maximum possible production, based on an examination of current worldwide agricultural statistics, a much lower estimate. We will examine all three. A_u values range widely, depending on whether we look at old, very dense urban populations or new populations with a strongly suburban living pattern. Considerable discussion is required by each of these area variables. In each case, we will work in both acres and hectares, because although many statistics are recorded in hectares, most students will have a greater intuitive feeling for the meaning of an acre. One acre is equal to 0.405 hectare.

Somewhat different estimates can be found in the literature for A_t. Hendricks (1969) states that there are 32 billion acres in the world (13×10^9 hectares), of which about a fourth are potentially arable (8 billion acres or 3.25×10^9 hectares). Other documents give this figure as about 4.6×10^9 hectares. The most careful assessment I have seen is by de Wit (1967), who computes the arable land in each $10°$ latitudinal band of the planet separately (Table 11-4). From his figures it would appear that the amount of agricultural land in the world is 2.29×10^9 hectares.

An extremely difficult figure to estimate accurately is A_f, the land area required to feed one person. Our figures are obtained from the various estimates in Chap. 11 and the Production Yearbooks of the Food and Agriculture Organization, United Nations.

To complete our calculation, the last estimate we need is A_u, the total amount of agricultural land converted to other uses for each person added to the population. This figure also varies considerably, depending on whether we are talking about new cities in the Western world, with a great deal of urban sprawl, or dense old cities. From Kyllonen (1967) it appears that a reasonable range of values for population density in cities is from 4,000 per square mile in new, sprawling cities to 10,000 per square mile in old, dense cities. These figures correspond to 0.16 acre per person and 0.064 acre per person (0.065 and 0.026 hectare per person) respectively. De Wit allows 0.075 hectare per person for urban and recreational needs.

We now have the necessary information to compute ranges of values for the maximum number of people the world can support. In all cases, we assume that as in the past, expansion of cities will not occur in deserts, on the ocean, on the side of steep hills, or in swamps, but rather on prime agricultural land. The reason for this is that prime agricultural land is flat, and it is adjacent to an important source of income for a growing village, which then evolves into a town and finally a sprawling modern city. Thus, the additional population X can be expected to remove XA_u hectares from agricultural use. From this fact, it follows that

$$N + X = \frac{A_t - XA_u}{A_f}$$

Consequently, the maximum number of people that can be added to the present world population is given by

$$X = \frac{A_t - NA_f}{A_u + A_f}$$

To give this calculation the widest possible significance, values for A_f and A_u are such that they represent very wide variations in the whole pattern of life on earth. For the low estimates of A_u, we utilize the figures 0.075 hectare of de Wit and 0.026 hectare for dense old cities. To get the highest possible estimate for agricultural land taken out of production per person added to the population, we can reason as follows. Suppose that all over the world, people were to demand the same amount of urban land, highway and airport land, reservoirs and flood control land, wilderness parks, and recreation and wildlife land per person that was used per new person added to the United States population in 1969. Suppose, further, that all such land were to be taken out of land that would otherwise be potentially available for agricultural use. We can obtain such an estimate from the data given in Council on Environmental Quality, *First Annual Report to the President* (1970). Combining these figures with an estimate of the people added to the United States population while acreage in various categories was being subtracted from the land pool, we can get an estimate of the subtraction rate per person. Assuming that the number of people added to the population in the year in question was about 2.031×10^6, we can construct the following table, which shows what land was lost to nonagricultural uses in the United States during 1969:

use	total acres	acres per person added to population	hectares per person added
Urban development	4.2×10^5	0.21	0.084
Highways and airports	1.6×10^5	0.079	0.032
Reservoirs and flood control	4.2×10^5	0.21	0.084
Wilderness parks recreation and wildlife	10.0×10^5	0.49	0.199
Total			0.399

Three widely different values of A_f are used. The first is based on Kleiber's (1961) estimate of 2×10^8 kilocalories per hectare per year, which we take to be a theoretical upper limit to crop productivity (it falls between the estimates of de Wit and Odum). For an intermediate value we take 1.6×10^7 kilocalories per hectare per year, that obtained for the Netherlands for wheat in 1964. For a low estimate we take 1.01×10^7 kilocalories per hectare per year, the energy content of the 1967 wheat crop in Mexico. This yield was the product of a long breeding program, and it probably represents a reasonable estimate of the wheat yields that can be obtained in large countries with a heroic effort. This was, for example, 3.1 times the energy content per acre of the 1967 Indian wheat yield, and 60 percent higher than the energy content per acre of the 1967 United States wheat yield. Production figures are converted to the number of hectares required to feed an average person by the equation

Hectares required to support one man
$$= \frac{\text{annual energy requirement for average man, in kilocalories}}{\text{energy content of crop yield per hectare, in kilocalories}}$$

The additional world human population possible under different combinations of values for $A_t, A_u,$ and A_f is displayed in Table 15-2. Two overriding generalizations come out of this table. First, the ability of the planet to support people will be extraordinarily sensitive to the energy input to agriculture in the form of equipment, fertilizers, and other technology. This demonstrates that the central question for the future of man on this planet is energy availability. Second, an even more important determinant of possible population size is the quality of life demanded,

TABLE 15-2 Effect of amount of arable land, living and recreation space requirements, and agricultural yields on maximum possible additional world population over 3.6 billion

area required to feed average person, hectares	total arable land = 2.29×10^9 hectares			total arable land = 4.6×10^9 hectares		
	total arable land required per person for urban and recreation uses, hectares					
	0.399	0.075	0.026	0.399	0.075	0.026
	additional population in billions					
0.11	3.7	10	14	8.3	23	31
0.069	4.4	14	22	9.3	30	46
0.0055	5.7	28	72	11.6	57	145

or tolerated, by the population. If people are willing to do without recreational space, then more people can be accommodated on the planet. However, Table 15-2 reveals some of the limitations on the carrying capacity of the planet for people. One might argue that the high estimate of 0.399 for A_u is unreasonable. Is it? If very high yields per acre are desired, then a tremendous energy subsidy into agriculture will be required, and some of this must be obtained by taking away even more agricultural land per person for such uses as flooding additional valleys to make impoundments for hydroelectric power generating plants.

The insignificance of the oceanic contribution to world caloric needs is clear from comparison of the previous discussion on the oceans with Table 15-2. If the maximum caloric contribution we can expect from the ocean is 10 percent of the required support for 3.6 billion people, or the support for 0.36 billion, then this is only about one-twentieth of the poorest we can expect from cereal grains. In short, the oceans are nutrient-limited. If we depend on the upwelling of nutrients, however, we endanger stability (Rosenzweig, 1971).

The reader should note also that this entire discussion assumes that the cost of increasing populations on the planet will be for each person to give up all food other than cereal grains. Perhaps some people will not think that very high populations are a sufficient benefit to justify this cost.

MINERALS   There have been a number of careful recent assessments of the supplies of minerals relative to the demand. At first glance, it would appear from the writings of Cloud (1969, 1971) and Lovering (1969) that we will shortly run out of an increasing number of minerals. However, the technological optimists counterargue that this statement has no meaning, because if we have unlimited atomic energy, the energy can be used to extract minerals from the ocean or from very low-grade ore bodies, or to manufacture substitutes. Since we have seen that energy is also the crucial issue in food production and supply/demand trends for land, we will turn immediately to a consideration of the status of fossil and nuclear fuels.

There is no need to examine the situation for each of the fossil fuels, because the essential nature of the problem is the same for coal and natural gas as for crude oil; only the time constants are different. The nature of our predicament with respect to crude oil supply and demand is exposed by Figs. 15-2 and 15-3. Figure 15-2 has been calculated by using integral calculus to compute the cumulative crude oil production for the world to and including a given year, assuming that the present trend in production continues. That is, we assume that worldwide production keeps growing at about 6.9 percent per year. On the same graph two horizontal lines are plotted, for low and high estimates of the total amount of crude oil that could be ultimately recovered in the world. The estimates given are for 1,350 and 2,100 billion barrels (Hubbert, 1969). The startling fact about this graph is that if present trends in production continue, all the crude oil in the world will be gone in 1996 or 2002. Even if Hubbert's high estimate is too low by a factor of 2, the cor-

Cumulative world production of crude oil, billions of barrels

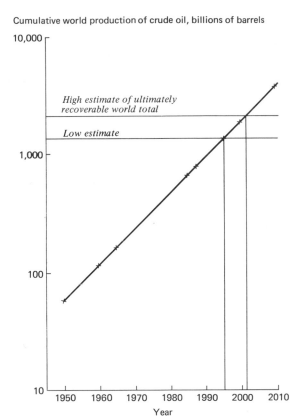

World crude oil production, barrels per person per year

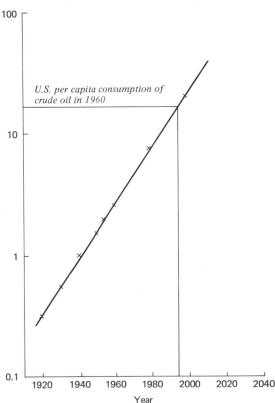

FIGURE 15-2 (Left) Trend in cumulative world production of crude oil, with projection assuming that present rate of increase in production per year continues to A.D. 2010. Computed by fitting trend line to production data in Hubbert (1969) and projecting cumulative production through integration.

FIGURE 15-3 (Right) Projection of world crude oil production per person (worldwide average). Computed by projecting trend line for worldwide crude oil production and using United Nations Medium variant population projections. Estimate of United States per capita crude oil consumption in 1960 computed from data in Landsberg et al. (1963).

rected figure would extend the life of world crude oil resources only 10 years if present trends will continue.

The central problem here is the likelihood that present trends will continue. The most critical issue bearing on that question is brought out in Fig. 15-3. In that graph average per person consumption of crude oil in the world is plotted against year. To give an idea of how realistic it is that this trend should continue, a horizontal line has been plotted to indicate United States per capita consumption of crude oil in 1960. It will be 1996 before the world average per person consumption of crude oil catches up to the 1960 level of consumption per person in the United States. But at that time, the world supply of crude oil will be just about exhausted. Is it reasonable to expect that consumption of crude oil per person in the rest of the world will continue to rise rapidly? In reality, every possible advertising effort is being expended by business to ensure that this will happen. Precisely because advertising is being used to encourage people of all countries to copy our life style, they are trying to copy it—and unconsciously hastening the expected consequences for crude oil consumption.

The full significance of the impending depletion of crude oil and gas stocks becomes clear when we study the relation between the energy produced by agriculture and the energy consumed by agriculture (Chap. 11). Production of fertilizer and farm machinery, and operation of farm machinery in turn, is simply a matter of having the energy with which to produce it. India lacks abundant supplies of cheap power and consequently has a low level of agricultural productivity.

To determine the future of resource availability in the world, it is necessary to turn directly to a consideration of nuclear sources, since growing public opposition to strip mining may limit the availability of coal. There are three means of obtaining atomic power: nonbreeder fission reactors, breeder fission reactors, and fusion reactors. Fusion reactors have not yet been able to generate power for even as long as a few seconds, even in a laboratory setting; it is unlikely that they will play a role in human affairs before 2010. Breeder reactors produce fissionable material while they burn fissionable material; consequently they stretch out world supplies of fissionable material enormously, and quite literally represent the salvation of mankind. Breeders are able to make use of a vastly greater amount of fuel than nonbreeders. In addition to uranium 235, which is fissionable under ordinary conditions, breeders can use uranium 238 and thorium 232, which are converted into fissionable isotopes in breeder reactors. However, breeders are not yet operational on a large scale and probably will not be for some decades. Consequently it is interesting to estimate how long world stocks of uranium oxides will last in conventional, nonbreeder reactors, which burn up uranium as if it were wood or coal.

Hubbert (1969) and Cloud (1971) have compared estimated availability of uranium oxide with estimated demand. A constraint is that, given the current economics of nuclear power generation, the uranium oxide must be available at $10 a pound or less. Worldwide demand for uranium oxide at this price, to and including the year 2000, will be about 650,000 to 675,000 tons. But assured available supply in an acceptable price range is only about 310,000 tons, plus a possible additional 350,000 tons that *may* be available. (See M. Davis, 1968.)

These facts by themselves would invite some concern, but an additional factor suggests that the seriousness of the situation may not yet be appreciated by many authorities. The whole point of Fig. 15-2 is that when demand for a fixed total amount of some stock resource is growing exponentially, the cumulative total amount used grows by a tremendous absolute amount each year after the process has been running for some time. This is bad enough if the compound annual growth rate is only 6.9 percent per year, as in the case of world demand for crude oil. However, Hubbert mentions that the current estimate of mean exponential growth rate per annum of nuclear power generation capacity in the United States is 31 percent per year. This means that the total demand for nuclear fuel will double every 2.4 years. Crude oil production, by contrast, doubles only every 10 years.

Since real problems about power availability are developing in our society, it is clearly mandatory that we begin to discover why we have them and what we must do about them. One possibility is to try to discover a new source of power. Another is to cut down sharply on our power consumption by eliminating waste in power use and increasing the overall energy efficiency of society. Clearly, we must do both. Basic and developmental research on breeder and fusion power generators, geothermal power, and all other sources must be pushed with even greater vigor than at present. However, we must also learn how to make much more efficient use of energy in all aspects of societal functioning: transportation, agriculture, manufacturing, and even leisure. It is this need for greater efficiency in the use of energy which provides the motivation for scientific management of resources discussed in other chapters.

Our use of energy has important implications for international relations, and will to a considerable extent determine whether underdeveloped countries will ever be able to improve their standards of living significantly. One example will illustrate how an input of energy, small compared with our comsumption, could have an enormous impact on the world. Stout (1968) has shown that 6,000 megawatts of nuclear power generating capacity installed by 1990 in the Gangetic plain of India would have a tremendous effect on the life of the region. Among other functions, this amount of power capacity could be used to produce the nitrogenous fertilizer and pump the water needed to have a tremendous effect on agricultural production, and close India's food gap. To illustrate how small an amount of power this is compared with that used in the United States, United States peak load electrical power generation capacity in 1968 was 244,000 megawatts. It is expected to be 1,056,000 megawatts by 1990.

ENVIRONMENTAL EFFECTS ON HUMAN BEHAVIOR AND HEALTH

One overriding fact stands out in any careful inspection of death rate tables on modern technological society. Every category of diseases which cause death in man has, with three exceptions, shown dramatic decline or a leveling off in the last two decades. The exceptions are respiratory cancer, emphysema, and arteriosclerotic heart disease, including coronary disease. Each of these shows a highly significant increase in incidence during the last two decades, the magnitude of which is indicated by Fig. 15-4. This change tells us a great deal about what is happening to the environment and man's position in it. The traditional great enemies of mankind, the infectious diseases, are not presently of much significance in modern technological society. Plague, typhus, smallpox, cholera, influenza, and the mosquito-borne malaria and yellow fever are all under control in advanced societies. However, it is becoming apparent that they are being replaced by a new set of diseases which are evidence of man's new environmental problems. These are the respiratory cancers and emphysema, the diseases caused by air pollution and smoking, and coronary heart disease, the best single measure of stress in our death rate statistics. In short, certain public health hazards of modern society

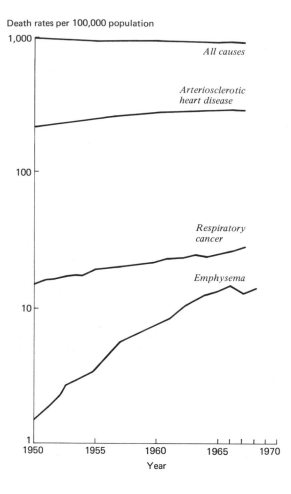

Death rates per 100,000 population

FIGURE 15-4 Trends in incidence of selected diseases in modern technological society. All incidences are expressed in death rates per 100,000 population per year. The top two lines are for the entire United States population, and the bottom two are for California. Data for the top two lines are from the U.S. Public Health Service, and for the bottom two lines are from the California State Bureau of Vital Statistics.

illustrate Principle 4: probability of survival is diminished as the intensity of these variables increases beyond an optimal or sufficient level.

Figure 15-4 shows that while heart disease, respiratory cancer, and emphysema are all increasing in importance, the number of people killed per 100,000 per year is greatest for heart disease, less for respiratory cancer, and least for emphysema, but the greatest rate of increase is found for emphysema. What is the significance of these patterns, and what causes them?

The important question to ask about rapidly rising death rates is: Are these increases in mortality due to particular diseases large enough to have any effect on life expectancy? The expectation of life in years is now about 71 when a person is born in the United States. If the population were stable and had a stable age distribution, then with this life expectancy, about one seventy-first of the population would die each year and the death rate per 100,000 would be 100,000/71, or 1,408 per year. The death rate from all causes in the United States is now hovering around 930 to 970 deaths per 100,000 per year, because the population is unusually young, owing to rapid recent growth. However, when the population stops growing, if life expectancy stays at 71 years, death rates will stabilize at 1,408. Will the diseases shown growing so rapidly in Fig. 15-4 add to the overall death rate, and hence reduce our life expectancy? The answer depends on whether such a disease kills people at an age beyond which they would not have survived in any case, or whether it kills prematurely people who would not otherwise have died at that age. Interesting evidence is accumulating on this point.

One might argue, for example, that a certain type of person is congenitally susceptible to lung diseases, and that if one of the lung diseases does not kill him, one of the others will. A way of exploring this possibility is to look at the death rates due to all the lung diseases separately, together with the death rates for all of them combined. In this way we can see whether emphysema and respiratory cancer are merely killing people who in earlier times would have died of tuberculosis, for example. Table 15-3 illustrates the situation.

Until 1955 in California, any increase in the death rate due to emphysema and respiratory cancer was more than compensated for by a drop in the death rate for other respiratory diseases. Consequently, it could be argued that these diseases were killing only that genetic group with an unusual susceptibility to fatal respiratory diseases. Also, it could

TABLE 15-3 Relative contribution to the overall respiratory disease death rate by various diseases in California, 1950 to 1967. All rates are per 100,000 persons per year.

year	death rate due to respiratory tuberculosis	death rate due to emphysema	death rate due to respiratory cancer	total death rate from respiratory diseases*
1950	19.9	1.5	15.2	43.2
1951	17.8	1.8	16.4	42.8
1952	14.2	2.2	16.5	40.6
1953	10.5	2.8	17.3	38.2
1954	8.9	3.1	17.8	37.4
1955	7.6	3.5	19.4	37.5
1956	6.9	4.4	19.6	38.4
1957	5.8	5.7	20.2	39.2
1958	5.5	6.3	20.9	40.8
1959	5.0	6.9	21.3	40.8
1960	4.7	7.7	22.3	43.2
1961	3.4	8.3	23.1	41.9
1962	3.5	10.1	23.6	44.8
1963	3.1	11.2	24.8	46.1
1964	3.0	12.7	24.9	47.2
1965	2.9	13.3	25.8	47.8
1966	2.6	14.6	27.2	51.2
1967	2.3	13.2	28.5	50.2
1968	2.1	14.2	29.9	52.8
1969				
1970				

SOURCE: Data from California State Bureau of Vital Statistics.
* The last column is not the sum of the first three columns because it includes also death rates due to a number of minor respiratory diseases, such as asthma and bronchitis.

be argued that the great recent rise in the emphysema death rate was due to increased physician recognition of this disease, which had earlier been classified as something else. However, beginning in 1955, the combined death rate from emphysema and lung cancer rose faster than the combined death rates from all other respiratory diseases dropped. Consequently, the total death rate from respiratory disease in 1962 was higher than it had been for this group in 1950. Also, death rates from most other respiratory diseases have now dropped as low as they can go (about zero), and so if there is any further increase in the death rate due to emphysema and lung cancer, there can be no further compensation for this by dropping death rates of other respiratory diseases. Thus, it is a reasonable expectation that overall respiratory disease death rates will rise from now on, if present trends in these two diseases continue. This means that there will be an increase in the overall death rate, and a decrease in life expectancy, unless there is a compensating drop in the death rates due to a decrease in deaths from some other organ system.

Determining what causes the observed trends in respiratory diseases and coronary heart disease is an extremely complex and expensive research operation. The reason is that there are a multiplicity of causes, and it takes research of considerable sophistication to assess the effects on mortality rates of different causative agents. For example, heart disease is in part the product of stress, in the sense of social or occupational stress, but it is also associated with cadmium in the air, which implies a very different causal pathway. The respiratory diseases can be attributed in part to air pollution, but they can be explained to a greater extent by smoking, and to some extent by social and economic conditions. It has only been possible to arrive at quantitative statements about causation, or possible causation, therefore, on the basis of statistical analysis of very large samples of data.

Before discussing such analyses, however, it seems worthwhile to point out that there may often be rather simple ways of manipulating data which can be most revealing and may indicate whether it is worthwhile to proceed with a more complex analysis. We may illustrate by using data on lung cancer and emphysema from three counties in, or adjacent to, the Los Angeles basin. A postulated history of air pollutant concentrations and respiratory mortality rates for these three counties is depicted in Fig. 15-5.

Los Angeles was the first of these counties to become intensively urbanized; it already had a population of just over 5 million in 1954, and it grew rather slowly until 1967, by which time the population was just over 7 million. The population of the other two counties grew very rapidly in this period, that of Orange County increasing from 303,000 to 1,269,000 in this period, and that of Riverside County increasing from 208,000 to 440,000. However, one additional very important fact about these three counties is that Los Angeles County lies in a topographic bowl, surrounded by the Santa Monica Mountains, San Rafael Hills, Puente Hills, Chino Hills, and farther to the east, the San Gabriel Mountains. As the total volume of pollution built up in the air over Los Angeles County, it gradually spread out of the county through passes through the hills, toward Riverside County to the east and southwards into Orange

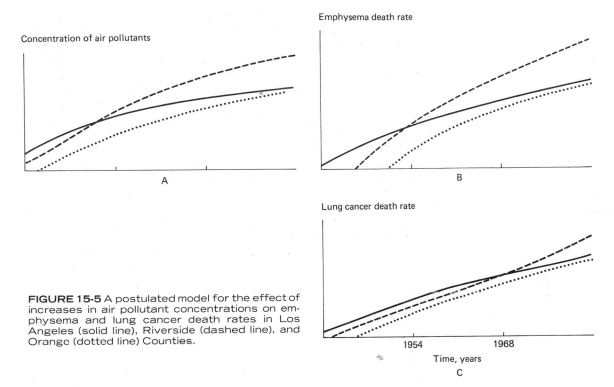

Concentration of air pollutants

A

Emphysema death rate

B

Lung cancer death rate

1954   1968

Time, years

C

FIGURE 15-5 A postulated model for the effect of increases in air pollutant concentrations on emphysema and lung cancer death rates in Los Angeles (solid line), Riverside (dashed line), and Orange (dotted line) Counties.

County. Although we do not have enough data to know exactly how much pollution there was in each county each year, Fig. 15-5A is a reasonable reconstruction of the pollution history of the area, and also is consistent with the facts the author does have. The pollution probably became more concentrated in Los Angeles first, because it urbanized first. Then, as urbanization developed in Riverside County, but also as the Los Angeles pollution began to blow east in greater quantities, the pollution level of Riverside County built up to the point where it reached and then surpassed that of Los Angeles County. The pollution level of Orange County is now building up rapidly, but it has not yet reached that of Los Angeles County. To indicate that this reconstruction is consistent with the facts, we can examine the published statistics of the State of California Bureau of Air Sanitation on air pollution for 1967 in the *Clean Air Quarterly* for December, 1967. During June, July, and August of that year, the number of hours in which state ambient air quality standards for oxidant were equalled or exceeded were 95 for Riverside, 19 for Los Angeles, and 6 for Anaheim, in Orange County. This is the order depicted at the point where the three air pollution lines intersect the line for 1968 in A of Fig. 15-5. B postulates that the trend lines for emphysema death rates in the three counties will respond with almost no lag to increase in air pollution concentration. C, following the theories of cancer development by such authors as Stocks (1966), assumes that there will be a lag of several years from the time that pollution builds up to the time that cancer death rates build up. If this is correct, then the crossing over of the pollution concentration lines for Riverside and Los Angeles Counties should be revealed much later in a crossing over of the lung cancer death rate lines.

We can now turn to comparison of the actual death rate statistics in Fig. 15-6 with the postulated model in Fig. 15-5. In the lung cancer death rate lines, the actual data, like the model, show the order from top to bottom: Los Angeles, then Riverside, then Orange County, with the Riverside line crossing the Los Angeles line twice — at the beginning and near the end of the period. In the case of the emphysema lines, the order is different: from top to bottom, Riverside, then Los Angeles, with Orange on the bottom. The fact that the Orange County emphysema line catches up to the Los Angeles emphysema line and the Riverside County lung cancer line gradually passes the Los Angeles lung cancer line suggests that air pollution is in fact a contributor to causing emphysema and lung cancer, along with smoking and social and economic factors. Statistical analyses bear this out. The most comprehensive attempt to assess the importance of air pollution as a determining factor in the incidence of various respiratory diseases, cancers, and infant and total mortality rates was conducted by Lave and Seskin (1970). This study represents a comprehensive review of all the data collected in a very large number of other studies. Their most important conclusions are as follows.

There is a statistically significant effect by air pollution, not only on

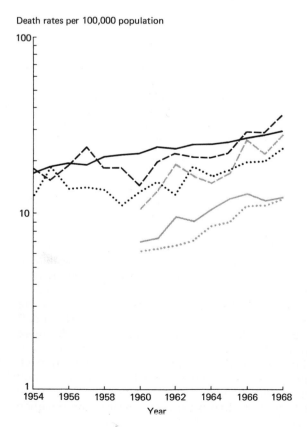

Death rates per 100,000 population

Year

FIGURE 15-6 Death rates per 100,000 population due to emphysema (shaded lines) and lung cancer (dark lines) in Los Angeles (solid), Riverside (dashed), and Orange (dotted) Counties. The death rates were computed from population and mortality data published each year by the California State Bureau of Vital Statistics.

the incidence of bronchitis, emphysema, pneumonia, lung cancer, and other respiratory diseases, but also on the incidence of stomach cancer, the total death rate, the infant death rate, and the fetal death rate.

The magnitude of this effect is startling. For example, statistical analysis of the effect of variations in air pollution and other factors in 114 metropolitan areas in the United States accounted for 80 percent of the variation in total death rate from city to city. Air pollutant concentrations played a highly significant role in this statistical relation. A 10 percent decrease in the minimum concentration of measured particulate air pollutants would decrease the total death rate by 0.5 percent. This statement can perhaps be made more meaningful by translating it into drop in life expectancy. In Fig. 15-7, the expectation of life at birth is plotted against death rate for the United States for the last 40 years. The graph

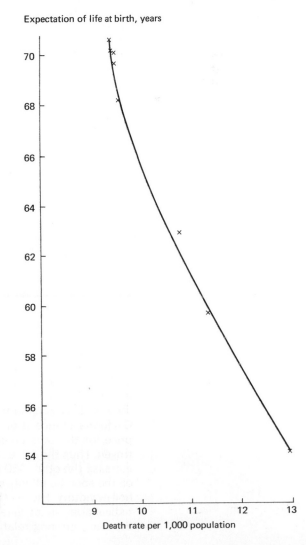

FIGURE 15-7 Empirical relation between expectation of life at birth and death rate per 1,000 population for the United States between 1920 and 1967, data from the U.S. Public Health Service, *Vital Statistics of the United States.*

shows that a change in death rate from 10 to 9.5 increases the life expectation at birth from 66.0 to 69.2 years. That is, a 5 percent drop in the death rate increases life expectation 3.2 years, or a 0.5 percent drop in the death rate increases life expectation 0.32 years, or about 17 weeks.

Lave and Seskin were able to translate the cost of air pollution into approximate dollar costs. For example, the dollar savings resulting from a 50 percent reduction in air pollution in major United States urban areas, causing reduction in incidence of respiratory diseases, would be 1,222 million dollars, each year. Suppose that there are 115 million cars in the United States. This cost would be equivalent to $11 per car per year or, say, $55 for an alteration that lasted for 5 years. It is amazing that society has not seen fit to spend this money. The dollar costs of pollution in terms of effect on public health are probably very low estimates. For example, Lave and Seskin estimate that if air pollution were reduced by 50 percent, the savings due to reduction in incidence of other diseases would be 468 million dollars a year for cardiovascular mortality and 390 million a year for all kinds of cancer. They estimate that a 50 percent reduction in air pollution in United States cities would have overall savings of 2,080 million dollars a year because of reduction in incidence of all diseases.

Another important point made by these authors is that the indices of air pollution they use are not those that make newspaper headlines when indices are usually high and the public is warned of possible danger from overexertion. Their indices measure average air pollution concentrations as encountered over long periods of time in urban areas. This type of air pollution has a cumulative, chronic effect on health which is probably much greater over a period of years than the more spectacular short-term very high concentrations which receive so much attention. Further, it probably affects incidence of a wide variety of diseases through increase in the mutation rate (Hickey, 1971).

It is of interest to consider how probable it is that the air pollution concentrations will level off or decline in big cities in the near future. Most readers would expect the new legislation controlling pollution emission to result in a decline very soon. This may not happen, because some new legislation may be having the surprising effect of sharply increasing total emission. The reason for this is that limitation of certain pollutants may depend on physical devices which limit certain pollutants at a price of sharply increasing the emission of others not mentioned in the legislation. In testimony presented to the U.S. House of Representatives in 1970, two cars were compared (Slade, 1970). One was 9 years old, and it emitted 500 parts per million of oxides of nitrogen and 500 parts per million of hydrocarbons. The second car was new in 1969. It met the California standard of 275 parts per million of hydrocarbons, but the price for this was emission of 2,800 parts per million of oxides of nitrogen. Thus the price for a 45 percent reduction in one pollutant was to increase the other 460 percent. Considering the best available evidence on the specific effects of nitrogen oxides on human health, this seems a hollow victory. Hickey (1971) and his associates have found through statistical analysis of data on a large number of United States cities that there are striking relations between city-to-city variation in concentrations of specific air pollutants and the incidence of specific diseases. For example, nitrogen dioxide concentrations account for a great deal of city-to-city variation in various cancers and heart disease. Three different studies have shown experimentally that nitrogen oxides will produce emphysema in laboratory animals (Kleinerman and Wright, 1962; Freeman and Haydon, 1964; Haydon et al., 1965).

In summary, it is not yet clear that control of pollution emission will be effective as a means of limiting the health hazard. This suggests the more fundamental approach of decreasing pollution by using less energy, and using it more efficiently.

The incidence of coronary heart disease is increased by social and psychological stress as well as exposure to certain chemicals in air or water. Hinkle et al. (1968) conducted an elaborate study of the factors related to the incidence of coronary heart disease in 270,000 men of the Bell System Operating Companies. They found that the incidence of this disease did not increase as rank in the company increased, for men of a given age, but, in general, decreased. Also, at almost every rank, the incidence of the disease was lower in men who had gone to college than in men who had not. The interpretation to be placed on these data is that an objective measure of stress is not what determines the incidence of coronary heart disease, but rather the ability of the individual to cope with stress. Further, differences between individuals in their ability to cope with stress are probably determined at a very early age. Whether a person goes to college may not be so important as whether he came out of a social and economic background which made it easy for him to go to college. This is another interesting illustration of Principle 12. A certain type of social and economic background accustoms the person to the stress involved in attempting a high level of achievement. Consequently, for such people, on the average, the unit impact of a stressful institutional environment (college, business, industry, or government) is less in later life. This, in turn, may mean that a population has a motive for upgrading itself through education in addition to the more commonly recognized social motives. Consequently, we are now led to consider the impact of population growth on the ability of a population to educate itself adequately.

THE EFFECTS OF DENSE POPULATIONS AND HIGH RATES OF POPULATION INCREASE ON SOCIAL EFFICIENCY

In any evaluation of alternative future strategies for mankind, the central question to be asked from an economic point of view is: What does each strategy do to the ratio of gross income per capita to gross cost of living per capita? Also, to make this ratio as revealing as possible, a host of hidden measures of income and costs must be included in the numerator and denominator, respectively. For example, being able to holiday at an uncrowded beach is a form of income, whereas losing 3 hours going to and from work every day because of traffic congestion is a contribution to cost of living.

The best strategy for society is one which increases the ratio of gross income per capita to gross living cost. A neutral strategy keeps the ratio constant. A bad strategy decreases gross income per capita and increases the gross cost of living. It is assumed by most people that population increase is a good social strategy for mankind, because it will lead to an increase in gross per capita income, due to larger markets and a more active economy, and a gross cost of living that will certainly increase less than gross income and might even level off or decline. In effect, then, most people assume that a tenfold increase in the population will increase gross national income more than ten times, and will increase gross national costs no more than ten times, and possibly less than that.

This is a proposition that can be examined objectively, either by using data or by arguing from first principles. The central problem is: Are there any mechanisms at work in society which cause per capita assets to decline with increasing population density, and cause per capita cost of living to increase with increasing population density?

In fact there are a large number of such mechanisms.

One group of mechanisms which will decrease per capita income and increase per capita costs has to do with the availability of raw materials. Raw materials of all types are being depleted, and the most accessible minerals and fuels have already been discovered and used up. Those materials left are increasingly more difficult to obtain, such as oil shale, which is covered with a deep rock overburden, and offshore oil deposits, which require expensive drilling rigs. Their inaccessibility has two important implications. First, obtaining the remaining minerals and fuels will depend more and more on high-technology extractive processes, in which the demand for unskilled labor will be low. Extraction will require very complex heavy equipment, rather than a lot of labor. Consequently, it is unlikely that this activity will require an increase in the labor force proportional to the population increase. (In general, the demand for blue-collar workers in a society declines as the technological level increases.) The same problem will occur in farming and forestry, because farming will become more mechanized and forests will become depleted. In summary, if the world moves to a high human population density based on a nuclear power technology, there will be real problems in providing enough jobs for the people who in a more primitive society would earn a living in resource acquisition.

Another problem, pointed out by Lovering (1969) is that as the cumulative mineral production increases and the less accessible minerals are depleted, the energy cost to produce still more minerals goes up, causing the cost per pound of minerals to go up. This point is illustrated by Table 15-4; it is an example of the impending depletion of Principle 4.

If shortages develop for an increasing range of raw materials, they will have an important effect on wealth in society because raw materials generate capital.

Per capita costs of food must increase with increasing population density, because there will be fewer acres on which to produce food, and it will consequently be necessary to produce more food per acre. This means that more fertilizer and other inputs of agricultural technology will be required, and the cost per unit of food produced will increase. In general, the energy subsidy into agriculture, with its attendant costs, will increase.

As the population density increases, the competition for land will intensify, and the cost of land will increase.

Increasing population densities will increase pollution. Further, the total amount of pollution will increase more rapidly than population density, because as population density increases, the energy cost per capita to obtain everything will increase, and rate of production of pollution is a consequence of rate of energy expenditure. Consequently, the public health costs of pollution per person will increase.

Several different kinds of transportation problems arise as population density increases. First, the average distance between the site of a demand and the site of a supply increases. The simplest example is the effect on the average distance from home to work when a man gets promoted and moves from a metropolitan area of 50,000 people to one of 5 million. As indicated in Chap. 6, his transportation overhead in time and money will be greater—on the average ten times as great, assuming equivalent urban densities. This means that the demand per capita for space, matter, energy, time, and money for use in transportation must increase. Because, in many cases, it will be impossible to meet the demand for extra space for transportation as population density increases, per capita time lost to waiting and queuing because of congestion will increase. For example, New York City and San Francisco have both run out of possible places to build a large new airport close to the city; in many crowded urban areas, all possible places where a new freeway system could be built are already occupied.

International transportation costs must increase per capita as the number of nations self-sufficient in resources declines still further. This means that the amount of long-distance shipment of raw materials and finished products is bound to increase as population density increases, making all these items more expensive because of the increased transportation overhead. Even worse, a variety of international tensions will intensify, as developing nations begin to realize that their chances of ever developing will be blocked if they keep exporting raw materials that they need for their own industrialization.

As pointed out in the preceding chapter, increasing population size results in an increased cost per taxpayer for all services—police, fire, education, for example—and the distorted age distribution leads to a

TABLE 15-4 The energy cost of producing minerals as the more accessible deposits are depleted

year	thousands of horse-power, equipment installed in U.S. mines	value of all U.S. mineral production, millions of dollars	dollars worth of mineral production per horsepower
1940	7,332	5,840	797
1950	22,000	14,285	649
1955	30,768	17,557	571
1960	34,700	18,032	519
1965	40,300	21,524	534
1968	43,400	24,978	576

SOURCE: Data on horsepower installed are from Twentieth Century Fund, J. F. Dewhurst and Associates, America's Needs and Resources, A New Survey; data on value of production are from U.S. Department of the Interior, Bureau of Mines, Minerals Yearbooks (includes metals, nonmetals, and fuels; before 1960, value is annual average for 5-year period beginning year after stated year).

decreased capital-generating ability within society with which to pay for the increased costs.

Further, there are mechanisms which will lead to a decrease in the number of jobs available. Market saturation will diminish the number of jobs available for people who make products to satisfy market demands, and increased communication will mean that once any person anywhere has solved a problem, his solution will be available to everyone, everywhere. This will decrease the per capita demand for problem solvers.

The per capita cost of controlling pollution becomes greater with increasing population density, because of the increased energy required per person, and hence the increased pollution per person. The environment can disperse or degrade the pollution produced by a very small number of people, but as the population increases, it becomes more difficult to do this, and as the environment becomes saturated with pollution, per capita costs of disposal become tremendous.

In short, it would appear that those people who have seen constantly increasing human population densities as an economic blessing may have overlooked some problems.

THE EFFECT OF POPULATION DENSITY ON THE PHYSICAL AND CHEMICAL STATE OF THE PLANET

A great deal has been written about the effect that man is having on the physical and chemical properties of the entire planet, and the ability of the planet to support life. Entire books are now being devoted to consideration of the way in which man is beginning to interfere with the gas balance of the planet, the nitrogen cycle, and other components of the normal dynamics of the planet (see, for example, Singer, 1970).

Rather than reviewing the massive body of literature on this subject, it seems more useful as a means of increasing the student's understanding of the nature of our present situation to examine one reasonably well-understood mechanism in some detail. The essential nature of the argument is the same for other phenomena.

The phenomenon we have selected is world temperature. This phenomenon has been studied for a long time, an immense amount of high-quality data is available, and the theory is well understood. It is clear that three factors are more important than all others in determining the temperature of the world: variations in solar radiation intensity, as measured by sunspot numbers, the carbon dioxide concentration of the air, and the effect of small particles on the transparency of the air to sunlight. Bryson and Wendland (1970) have shown recently that they can account for temperature changes of the world over the last 80 years with the following simple formula, derived from statistical analysis:

$$\partial T \, (°C) = -3.546 + 0.012 \, CO_2 - 0.002 \, dust + 0.006 \, sunspots$$

This equation states that the world temperature T is increased by increasing concentrations of carbon dioxide, decreased by increasing dust concentrations, and increased by increasing sunspots. Carbon dioxide increases the temperature through the greenhouse effect, which prevents the reradiation of solar energy out to space; dust and fine particles decrease the world temperature by decreasing atmospheric transparency so that incoming energy bounces back into space before it can hit the surface of the planet. It is particularly noteworthy that this equation accounts for world temperature very precisely in the last two decades.

Now what can we tell from this equation about the future temperature of the world? First, the effect of dust and fine particles is beginning to nullify the effect of carbon dioxide. The equation states that a unit increase in dust will have only one-sixth the effect of a unit increase in carbon dioxide (the regression coefficient for dust is $-.002$, whereas it is .012 for carbon dioxide). However, between 1950 and 1960, the carbon dioxide concentration increased only about 7 percent, whereas the dust content of the air tripled (Bryson and Wendland, 1970). Thus, the net effect of dust and fine particles is increasing more than four times as fast as the effect of carbon dioxide. Second, if dust and fine particles in the air are related to industrial production, then we can expect them to increase in concentration very fast indeed if present trends continue. Bryson and Wendland show that between 1930 and 1965 the dustfall in the Caucasus Mountains increased about twelvefold. This is extremely interesting, because world crude oil production has increased about 6.9 percent a year during that period; this rate would result in about a twelvefold increase over a 35-year period. Consequently, dustfall seems to be increasing at about the same rate as one of the best measures of worldwide industrial activity. Now suppose that industrial activity keeps increasing at the present rate for another 35 years. Suppose, further, that the amount of dust increases proportionately. What will be the effect on world temperature? Looking at only that component of the equation due to dust and fine particles, and assuming a continuation of present trends, we get the following numbers:

year	dustfall, milligrams per liter	number of degrees drop in world temperature due to dust and fine particles, °C
1930	20	0.04
1965	240	0.48
2000	2,880	5.76

The only rational conclusion we can arrive at is that it is simply not possible for industrial activity to keep expanding at the present rate unless the degree of pollutant emission control becomes at least ten times as effective per unit of pollution as at present. The cost implied by this degree of control will add significantly to the cost of all transportation and manufacturing activity.

OVERVIEW OF THE PESSIMISTIC ARGUMENT Adding up all the items in our arguments that contribute to the costs of population growth, we see that the sum of the costs is so great that further population growth appears unattractive. Consequently, there is ample motive to develop some overall global strategy for mankind that will allow us to get out of our present pattern of living, in which the emphasis is on growth and more growth. The transition should of course be made with minimal disruption of existing conditions.

THE APPROPRIATE GOALS FOR MANKIND

The logical approach to determining a global strategy for mankind is to set certain goals, and then decide what steps must be followed to attain them.

An appropriate goal for our species is to adjust the population-environment balance as quickly as possible so that the quality of life will be optimal for as far into the future as we can see. A second goal is the systematic avoidance of any activity that will produce large-amplitude instability in the population-environment system at any time in the future.

These two goals seem rather straightforward and simple, but when all their implications are recognized, it appears that they amount to an elaborate, ambitious action program for which the political support at the moment is sadly lacking.

In broad outline, this action program has the following six goals:

1 Provide strong, governmentally supported incentives for limiting family size. Determine the optimum population size for the world, each country, and each region, together with the optimum age distribution and geographic distribution. Then try to attain this demographic situation as quickly as possible.

2 Make wise, efficient use of all natural resources, including food, land, energy, wood, soil, minerals, and human time.

3 Develop an economic theory based on equilibrium, rather than growth economies, and apply the theory to social management as quickly as possible.

4 Routinely monitor the physical and chemical state of the planet, and rigidly control any activities degrading it.

5 Legislate rigidly and effectively against any activity which can cause large-amplitude environmental instability.

6 Guarantee all citizens a basic set of environmental rights.

COMPUTATION OF THE OPTIMUM POPULATION

One interesting approach to this problem is to ask the question: How long do we want humanity to survive on this planet under any given set of conditions? The answer to this question will depend on the number of people in the world and their life-style. For each life-style the length of time different numbers of people can be supported can be computed, and then the electorate can be asked to vote on the population size and the life-style they desire, and the appropriate strategies can be pursued to attain this end.

The calculations can be illustrated by considering any resource in short supply in relation to long-term demand, such as the metal chromium, the essential ingredient of stainless steel. A reasonable outside estimate for the entire world supply of chromium in the form of chromic oxide is $1,200 \times 10^6$ tons. In the calculations, two different use rates have been assumed: the present United States use rate and the present use rate in the non-Communist countries other than the United States. The first rate is 375 tons per 100,000 people per year, and the second is 14 tons per 100,000 people per year. From these figures, we can compute how long chromium will last in the world, for different world population sizes. The graph which results is Fig. 15-8. This graph, like all similar ones that can be made for various resources, brings out very dramatically

World population, billions

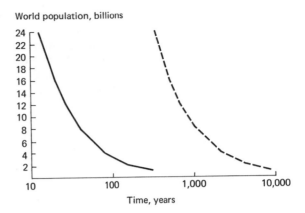

FIGURE 15-8 The trade-off involved in life-style, world population size, and the number of years the life-style can be maintained. The graph was computed by using the stock of chromic oxide as a limited resource, and the two life-styles were based on the way two populations used chromic oxide in 1970: the United States use rate (solid line) and the other non-Communist countries (dashed line).

Time, years

the trade-offs that can be made between world population size, life-style, and the length of time the life-style can be maintained. Thus, if the world population rose to 24 billion people, about six times the total expected in 1975, and the average rate of chromium consumption was what it has been recently in the United States, then world resources of chromium would last 13 years. At the other extreme, if the world population fell to 1 billion, less than a third of the current population, then if a life-style like that of the average non-American and non-Communist person of 1970 were adopted, chromium would last almost 9,000 years. This would be ample time to develop a sophisticated recycling technology. It would be extremely interesting to observe the results if a number of graphs, on different resources, were presented to the electorate so that they could vote for their life-style and population size. It should be noted that the time could be made indefinite with proper planning and management, but only in this way. The present unplanned drift leads inexorably to disaster.

After determination of the optimum population, which the electorate would probably decide should be 1 billion or less if they were well-informed, there would remain the problem of determining the optimal demographic characteristics for the population. This could be arrived at by computer simulation, as described in Chap. 6. Unfortunately, as indicated earlier in this chapter, it would take a long time to reduce the population to the desired level. The more numerous the world population becomes, the longer it will take.

USE OF NATURAL RESOURCES

Several strategies are available to modern technological society for sharply increasing the efficiency of energy use.

The transportation overhead involved in all man's activities should be cut to a minimum. Too high a proportion of the incident solar radiation and the energy stored in fossil fuels is being dissipated in moving food over large distances, and in moving man over large distances. We are all familiar with such operations as the shipment of Florida citrus fruits to California, and the shipment of shrimps in cans from California to other parts of the world surrounded by shrimps, but thousands of miles away. The shipment of canned shrimps from California to Trinidad, about

4,600 miles away, is an example. That a man makes a round trip of 100 or more miles to work every day is the strongest kind of evidence that the area adjacent to his work is horribly unattractive, and this problem needs to be rectified. It is beside the point that the long trip is easy, necessary, pleasant, or healthy. Cities should be designed to go up, rather than out, and they should be designed to be pleasant and interesting. In fact, we must try to minimize transportation overhead by striving for local, regional, and national self-sufficiency in raw materials and goods, because soon we will not have any choice. The present long-distance exchanges are possible only because the underdeveloped countries need money and manufactured products more than they need their own raw materials. As their own industrial base expands, however, they will need their own raw materials and may no longer export them.

In general, of course, we will minimize the probability of war and civil insurrection in the future if we try to even out resource use over racial, social, and national groups. Thus, we can deal with two problems at once by minimizing transportation overhead: energy inefficiency and political tension.

We must conserve mineral use by recycling and by dispensing with planned obsolescence, which increases the rate at which minerals are being lost to rust in the form of abandoned cars and other equipment, much of which is not melted down for reuse.

Man can maximize the efficiency of utilization of solar radiation and minerals at any farm site on the earth's surface by cropping those species of plants and animals that make the most efficient use of the site's resources. In more instances than have been hitherto suspected, this will be achieved by cropping those species of plants and animals which occur naturally at a particular place, rather than by cropping imported species or strains which have evolved or been bred elsewhere.

Too high a proportion of our food is being grown on soil that is suboptimal for the kind of food raised on it. Cities should be built on suboptimal agricultural soil. The present practice, building them on highest-quality agricultural soil and then expanding them outwards over progressively poorer agricultural soil, makes no sense. Crops, not man, should be grown on the very best agricultural soil. A particularly glaring example is the replacement of citrus groves by man in the Pasadena portion of the Los Angeles basin. The groves were replaced by new groves set out in virtually pure sand in such places as the southeast corner of California.

EQUILIBRIUM ECONOMIC THEORY

Much of present economic theory is predicated on the assumption that a fairly constant annual percentage increase in gross national product is necessary, desirable, and possible. In fact, three different economic approaches to operating nations and the world are worth comparing: the growth economy just mentioned, a completely non-growth economy, or a growth economy in which there is a tremendously rapid shift in the mix of gross national product away from activities based on consumption of matter and energy, toward activities providing services. It is hoped that this book has made it clear that a growth economy based on anything like the present mix of gross national product is impossible, for several reasons. We lack the resource base, we would pollute ourselves into extinction, and even if we get the breeder and fusion nuclear power generating plants operational on a vast scale, there would be insuperable social problems related to supply and demand for capital, tax money, and composition of the labor force, and so on.

On the other hand, man is stimulated by growth and the potential for growth. It would appear, then, that what we need is a new type of economic theory which allows for constant growth in the gross national product, but constantly depresses, and very rapidly, the proportion of that mix based on use of matter and energy. This means that we must very quickly shift to a society based on realization of the potential of the human animal, as opposed to a society in which some people become extraordinarily wealthy by exploitation of others. Thus, we must shift very rapidly to a society in which a much higher proportion of gross national product goes into education, libraries, research, culture, communication, entertainment, leisure, health service, medical research, and other social services. Design and development of regions, cities, and transportation systems should involve a much greater proportion of our total national and international effort. The fantastic inefficiencies and congestion developing in these now is mute testimony to the small amount of imaginative thinking and planning presently expended on them.

THE PHYSICAL AND CHEMICAL STATE OF THE PLANET

Man is now able to have an effect on various geophysical characteristics of the planet, such as the dust in the atmosphere, large enough to diminish or terminate the usefulness of the planet as an environment for life. The time would appear to be overdue for the monitoring and management of geophysical parameters on a cooperative international basis. There should be an international legal basis for making possible sanctions against any country polluting air or water to an extent hazardous to the maintenance of stable and salubrious weather conditions for all countries.

THE PREVENTION OF INSTABILITY

This book suggests several specific global strategies for minimizing the probability of large-amplitude instability in human society.

First, keeping human population densities low minimizes the probability of explosive eruptions of pandemic disease, for the reasons pointed out in Chap. 13 in connection with the Kermack-McKendrick threshold theorem. It also minimizes the probability of starvation, and of mass social instability due to exhaustion or depletion of resources, or large inequities in the difference between per capita availability or use of resources between different countries, regions, or social and racial groups. (Peccei, 1969).

Man can maximize economic and social stability by departing from monoculture of large land tracts insofar as possible, so that complexity of trophic food webs is maximized (Principle 13). The penalty for going to the extreme in monoculture is well illustrated by the Irish potato famine after Ireland became largely dependent on this one crop plant. Between 1785 and 1845, the Irish population increased from 2.8 to 8.3 million; during this period Irish agriculture depended more and more on growing potatoes. A potato blight between 1845 and 1851 resulted in about 1 million of the Irish starving to death and another 2 million emigrating. As indicated elsewhere in this book, the probability that blight will spread quickly through a nation's entire food supply is heightened by planting enormous tracts of land out to the same crop.

Another means of minimizing instability is intensive programs of international monitoring, research, and management of such large resources as the air, the oceans, the deserts, and the arctic and antarctic wastes as catastrophe-warning devices, and to ensure that resources are being used in the most imaginative and wisest possible fashion. The deserts, the ice caps, and the oceans should be made as productive as possible, in a way consistent with the goal of maintaining stability elsewhere. Artificial deserts should be made to retreat, not expand, and we should ensure that we do nothing to poison the oceans.

A very important means of maintaining stability is to ensure that there is considerable diversity in land use practices (Principle 13). Consequently, we should not attempt intensive cultivation everywhere, because if we do, and something goes wrong that we haven't expected, then we will have left ourselves without a life raft. Therefore, at least while we are learning how to plan rationally for land use, we must set aside vast tracts of wilderness in all the world's diverse habitats. We should not try to manage these, only protect them.

Further, we must experiment with various land use strategies. In this way we can learn empirically what can be done to manage each habitat so that civilization can be preserved in a reasonably natural setting. It is noteworthy that such experiments will require a type of patience not yet exhibited by any civilization.

If we do not adopt such prudent measures, then our civilization is in the position of an explorer that unwittingly destroys his path behind him, so that if he is going the wrong way, there is no way back.

THE TACTICS AND STRATEGY OF CORRECTIVE ACTION

Analysis of the errors committed in dealing with environment-population problems in the past indicates that there have been two main causes for them: defects in judgment on the part of managers, planners, executives, or politicians, and lack of political support for wise policies, which would prevent the exploitation of the masses for personal gain. Two techniques are available for dealing with these problems.

In short, these two techniques involve linking two modern technologies: computer simulation and television. More penetrating historical analysis of the costs and benefits associated with various social changes, with the results used to construct computer simulation models, provides us with a tool for playing games that can be used to sharpen the judgment of managers. Taking the next step of linking up this tool with television, so that the electorate can see the consequences of choosing various alternatives for future costs and benefits, provides us with a tool for developing a political mandate for rational social decision making.

The idea behind computer simulation is that before any major social change is effected, a realistic cost-benefit analysis can predetermine what net social benefit would result. Unfortunately, it appears to be a natural human failing to perceive most readily the benefits of various changes while remaining quite blind to the real costs of change. For example, it is clear to the people involved that there are short-term benefits accruing from cutting down redwood trees in California at a rate that is too high, compared with their growth rate, or indefinitely expanding the population of southern California, or wiping out the world supply of whales or petroleum. What has not been widely perceived is the entire range of long-term deleterious effects from each of these moves. Thus, it appears that there is a real need to dramatize, for public benefit, the costs as well as the benefits from such ventures, over a long time period. This can be done by constructing computer simulation games which can be played on television, with graphic computer output displaying the long-term trends in various critical social variables resulting from following each of several different alternative strategy options.

The data base for such models should be not only those parts of the world recently subjected to intensive development, but also those parts which have been intensively cultivated for very long periods of time, such as the Nile Valley, Mesopotamia, and India. By using such a data base, not only the short-term effects, but also the long-term effects of various prospective changes in newer parts of the world could be built into computer simulation models.

Another motive for using computer simulation as a training device is clear from study of the present and past records of resource management. There have been two types of defects in resource management. Techniques for solving problems and making objective comparisons of different management strategies have been either inadequate or completely unavailable. Even where inadequate techniques were available, the judgment in individuals and societies in applying such techniques has been faulty. There have been errors in determining when, where, and how much various techniques of management should be used, and in the mode of use. For example, the judgment of local irrigation experts in determining what steps should be taken to prevent drought, waterlogging, or salinization has been incorrect in many instances. The record of local pest management schemes is very spotty. Sharpening up judgments by playing games repeatedly with realistic computer simulation models, which compress the time required to gain a great deal of trial-and-error experience through witnessing the costs and benefits of various strategies, would appear to be a useful pedagogic device.

A breakthrough in this type of activity is illustrated by the work of Forrester (1971) in connection with the Club of Rome Project on the Predicament of Mankind. Forrester used computer simulation to demonstrate the effect of various global strategies on a number of critical-state variables: worldwide population, pollution, natural resource stocks, capital investment, and quality of life.

In summary, although there is much evidence that the future for mankind will be bleak if present practices and trends continue, there is much reason for hope because of the availability of new techniques with which we can learn to deal with our problem.

SUGGESTIONS FOR INDIVIDUAL AND GROUP PROJECTS

Update information on world supplies of crude oil, gas, coal, and uranium ore. Update information on worldwide rates of production of these materials. Is the general picture conveyed in this chapter still realistic?

Read history books for evidence of the relation between irrigation systems, forest management practices, and agricultural productivity. See, for example, T. Jacobsen and R. M. Adams, Salt and Silt in Ancient Mesopotamian Agriculture, *Science*, **128**:1251–1258 (1958).

Using the FAO Yearbook of Fishery Statistics, United Nations, update information on worldwide fish harvests. Make graphs of yields against year for both total catch and particular stocks (North Sea species, Pacific tuna, Peruvian anchovies, etc.). Does it appear that we can increase catches much more?

Can you think of legal or legislative devices for combating pollution and conserving natural resources that would be more effective than those now available?

REFERENCES

Bryson, R. A., and W. M. Wendland: Climatic Effects of Atmospheric Pollution, in S. F. Singer (ed.), "Global Effects of Environmental Pollution," pp. 130–138, D. Reidel, Dordrecht, Holland, 1970.

Clark, C.: "Population Growth and Land Use," Macmillan, London and Basingstoke, 1967.

Cloud, P.: Mineral Resources from the Sea, in National Academy of Sciences, National Research Council, "Resources and Man," pp. 135–155, Freeman, San Francisco, 1969.

————: Mineral Resources in Fact and Fancy, in W. W. Murdoch (ed.), "Environment," pp. 71–78, Sinauer Associates, Stamford, Conn., 1971.

Council on Environmental Quality: *First Annual Report to the President*, 1970.

Davis, K.: Population Policy: Will Current Programs Succeed? *Science*, **158**:730–739 (1967).

Davis, M.: Future Uranium Demands, *Brit. Nucl. Energy Soc. J.*, **7**:159–162 (1968).

de Wit, C. T.: Photosynthesis: Its Relationship to Overpopulation, in A. San Pietro, F. Greer, and T. J. Army (eds.), "Harvesting the Sun" pp. 315–320, Academic, New York, 1967.

Forrester, J. W.: "World Dynamics," Wright-Allen Press, Cambridge, Mass., 1971.

Freeman, G., and G. B. Haydon: Emphysema after Low-level Exposure to NO_2, *Arch. Environ. Health*, **8**:125–128 (1964).

Frejka, T.: Reflections on the Demographic Conditions Needed to Establish a U.S. Stationary Population Growth, *Population Studies*, **22**:379–397 (1968).

Haydon, G. B., G. Freeman, and N. J. Furiosi: Covert Pathogenesis of NO_2 Induced Emphysema in Rat, *Arch. Environ. Health*, **11**:776–783 (1965).

Hendricks, S. B.: Food from the Land, in Committee on Resources and Man, National Academy of Sciences, National Research Council, "Resources and Man," pp. 65–85, Freeman, San Francisco, 1969.

Hickey, R. J.: Air Pollution, in W. W. Murdoch (ed.), "Environment," pp. 189–212, Sinauer Associates, Stamford, Conn., 1971.

Hinkle, L. E., Jr., L. H. Whitney, E. W. Lehman, J. Dunn, B. Benjamin, R. King, A. Plakun, and B. Flehinger: Occupation, Education, and Coronary Heart Disease, *Science*, **161**:238–246 (1968).

Hubbert, M. K.: Energy Resources, in National Academy of Sciences, National Research Council, "Resources and Man," pp. 157–242, Freeman, San Francisco, 1969.

Keyfitz, N.: The Numbers and Distribution of Mankind, in W. W. Murdoch (ed.), "Environment," pp. 31–52, Sinauer Associates, Stamford, Conn., 1971.

Kleiber, M.: "The Fire of Life," Wiley, New York, 1961.

Kleinerman, J., and G. W. Wright: Experimental Production of a Lesion Resembling Human Microbullous Emphysema in Chronic Bronchitis, *Federation Proc.*, **21**:439 (1962).

Kyllonen, R. L.: Crime Rate Vs. Population Density in United States Cities: A Model, *Gen. Systems*, **12**:137–145 (1967).

Landsberg, H. H., L. L. Fischman, and J. L. Fisher: "Resources in America's Future," Johns Hopkins, Baltimore, 1963.

Lave, L. B., and E. P. Seskin: Air Pollution and Human Health, *Science*, **169**:723–733 (1970).

Lovering, T. S.: Mineral Resources from the Land, in National Academy of Sciences, National Research Council, "Resources and Man," pp. 109–134, Freeman, San Francisco, 1969.

Odum, H. T.: "Environment, Power and Society," Wiley, New York, 1971.

Peccei, A.: "The Chasm Ahead," Macmillan, New York, 1969.

President's Science Advisory Committee: "The World Food Problem," vol. III, the White House, Washington, 1967.

Ricker, W. E.: Food from The sea, in National Academy of Sciences, National Research Council, "Resources and Man," pp. 87–108, Freeman, San Francisco, 1969.

Rosenzweig, M. L.: Paradox of Enrichment: Destabilization of Exploitation Ecosystems in Ecological Time, *Science*, **171**:385–387 (1971).

Singer, S. F. (ed.): "Global Effects of Environmental Pollution," D. Reidel, Dordrecht, Holland, 1970.

Slade, A.: in "Technology Assessment," Hearings before the Subcommittee on Science, Research and Development of the Committee on Science and Astronautics, H. R. 17046, Part II, 91st Cong., 2d Sess., 1970, pp. 526–529.

Stocks, P.: Recent Epidemiological Studies of Lung Cancer Mortality, Cigarette Smoking and Air Pollution, with Discussion of a New Hypothesis of Causation, *Brit. J. Cancer*, **20**:595–623 (1966).

Stout, P. R.: "Potential Agricultural Production from Nuclear Powered Agro-industrial Complexes Designed for the Upper Indo-Gangetic Plain," ORNL-4292 UC-80-Reactor Technol., Oak Ridge National Laboratory, Oak Ridge, Tenn., 1968.

name index

subject index

McGRAW-HILL SERIES IN
POPULATION BIOLOGY

Consulting Editors
Paul R. Ehrlich, Stanford University
Richard W. Holm, Stanford University

BRIGGS: Marine Zoogeography
EDMUNDS and LETEY: Environmental Administration
EHRLICH, HOLM, and PARNELL: The Process of Evolution
GEORGE and McKINLEY: Urban Ecology
HAMILTON: Life's Color Code
POOLE: An Introduction to Quantitative Ecology
STAHL: Vertebrate History: Problems in Evolution
WATT: Ecology and Resource Management
WATT: Principles of Environmental Science
WELLER: The Course of Evolution

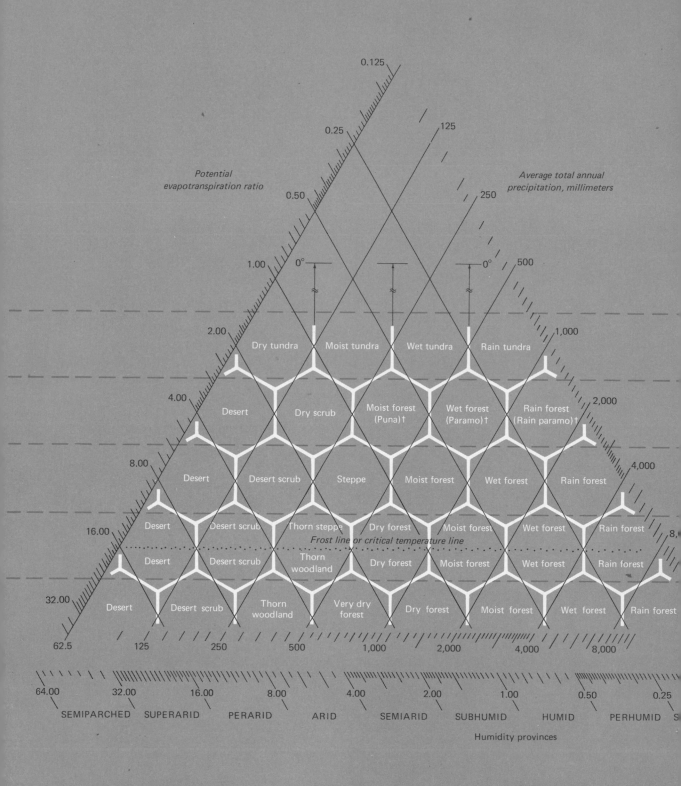